Radiochemical Syntheses

VOLUME 2

WILEY SERIES ON RADIOCHEMICAL SYNTHESES

Peter J. H. Scott and Michael R. Kilbourn, Series Editors

Radiochemical Syntheses, Volume 1-Radiopharmaceuticals for Positron Emission Tomography
Edited by Peter J. H. Scott and Brian G. Hockley

Radiochemical Syntheses, Volume 2: Further Radiopharmaceuticals for Positron Emission Tomography and New Strategies for Their Production
Edited by Peter J. H. Scott

Radiochemical Syntheses

VOLUME 2

FURTHER RADIOPHARMACEUTICALS FOR POSITRON EMISSION TOMOGRAPHY AND NEW STRATEGIES FOR THEIR PRODUCTION

Edited by

PETER J. H. SCOTT

WILEY

Published by John Wiley & Sons, Inc., Hoboken, New Jersey
Published simultaneously in Canada

Library of Congress Cataloging-in-Publication Data:

Further radiopharmaceuticals for positron emission tomography and new strategies for their production / edited by Peter J. H. Scott.
pages cm. – (Radiochemical syntheses ; volume 2)
Includes bibliographical references and index.
ISBN 978-1-118-23784-7 (cloth : alk. paper)
1. Radiopharmaceuticals–Synthesis. 2. Tomography, Emission. I. Scott, Peter J. H., editor.
RS431.R34R326 2015
615.8′42–dc23

2015001063

Set in 10.25/12pt Times by SPi Global, Pondicherry, India

10 9 8 7 6 5 4 3 2 1

1 2015

MIX
Paper from responsible sources
FSC
www.fsc.org
FSC® C013604

CONTENTS

CONTRIBUTORS

Antero J. Abrunhosa
Institute for Nuclear Sciences Applied to Health (ICNAS), Institute for Biomedical Imaging and Life Sciences (IBILI), Faculty of Medicine, University of Coimbra, Coimbra, Portugal

Franklin I. Aigbirhio
Wolfson Brain Imaging Centre, Addenbrooke's Hospital, University of Cambridge, Cambridge, UK

David Alagille
Molecular NeuroImaging, LLC, New Haven, Connecticut and Institute for Neurodegenerative Disorders, New Haven, Connecticut, USA

Vítor H. Alves
Institute for Nuclear Sciences Applied to Health (ICNAS), Institute for Biomedical Imaging and Life Sciences (IBILI), Faculty of Medicine, University of Coimbra, Coimbra, Portugal

Vadim Bernard-Gauthier
McConnell Brain Imaging Centre, Montreal Neurological Institute, McGill University, Montreal, Canada

Denis R. Bouvet
GE Healthcare, The Grove Centre, Amersham, Buckinghamshire, UK

Allen F. Brooks
Department of Radiology, University of Michigan School of Medicine, Ann Arbor, Michigan, USA

Lynda J. Brown
Department of Chemistry, University of Southampton, Southampton, UK

Richard C. D. Brown
Department of Chemistry, University of Southampton, Southampton, UK

Stephen M. Carlin
Athinoula A. Martinos Center for Biomedical Imaging, Department of Radiology, Massachusetts General Hospital, Harvard Medical School, Charlestown, Massachusetts, USA

Garret M. Carpenter
Department of Radiology, University of Michigan School of Medicine, Ann Arbor, Michigan, USA

Miguel Castelo-Branco
Institute for Nuclear Sciences Applied to Health (ICNAS), Institute for Biomedical Imaging and Life Sciences (IBILI), Faculty of Medicine, University of Coimbra, Coimbra, Portugal

Sue M. Champion
GE Healthcare, The Grove Centre, Amersham, Buckinghamshire, UK

Ashok Chaudhary
Siemens Molecular Imaging, Culver City, California, USA

Kai Chen
Molecular Imaging Center, Department of Radiology, Keck School of Medicine, University of Southern California, Los Angeles, California, USA

Thomas Lee Collier
Division of Nuclear Medicine and Molecular Imaging, Massachusetts General Hospital, Boston, Massachusetts, USA;
Department of Radiology, Harvard Medical School, Boston, Massachusetts and Advion Inc., Ithaca, New York, USA

Peter S. Conti
Molecular Imaging Center, Department of Radiology, Keck School of Medicine, University of Southern California, Los Angeles, California, USA

David W. Dick
Department of Radiology, University of Iowa Hospitals & Clinics, Iowa City, Iowa, USA

Maria V. Fawaz
Department of Radiology, University of Michigan School of Medicine, Ann Arbor, Michigan, USA

Umesh Gangadharmath
Siemens Molecular Imaging, Culver City, California, USA

Alex M. Gibson
GE Healthcare, The Grove Centre, Amersham, Buckinghamshire, UK

Bradford D. Henderson
Department of Radiology, University of Michigan School of Medicine,
Ann Arbor, Michigan, USA

Brian G. Hockley
Department of Radiology, University of Michigan School of Medicine,
Ann Arbor, Michigan, USA

Jacob M. Hooker
Athinoula A. Martinos Center for Biomedical Imaging, Department of Radiology,
Massachusetts General Hospital, Harvard Medical School, Charlestown,
Massachusetts, USA

Naoko Ichiishi
Department of Chemistry, University of Michigan, Ann Arbor, Michigan, USA

Alexander Jackson
GE Healthcare, The Grove Centre, Amersham, Buckinghamshire, UK

Raul N. Jackson
Division of Nuclear Medicine and Molecular Imaging, Massachusetts General
Hospital, Boston, Massachusetts, USA

Keunsam Jang
Department of Radiology, University of Michigan School of Medicine,
Ann Arbor, Michigan, USA

Klaus Jurkschat
Lehrstuhl für Anorganische Chemie II, TU University of Dortmund, Dortmund,
Germany

Imtiaz Khan
GE Healthcare, The Grove Centre, Amersham, Buckinghamshire, UK

Hartmuth C. Kolb
Siemens Molecular Imaging, Culver City, California, USA

Alexey Kostikov
McConnell Brain Imaging Centre, Montreal Neurological Institute,
McGill University, Montreal, Canada

Steven H. Liang
Division of Nuclear Medicine and Molecular Imaging, Massachusetts General
Hospital, Boston, Massachusetts and
Department of Radiology, Harvard Medical School, Boston, Massachusetts, USA

Eli Livni
Division of Nuclear Medicine and Molecular Imaging, Massachusetts General Hospital, Boston, Massachusetts and
Department of Radiology, Harvard Medical School, Boston, Massachusetts, USA

Shuiyu Lu
Molecular Imaging Branch, National Institute of Mental Health, Bethesda, Maryland, USA

Nianchun Ma
Department of Chemistry, University of Southampton, Southampton, UK

Philip W. Miller
Department of Chemistry, Imperial College London, South Kensington, London, UK

Nicolas Millot
GE Healthcare, The Grove Centre, Amersham, Buckinghamshire, UK

Vani P. Mocharla
Siemens Molecular Imaging, Culver City, California, USA

Christian K. Moseley
Athinoula A. Martinos Center for Biomedical Imaging, Department of Radiology, Massachusetts General Hospital, Harvard Medical School, Charlestown, Massachusetts, USA

Fanrong Mu
Siemens Molecular Imaging, Culver City, California, USA

Jan Passchier
Imanova Centre for Imaging Sciences, Imperial College London, Hammersmith Hospital, London, UK

Victor W. Pike
Molecular Imaging Branch, National Institute of Mental Health, Bethesda, Maryland, USA

Christophe Plisson
Imanova Centre for Imaging Sciences, Imperial College London, Hammersmith Hospital, London, UK

Waqas Rafique
Kjemisk Institutt, Universitetet i Oslo, Oslo, Norway

Joaquim Ramada-Magalhaes
Imanova Centre for Imaging Sciences, Imperial College London, Hammersmith Hospital, London, UK

Hong Ren
Athinoula A. Martinos Center for Biomedical Imaging,
Massachusetts General Hospital, Harvard Medical School, Charlestown, Massachusetts, USA

Peter A. Rice
Division of Nuclear Medicine and Molecular Imaging, Department of Radiology, Massachusetts General Hospital, Boston, Massachusetts, USA

Patrick J. Riss
Kjemisk Institutt, Universitetet i Oslo, Oslo, Norway and
Wolfson Brain Imaging Centre, Addenbrooke's Hospital, University of Cambridge, Cambridge, UK

Melissa E. Rodnick
Department of Radiology, University of Michigan School of Medicine, Ann Arbor, Michigan, USA

Melanie S. Sanford
Department of Chemistry, University of Michigan, Ann Arbor, Michigan, USA

Nathan J. Schauer
Athinoula A. Martinos Center for Biomedical Imaging, Department of Radiology, Massachusetts General Hospital, Harvard Medical School, Charlestown, Massachusetts, USA

Ralf Schirrmacher
McConnell Brain Imaging Centre, Montreal Neurological Institute, McGill University, Montreal, Canada

Esther Schirrmacher
McConnell Brain Imaging Centre, Montreal Neurological Institute, McGill University, Montreal, Canada

Peter J. H. Scott
Department of Radiology, University of Michigan School of Medicine, Ann Arbor, Michigan, USA

Xia Shao
Department of Radiology, University of Michigan School of Medicine, Ann Arbor, Michigan, USA

Megan N. Stewart
Department of Radiology, University of Michigan School of Medicine, Ann Arbor, Michigan, USA

Gilles Tamagnan
Molecular NeuroImaging, LLC, New Haven, Connecticut and Institute for Neurodegenerative Disorders, New Haven, Connecticut, USA

Sanjay Telu
Molecular Imaging Branch, National Institute of Mental Health, Bethesda, Maryland, USA

Joseph J. Topczewski
Department of Chemistry, University of Michigan, Ann Arbor, Michigan, USA

Neil Vasdev
Division of Nuclear Medicine and Molecular Imaging, Massachusetts General Hospital, Boston, Massachusetts and
Department of Radiology, Harvard Medical School, Boston, Massachusetts, USA

Harry J. Wadsworth
GE Healthcare, The Grove Centre, Amersham, Amersham, Buckinghamshire, UK

Joseph C. Walsh
Siemens Molecular Imaging, Culver City, California, USA

Björn Wängler
Molecular Imaging and Radiochemistry, Department of Clinical Radiology and Nuclear Medicine, Medical Faculty Mannheim, Heidelberg University, Mannheim, Germany

Carmen Wängler
Biomedical Chemistry, Department of Clinical Radiology and Nuclear Medicine, Medical Faculty Mannheim, Heidelberg University, Mannheim, Germany

G. Leonard Watkins
Department of Radiology, University of Iowa Hospitals & Clinics, Iowa City, Iowa, USA

Daniel L. Yokell
Division of Nuclear Medicine and Molecular Imaging, Massachusetts General Hospital, Boston, Massachusetts, USA

Zhihong Zhu
Siemens Molecular Imaging, Culver City, California, USA

EDITORIAL PREFACE

Our intention for founding *Radiochemical Syntheses* was to make readily available a series of monographs that offer detailed procedures for the synthesis of radiolabeled compounds. These volumes (available as books, as pdf files through Wiley Online Library, and also as convenient ebooks suitable for smartphone and tablet use) are intended to offer experimental details that are beyond the capacity of the experimental section of a typical journal article but that are often critical for successful implementation of a given radiochemical synthesis. This is perhaps particularly true in the land of radiochemistry, even more so than in organic chemistry for which Organic Syntheses was founded almost a century ago to address similar issues. We believe that this goal was accomplished with Volume 1 of *Radiochemical Syntheses*, published in 2012, judging by the positive feedback that has been received from radiochemists utilizing the book all over the world.

Volume 2 aims to continue the momentum of the first volume, and the first half of the book presents methods for the synthesis of another group of radiopharmaceuticals for positron emission tomography (PET) imaging. We have continued our theme of selecting only those radiopharmaceuticals in clinical use at two or more PET Centers around the world. An alternative synthesis of fluorocholine is presented from our laboratory, utilizing fluoromethyl tosylate instead of the synthesis using fluorobromomethane reported in Volume 1 as we believe both methods are useful and should be available to the radiochemistry community. Building on this notion, it is hoped and encouraged that if any radiochemists have alternative methods for synthesizing any of the radiopharmaceuticals considered or if improvements to the synthetic methods described in this or the previous volume are discovered, they will consider furnishing monographs to future volumes of *Radiochemical Syntheses*. This way, the series will stay current, and the best synthetic techniques can be put in the hands of the global radiochemistry community.

The second half of the book takes a different direction and has more of a feel of Organic Syntheses about it. We regularly read about new methods for incorporating positron-emitting radioisotopes into bioactive molecules in the literature, but frequently, those reports are the first and last time we hear about such methodology. One reason for this is that such papers (which are frequently two- or three-page letters) often lack the details really needed to implement the complex radiochemistry for new substrates. We wish to address this issue going forward and hope to see

Radiochemical Syntheses grow into the peer-reviewed forum for communicating reliable radiochemical methodology to the field. With this goal in mind, potential authors and guest volume editors are encouraged to obtain information about preparing monographs or future volumes from the Radiochemical Syntheses Editorial Office (pjhscott@umich.edu).

Finally, *Radiochemical Syntheses* would not be possible without the dedication and enthusiasm of the authors, the members of the Editorial Advisory Board who peer-review all of the submitted manuscripts, and the tireless efforts of Jonathan Rose at Wiley for championing our cause and demonstrating exhaustive patience when dealing with the editors' special talent for missing deadlines!

Peter J. H. Scott, PhD
Michael R. Kilbourn, PhD

The University of Michigan
Ann Arbor, Michigan
March 2015

ABBREVIATIONS

μA	Microampere
μg	Microgram
μl	Microliter
μm	Micrometer
μmol	Micromole
18-cr-6	18-Crown-6; 1,4,7,10,13,16-hexaoxacyclooctadecane
Ac	Acetate
AChE	Acetylcholinesterase
AD	Alzheimer's disease
BEMP	2-*tert*-Butylimino-2-diethylamino-1,3-dimethylperhydro-1,3,2-diazaphosphorine
Bn	Benzyl
Boc	*tert*-Butyloxycarbonyl protecting group
Bq	Becquerel
CB1	Cannabinoid receptor 1
CB2	Cannabinoid receptor 2
cGMP	Current good manufacturing practice
Ci	Curie
CNS	Central nervous system
CT	Computed tomography
D_2	Dopamine D_2 receptor
D_3	Dopamine D_3 receptor
DABCO	1,4-Diazabicyclo[2.2.2]octane; triethylenediamine
DASB	3-Amino-4-[2-[(di(methyl)amino)methyl]phenyl] sulfanylbenzonitrile
DBN	1,5-Diazabicyclo[4.3.0]non-5-ene
DBU	1,8-Diazabicyclo[5.4.0]-undec-7-ene
DEAE-C	Diethylaminoethyl cellulose
DMAE	Dimethylaminoethanol
DMAP	4-(Dimethylamino)pyridine
DMF	*N,N*-Dimethylformamide
DMSO	Dimethyl sulfoxide
DNA	Deoxyribonucleic acid
DTBZ	Dihydrotetrabenazine
EOB	End of bombardment

EOS	End of synthesis
EP	European Pharmacopeia
Et_2O	Diethyl ether
EtOH	Ethanol
EU	Endotoxin units
FCH	Fluorocholine
FDG	2-Fludeoxyglucose
FDOPA	Fluorodopa
FEOBV	Fluoroethoxy benzovesamicol
FIAC	2′-Fluoro-5-iodo-1-β-D-arabinofuranosyl-cytosine
FIAU	2′-Fluoro-5-iodo-1-β-D-arabinofuranosyl-uracil
FLBT	Flubatine
FLT	Fluorothymidine
FMAU	2′-Deoxy-2′-fluoro-5-methyl-1-β-D-arabinofuranosyluracil
Fmoc	Fluorenylmethyloxycarbonyl
FPEB	3-Fluoro-5-(2-pyridinylethynyl)benzonitrile
FTM	Fluid thioglycolate media
GBq	Gigabecquerel
GC	Gas chromatography
h	Hours
HED	meta-Hydroxyephedrine
HMDS	Hexamethyldisilazane
HPLC	High-performance liquid chromatography
i.d.	Internal diameter
ICH	The International Conference on Harmonisation of Technical Requirements for Registration of Pharmaceuticals for Human Use
ICP-MS	Inductively coupled plasma mass spectrometry
K222	Kryptofix-2.2.2; Crypt-222; 2.2.2-cryptand; 4,7,13,16,21,24-hexaoxa-1,10-diazabicyclo[8.8.8]hexacosane
keV	Kiloelectron volt
LAL	Limulus amebocyte lysate
LC	Liquid chromatography
M	Molar
MBq	Megabecquerel
MCA	Multichannel analyzer
mCi	Millicurie
MeCN	Acetonitrile
MeOH	Methanol
MeOTf	Methyl triflate
MES	Mesitylene; mesityl group
MeV	Megaelectron volt
mg	Milligram
mGluR5	Metabotropic glutamate receptor subtype 5
min	Minute
mL	Milliliter

mM	Millimolar
mmol	Millimole
MPOH	*N*-Methyl-4-piperidinol
nAChRs	Nicotinic acetylcholine receptors
NHC	*N*-Heterocyclic carbene
nm	Nanometer
NMR	Nuclear magnetic resonance
o.d.	Outer diameter
OMAR	1-(2,4-Dichlorophenyl)-4-cyano-5-(4-[^{11}C] methoxyphenyl)-*N*-(piperidin-1-yl)-1*H*-pyrazole-3-carboxamide
OTf	Triflate; trifluoromethanesulfonate; CF_3SO_3-
OTs	Tosylate; 4-toluenesulfonate; $CH_3C_6H_4SO_3$-
PBR	Peripheral benzodiazepine receptor
PD	Parkinson's disease
PET	Positron emission tomography
Ph. Eur.	European Pharmacopeia
PHNO	4-Propyl-3,4,4a,5,6,10b-hexahydro-2*H*-naphtho[1,2-b][1,4] oxazin-9-ol
PiB	Pittsburgh Compound B
PMP	*N*-Methyl-4-piperidinyl propionate
PN	Part number
ppb	Parts per billion
ppm	Parts per million
psig	Pounds per square inch gauge
QC	Quality control
QMA	Quaternary methyl ammonium
RCC	Radiochemical conversion
RCP	Radiochemical purity
RCY	Radiochemical yield
RGD	Arginine–glycine–aspartate
RLV	Resin–linker–vector
RRT	Relative retention time
sccm	Standard cubic centimeters per minute
SCDM	Soybean casein digest agar media
SFB	*N*-Succinimidyl 4-fluorobenzoate
SiFA	Silicon Fluoride Acceptors
SPECT	Single-photon emission computed tomography
$t_{1/2}$	Half-life
TCD	Thermal conductivity detector
TEA	Triethylamine
THF	Tetrahydrofuran
TLC	Thin-layer chromatography
TMEDA	*N,N,N′,N′*-Tetramethylethane-1,2-diamine
TMS	Trimethylsilyl
TSB	Tryptic soy broth

TSPO	18 kDa translocator protein
USP	United States Pharmacopeia
UV	Ultraviolet
v/v	Volume–volume concentration (percent)
VEGF	Vascular endothelial growth factor
w/w	Weight–weight concentration (percent)

PART I

FLUORINE-18 LABELED RADIOPHARMACEUTICALS

CHAPTER 1

SYNTHESIS OF (–)-[^{18}F]FLUBATINE ([^{18}F]FLBT)

Megan N. Stewart, Brian G. Hockley, and Peter J. H. Scott

Department of Radiology, University of Michigan School of Medicine, Ann Arbor, Michigan, USA

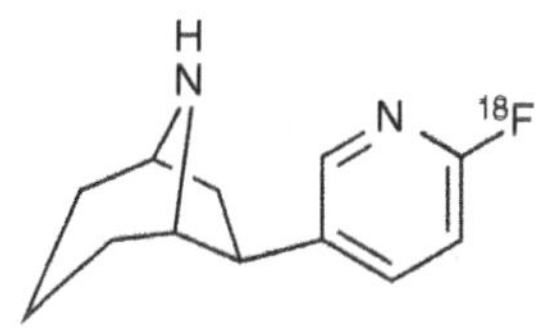

1 INTRODUCTION

Cognitive or depressive disorders and neurodegenerative diseases such as Alzheimer's disease (AD), dementia, and Parkinson's disease (PD) may be related to dysfunctional signaling through α4β2-nicotinic acetylcholine receptors (α4β2-nAChRs) [1, 2]. Alterations in the cholinergic system are also implicated in the progression of cognitive decline in the aforementioned neurodegenerative diseases, particularly AD [2–4]. The development of (–)-[^{18}F]flubatine as a high affinity and selective PET radiotracer with improved kinetics over the earlier developed ligands allows for noninvasive quantification of nAChRs [2, 5].

The first reported radiosynthesis of (–)-[^{18}F]flubatine, a derivative of epibatidine, utilized a norchloro-bromo-homoepibatidine (NCBrHEB) precursor that underwent a nucleophilic substitution with the bromine leaving group, then the enantiomers separated, and the product purified appropriately via HPLC [4–7]. However, due to low radiochemical yields, other candidate precursors were explored for radiolabeling and the trimethylammonium iodide-Boc-protected compound ((5-((1*R*,5*S*,6*S*)-8-*tert*-butoxycarbonyl)-8-azabicyclo[3.2.1]-octan-6-yl)-*N*,*N*,*N*-trimethylpyridin-2-aminium iodide, Boc-trimethylammonium homoepibatidine, **1**) was shown to give the best yields of

Radiochemical Syntheses, Volume 2: Further Radiopharmaceuticals for Positron Emission Tomography and New Strategies for Their Production, First Edition. Edited by Peter J. H. Scott.

approximately 60% and further adapted for fully automated synthesis [2, 8]. This precursor has since become commercially available, making [^{18}F]flubatine more accessible for clinicians and has been validated for clinical use in nonhuman primates [4].

2 SYNTHESIS PROCEDURES

CAUTION: All radiochemical syntheses must be carried out using appropriate equipment in a facility authorized for the use of radioactive materials. Personal protective equipment must be worn and all local radiation safety laws followed.

2.1 Production of [^{18}F]Fluoride

[^{18}O]H_2O (1.5 ml) [9] was loaded into the [^{18}F]fluoride target [10] of a General Electric Medical Systems (GEMS) PETtrace cyclotron [11]. The target was bombarded (60 µA beam for 30 min) to generate approximately 1.5 Ci (55.5 GBq) of [^{18}F]fluoride by the ^{18}O(p,n)^{18}F nuclear reaction.

2.2 Azeotropic Drying of [^{18}F]Fluoride

The [^{18}F]fluoride was delivered to a GEMS TRACERlab FX_{FN} synthesis module [11] as a solution in [^{18}O]H_2O (1.5 ml). This solution was passed through a Sep-Pak® QMA-Light cartridge [12] to trap the [^{18}F]fluoride and recycle the [^{18}O]H_2O. The [^{18}F]fluoride was then eluted into the TRACERlab FX_{FN} glassy carbon reaction vessel using a solution of aqueous potassium carbonate (3.5 mg in 0.5 ml H_2O) [13]. A solution of Kryptofix 222 (5 mg in 1 ml MeCN) [14] was added and the reaction mixture was azeotropically dried, initially at 80°C under vacuum for 4 min and subsequently at 60°C with both vacuum and argon flow for an additional 4 min.

2.3 Synthesis of (−)-[^{18}F]FLBT

A solution of (−)-FLBT precursor [15] (**1**, 0.5–1.0 mg) in anhydrous dimethyl sulfoxide (DMSO) [16] (0.6 ml) was added to the dried [^{18}F]fluoride, and the reaction was heated to 120°C with stirring for 10 min (Fig. 1). After this time, the reaction was cooled to 40°C, and 1.0 M aqueous hydrochloric acid (1 ml) was added. The reaction was stirred for 5 min at 80°C to hydrolyze the Boc protecting groups. The reaction mixture was neutralized with 0.5 M aqueous sodium hydroxide (2 ml) [17].

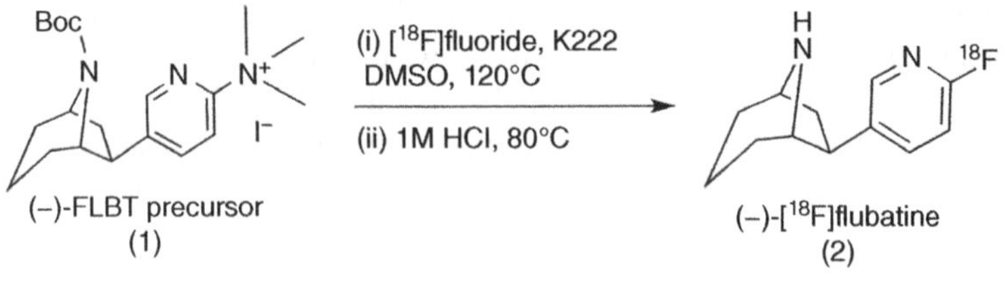

FIGURE 1 Synthesis of (−)-[^{18}F]FLBT.

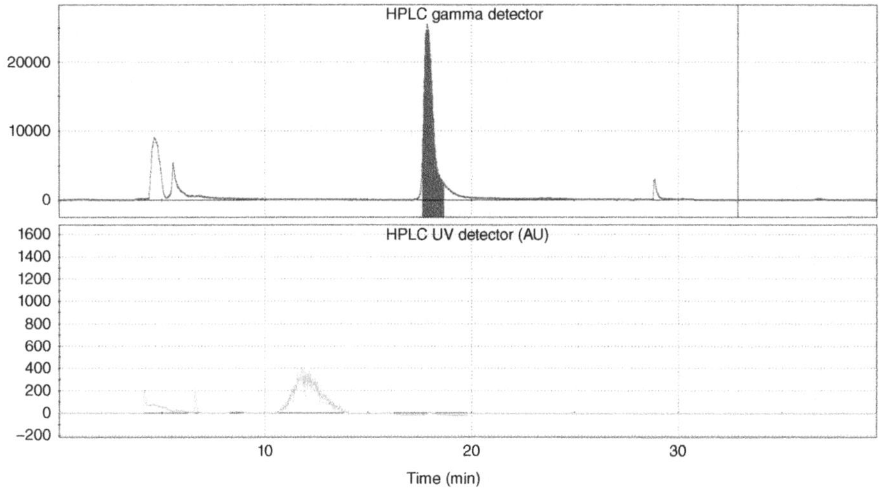

FIGURE 2 Semipreparative UV and radioactive HPLC traces for [^{18}F]FLBT.

2.4 Purification and Formulation of (−)-[^{18}F]FLBT

After hydrolysis, the crude reaction mixture was purified by semipreparative HPLC (Luna 10u C18(2) 250×10mm column [18], flow rate=4ml/min), and a representative HPLC trace is shown in Fig. 2.

The fraction corresponding to (−)-[^{18}F]FLBT (typically eluting between 20 and 25min) was collected for 1min into a vial precharged with 0.9% sodium chloride, USP [19] (6ml). The final formulation (10ml) was then passed through a sterile filter [20] into a sterile vial [21] to provide (−)-[^{18}F]FLBT (typically 50–115mCi (1.85–4.3GBq)) in an isotonic solution released for quality control. After synthesis was complete, the semipreparative HPLC column was flushed with 70% ethanol.

3 QUALITY CONTROL PROCEDURES

CAUTION: All radiochemicals produced for clinical use must have local regulatory approval (e.g., FDA, EMEA, MHRA, PFSB, etc.) prior to human use. Quality control procedures must be carried out by trained personnel, and each dose must meet all established QC criteria before release to the clinic.

Quality control (QC) procedures for (−)-[^{18}F]FLBT, based upon the current requirements for radiopharmaceuticals laid out in the US Pharmacopeia (USP) [22], are summarized in the following text. Complete QC data for three repeat batches of (−)-[^{18}F]FLBT produced using the method disclosed herein are summarized in Table 1. Each of the three doses met all of the established QC criteria

TABLE 1 QC Data for three Repeat Runs of (−)-[^{18}F]FLBT

QC Test	Release Criteria	Run 1	Run 2	Run 3
Yield/mCi (GBq)	N/A	79.7 (2.9)	101 (3.7)	65 (2.4)
Visual inspection	Clear, colorless	Clear, colorless	Clear, colorless	Clear, colorless
Radiochemical identity	RRT = 0.9–1.1	1.01	1.01	1.01
FLBT concentration	No limit established	0.84 μg/ml	0.66 μg/ml	0.33 μg/ml
Radiochemical purity	≥95%	99.9	100	100
Specific activity	No limit established	1.5 Ci/μmol	2.7 Ci/μmol	4.0 Ci/μmol
Residual solvent analysis	DMSO <5000 ppm	Pass	Pass	Pass
	Acetone <5000 ppm	Pass	Pass	Pass
	Acetonitrile <410 ppm	Pass	Pass	Pass
Dose pH	4.5–7.5	4.5	4.5	4.5
Residual Kryptofix 222	≤50 μg/ml	≤50 μg/ml	≤50 μg/ml	≤50 μg/ml
Sterile filter integrity test	>50 psi	>50 psi	>50 psi	>50 psi
Radionuclidic identity ($t_{1/2}$)	105–115 min	112	109	109
Endotoxin analysis	≤17.5 EU/ml	≤2 EU/ml	≤2 EU/ml	≤2 EU/ml
Sterility testing	No colony growth out to 14 days	Pass	Pass	Pass

3.1 Visual Inspection

The (−)-[^{18}F]FLBT dose is examined behind a PET L-block and must be clear, colorless, and free of particulate matter.

3.2 Radiochemical Identity

HPLC analysis of radiochemical identity was conducted using a Shimadzu LC-2010A$_{HT}$ Liquid Chromatograph [23] fitted with UV detectors and Bioscan γ-detectors [24] (column, Phenomenex Synergi Polar RP 150×4.6 mm [25]; mobile phase, 50% acetonitrile 50% water +0.1% acetic acid [26]; flow rate, 1 ml/min, λ=254 nm). The retention time of [^{18}F]FLBT is compared to that of the [^{19}F]FLBT reference standard [15] and must be ±10% (relative retention time (RRT) must be 0.9–1.1).

3.3 Radiochemical Purity

HPLC analysis of radiochemical purity was conducted using a Shimadzu LC-2010A$_{HT}$ Liquid Chromatograph [23] fitted with a UV detector and a Bioscan γ-detector [24] (column, Phenomenex Synergi Polar RP 150×4.6 mm [25]; mobile phase, 50% acetonitrile 50% water +0.1% acetic acid [26]; flow rate, 1 ml/min,

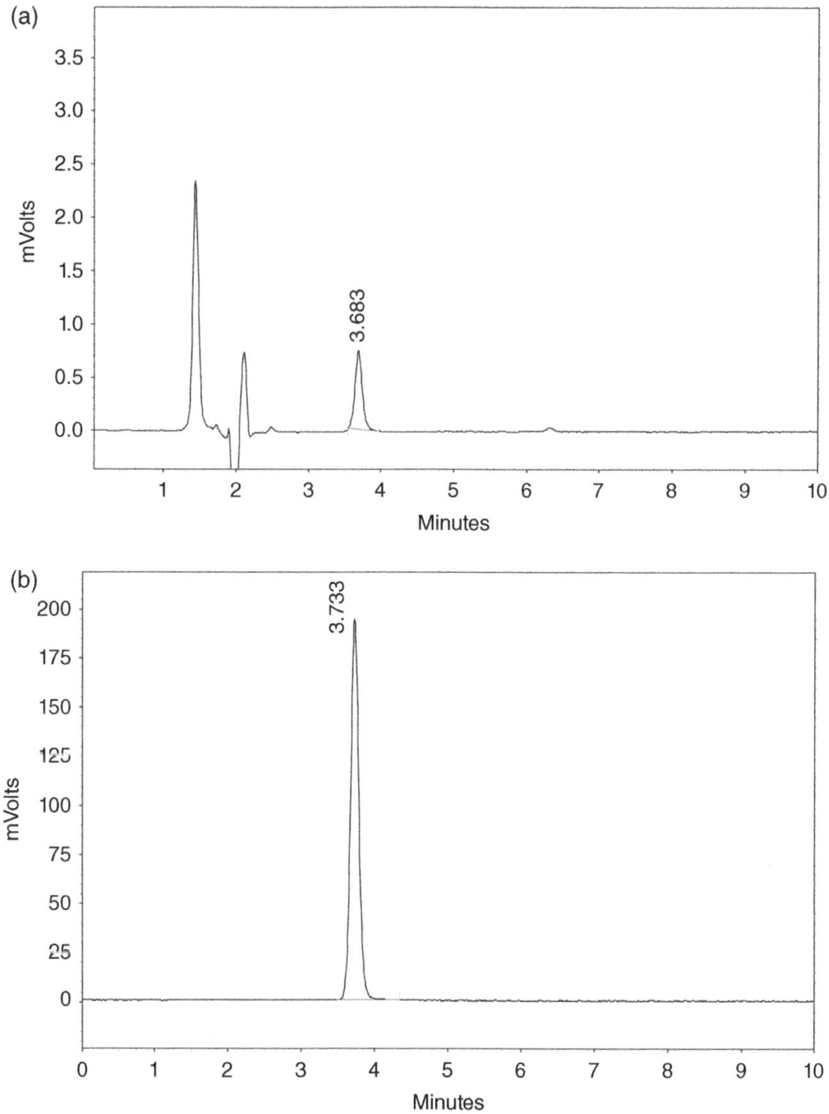

FIGURE 3 Analytical UV (a) and radioactive (b) HPLC trace for (−)-[^{18}F]FLBT.

$\lambda = 254$ nm). Radiochemical purity must be greater than 95%. Representative analytical HPLC traces are displayed in Fig. 3.

3.4 Specific Activity

There is no specific activity release limit for (−)-[^{18}F]FLBT in place at our institution. The injection volume is based on the injectable mass limit for (−)-FLBT and is calculated on a dose-by-dose basis by the administering study team.

3.5 Residual Solvent Analysis

Analysis of residual solvent levels (acetone (from TRACERlab FX_{FN} drying cycle), DMSO, and acetonitrile) in [^{18}F]FLBT doses was performed using a Shimadzu GC-2010 gas chromatograph (GC) equipped with a split/splitless inlet and flame ionization detector and a Restek Stabilwax GC column (35 m × 0.25 mm) using conditions outlined in the following text [23]. Limits imposed by the ICH Harmonised Tripartite Guidelines are <410 ppm/day for acetonitrile and <5000 ppm/day for class 3 solvents (DMSO, acetone) [27].

GC Conditions

Oven temperature profile:

1. Hold at 30°C for 3 min.
2. Increase 30–180°C at 60°C/min.
3. Hold at 180°C for 5 min.

Inlet temperature: 180°C
Column linear velocity: 18.8 cm/s
FID temperature: 250°C

3.6 Dose pH

The pH of the [^{18}F]FLBT dose was analyzed by applying a small amount of the dose to colorpHast® pH 2.0–9.0 nonbleeding pH-indicator strips [28] and determined by visual comparison to the scale provided. Dose pH must be 4.5–7.5.

3.7 Residual Kryptofix 222 Analysis

Residual Kryptofix 222 levels in (–)-[^{18}F]FLBT doses were analyzed using the established spot test [29]. Strips of plastic-backed silica gel TLC plates saturated with iodoplatinate reagent [29] were spotted with water (negative control), 50 µg/ml Kryptofix 222 standard (positive control) and (–)-[^{18}F]FLBT dose. If Kryptofix 222 is present in a sample, a blue-black spot appears. Spots for the three samples were compared, and a visual determination of residual Kryptofix 222 in the (–)-[^{18}F]FLBT dose was made. Less than 50 µg/ml is acceptable.

3.8 Sterile Filter Integrity Test

The sterile filter from the (–)-[^{18}F]FLBT (with needle still attached) was connected to a nitrogen supply via a regulator. The needle was submerged in water and the nitrogen pressure was gradually increased. If the pressure was raised above the filter acceptance pressure (50 psi) without seeing a stream of bubbles, the filter was considered intact. If a stream of bubble occurs at less than 50 psi, the test fails.

3.9 Radionuclidic Identity

Activities were measured using a Capintec CRC®-15R Radioisotope Dose Calibrator [30], and half-life was calculated using Equation 1. Calculated half-life must be 105–115 min:

$$T_{1/2} = -\ln 2\left(\frac{\text{time difference}}{\left(\ln\left(\text{ending activity / starting activity}\right)\right)}\right) \quad (1)$$

3.10 Endotoxin Analysis

Endotoxin content in doses of (−)-[^{18}F]FLBT was analyzed by a Charles River Laboratories Endosafe® Portable Testing System [31] and according to the USP. Doses must contain less than or equal to 17.5 endotoxin units (EU)/ml.

3.11 Sterility Testing

Samples of (−)-[^{18}F]FLBT were placed into fluid thioglycolate media (FTM) plates and soybean-casein digest agar media (SCDM) tubes. FTM tubes are used to test for anaerobes, aerobes, and microaerophiles, while SCDM tubes are used to test for nonfastidious and fastidious microorganisms [32]. (−)-[^{18}F]FLBT tubes were incubated along with positive and negative controls for 14 days. FTM tubes were incubated at 32°C and SCDM tubes were incubated at 22°C according to current USP guidelines [33]. Tubes were visually inspected on the 3rd, 8th, and 14th days of the test period and compared to the positive and negative standards. Positive standards must show growth (turbidity) in the tubes, and (−)-[^{18}F]FLBT/negative controls must have no culture growth after 14 days to be indicative of sterility.

WASTE DISPOSAL INFORMATION

All hazardous chemicals and toxic materials were disposed of according to Prudent Practices in the Laboratory (Washington, DC: National Academy Press, 1995).

CHEMICAL ABSTRACTS NOMENCLATURE (REGISTRY NUMBER)

Acetonitrile (75-05-8)
5-(((1*R*, 5*S*, 6*S*)-8-*tert*-butoxycarbonyl)-8-azabicyclo[3.2.1]octan-6-yl)-*N,N,N*-trimethylpyridin-2-aminium iodide (CAS-RN not assigned)
Carbonic acid, potassium salt (1:2) (584-08-7)
DMSO (67-68-5)
Ethanol (64-17-5)
(1*R*,5*S*,6*S*)-6-(6-Fluoropyridin-3-yl)-8-azabicyclo[3.2.1]octane (CAS-RN not assigned)
4,7,13,16,21,24-Hexaoxa-1,10-diazabicyclo[8.8.8]hexacosane (23978-09-8)
Hydrochloric acid (7647-01-0)
Sodium hydroxide (1310-73-2)

REFERENCES AND NOTES

For detailed supplier information, see Appendix 1.

1. P. M. Meyer, K. Strecker, K. Kendziorra, G. Becker, S. Hesse, D. Woelpl, A. Hensel, M. Patt, D. Sorger, F. Wegner, D. Lobsien, H. Barthel, P. Brust, H. J. Gertz, O. Sabri, *Arch Gen Psychiatr*, 2009, 66, 866.
2. S. Fischer, A. Hiller, R. Smits, A. Hoepping, U. Funke, B. Wenzel, P. Cumming, O. Sabri, J. Steinbach, P. Brust, *Appl Radiat Isot*, 2013, 74, 128.
3. R. Smits, S. Fischer, A. Hiller, W. Deuther-Conrad, B. Wenzel, M. Patt, P. Cumming, J. Steinbach, O. Sabri, P. Brust, A. Hoepping, *Bioorg Med Chem Lett*, 2014, 22, 804.
4. B. G. Hockley, M. N. Stewart, P. Sherman, C. Quesada, M. R. Kilbourn, R. L. Albin, P. J. H. Scott, *J Labelled Comp Rad*, 2013, 56, 595.
5. P. Brust, J. T. Patt, W. Deuther-Conrad, G. Becker, M. Patt, A. Schildan, D. Sorger, K. Kendziorra, P. Meyer, J. Steinbach, O. Sabri, *Synapse*, 2008, 62, 205.
6. J. T. Patt, W. Deuther-Conrad, K. Wohlfarth, D. Feuerbach, P. Brust, J. Steinbach, *J Labelled Comp Rad* 2003, 46, S168.
7. W. Deuther-Conrad, J. T. Patt, P. R. Lockman, D. D. Allen, M. Patt, A. Schildan, V. Ganapathy, J. Steinbach, O. Sabri, P. Brust, *Eur Neuropsychopharmacol*, 2008, 18, 222.
8. M. Patt, A. Schildan, B. Habermann, S. Fischer, A. Hiller, W. Deuther-Conrad, S. Wilke, R. Smits, A. Hoepping, G. Wagenknecht, J. Steinbach, P. Brust, O. Sabri, *Appl Radiat Isot*, 2013, 80, 7–11.
9. Virgin $[^{18}O]H_2O$ purchased from ABX, Amic, Rotem, or Medical Isotopes, and used as received.
10. GEMS silver high-yield $[^{18}F]$fluoride target.
11. GE Healthcare, United States.
12. Sep-Pak® QMA-Light cartridges were purchased from Waters (part no. WAT023525) and conditioned with 10 ml ethanol, 10 ml water, 10 ml 0.5 M sodium bicarbonate, and a further 10 ml water prior to use.
13. Potassium carbonate purchased from Sigma-Aldrich (part no. 209619) and used as received. Sterile water purchased from Hospira (part no. 0409-4887-50) and used as received.
14. Kryptofix 222 was purchased from Acros (part no. 29195-0010) and used as received. Anhydrous acetonitrile was purchased from Acros (part no. 61096-1000) and used as received.
15. (−)-$[^{18}F]$FLBT precursor and reference standard were purchased from ABX and used as received.
16. Anhydrous DMSO was purchased from Acros (part no. 61042-10000) and used as received. We notice a gradual decline in yield throughout the lifetime of a bottle of DMSO and stress the importance of using anhydrous solvents in this synthesis.
17. Sodium hydroxide 97% was purchased from Sigma-Aldrich (part no. 221465) and used as received.
18. Semi-preparative HPLC column: Phenomenex Luna C18(2), 10 μ, 250 × 10 mm (part no. 00G-4253-NO); flow rate = 4 ml/min; $\lambda = 254$ nm; semipreparative column was equilibrated with 200–300 ml of mobile phase prior to synthesis.
19. 0.9% Sodium Chloride, USP, was purchased from Hospira (part no. 0409-4888-50) and used as received.
20. Millex-GV sterile 0.22 μm filters were purchased from Millipore (part no. SLGV013SL) and used as received.
21. 10 ml sterile vials were purchased from Hollister-Stier (part no. 7515ZA) and vented with a sterile Millex-FG vent filter (part no. SLFG025LS).
22. US Pharmacopeia <823>. Radiopharmaceuticals for positron emission tomography-compounding. US Patent 32–NF 27. 2009.
23. Shimadzu Corporation, United States.
24. Bioscan, Inc., United States.
25. Analytical HPLC column was purchased from Phenomenex (part no. 00F-4336-E0); flow rate = 1.0 ml/min; $\lambda = 254$ nm; analytical column was equilibrated with mobile phase for 30–60 min prior to use.
26. Analytical mobile phase: 50% MeCN, 50% water + 0.1%acetic acid.
27. International Conference on Harmonisation of Technical Requirements for Registration of Pharmaceuticals for Human Use (ICH).

28. EMD Chemicals, Inc., United States (part no. 9578-3).
29. B. H. Mock, W. Winkle, M. T. Vavrek, *Nucl Med Biol*, 1997, 24, 193.
30. Capintec, Inc., United States.
31. Charles River Laboratories, United States.
32. Becton, Dickinson and Company, United States.
33. US Pharmacopeia <71>. Sterility tests. US Patent 32–NF 27. 2009.

CHAPTER 2

SYNTHESIS OF [^{18}F]-(−)FLUOROETHOXY BENZOVESAMICOL ([^{18}F]FEOBV)

BRIAN G. HOCKLEY, MEGAN N. STEWART, AND PETER J. H. SCOTT

Department of Radiology, University of Michigan School of Medicine, Ann Arbor, Michigan, USA

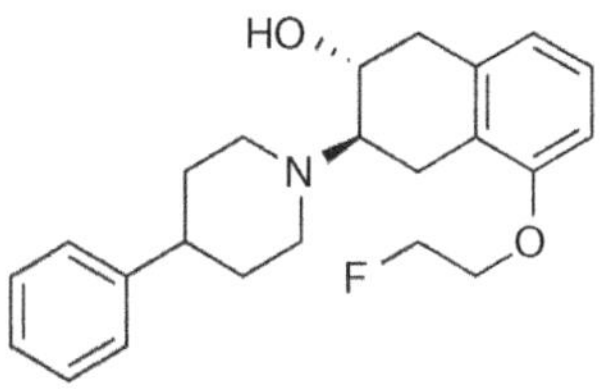

[^{18}F]FEOBV

1 INTRODUCTION

For more than a decade, [^{18}F]fluoroethoxy benzovesamicol ([^{18}F]FEOBV) has been considered a promising diagnostic PET radiopharmaceutical for imaging and quantification of the cholinergic system in normal aging, neurodegenerative disease, and cases of dementia. [^{18}F]FEOBV binds to the vesicular acetylcholine transporter protein in the presynaptic nerve terminal and has been shown to be sensitive to the subtle decreases in cholinergic innervation typical of early-stage Alzheimer's disease (AD) [1, 2]. Additionally, cholinergic imaging using [^{18}F]FEOBV was recently proposed to investigate mobility issues associated with Parkinson's disease (PD) [3]. Preclinical primate imaging, rodent imaging, rodent biodistribution, and postmortem analyses all strongly support the use of [^{18}F]FEOBV as a cholinergic imaging agent [2, 4–8]. The first human imaging studies using [^{18}F]FEOBV were conducted at the University of Michigan to demonstrate the tracer biodistribution, kinetics, and general imaging characteristics in healthy subjects. These studies found that [^{18}F]FEOBV imaging in humans was consistent

Radiochemical Syntheses, Volume 2: Further Radiopharmaceuticals for Positron Emission Tomography and New Strategies for Their Production, First Edition. Edited by Peter J. H. Scott.

with previous preclinical studies and that [^{18}F]FEOBV was well tolerated by all subjects, with no adverse events reported [9, 10].

2 SYNTHESIS PROCEDURES

CAUTION: All radiochemical syntheses must be carried out using appropriate equipment in a facility authorized for the use of radioactive materials. Personal protective equipment must be worn and all local radiation safety laws followed.

2.1 Production of [^{18}F]Fluoride

[^{18}O]H_2O (1.5 ml) [11] was loaded into the [^{18}F]fluoride target [12] of a General Electric Medical Systems (GEMS) PETtrace cyclotron [13]. The target was bombarded (60 μA beam for 30 min) to generate approximately 1.5 Ci (55.5 GBq) of [^{18}F]fluoride by the $^{18}O(p,n)^{18}F$ nuclear reaction.

2.2 Azeotropic Drying of [^{18}F]Fluoride

The [^{18}F]fluoride was delivered to a GEMS TRACERlab FX_{FN} synthesis module [14] as a solution in [^{18}O]H_2O (1.5 ml). This solution was passed through a Sep-Pak® QMA-Light cartridge [15] to trap the [^{18}F]fluoride and recycle the [^{18}O]H_2O. The [^{18}F]fluoride was then eluted into the TRACERlab FX_{FN} reaction vessel using a solution of aqueous potassium carbonate (3.5 mg in 0.5 ml H_2O) [16]. A solution of Kryptofix 222 (15 mg in 1 ml MeCN) [17] was added and the reaction mixture was azeotropically dried, initially at 80°C under vacuum for 4 min and subsequently at 60°C with both vacuum and argon flow for an additional 4 min.

2.3 Synthesis of [^{18}F]FEOBV

A solution of FEOBV precursor [18] (0.2–0.5 mg) in anhydrous dimethyl sulfoxide (DMSO) [19] (0.5 ml) was added to the dried [^{18}F]fluoride/K222 mixture, and the reaction was heated to 120°C with stirring for 10 min to prepare [^{18}F]FEOBV (Fig. 1). After this time, the reaction was cooled to 40–50°C, diluted with semi-preparative HPLC mobile phase, and transferred to the HPLC column.

2.4 Purification and Formulation of [^{18}F]FEOBV

After reaction, the crude reaction mixture was purified by semipreparative HPLC [20, 21], and a representative HPLC trace is provided in Fig. 2.

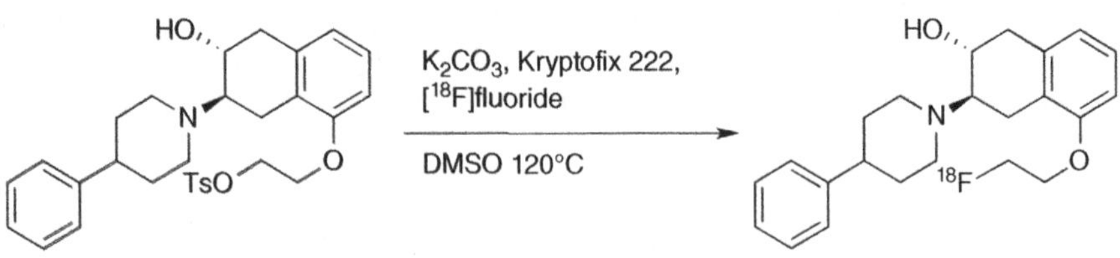

FIGURE 1 Synthesis of [^{18}F]FEOBV.

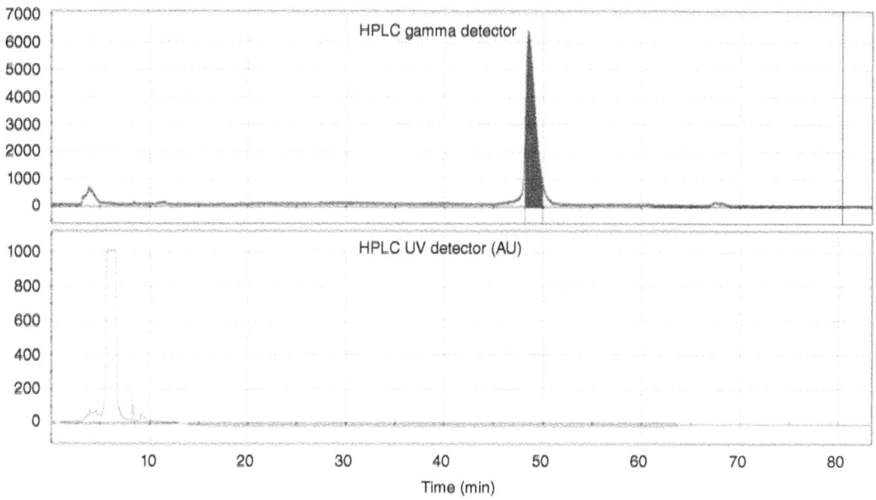

FIGURE 2 Semipreparative UV and radioactive HPLC traces for [^{18}F]FEOBV.

The fraction corresponding to [^{18}F]FEOBV (typically eluting between 20 and 25 min) was collected into a vial precharged with 50 ml Milli-Q water. This diluted solution was passed through a Sep-Pak C18 [22]. The Sep-Pak was washed with 10–12 ml of water; the product was then eluted from the Sep-Pak with 0.5 ml of 200 proof ethanol and diluted with 9.5 ml of normal saline, USP. The final formulation (10 ml) was then passed through a sterile filter [23] into a sterile vial [24] to provide [^{18}F]FEOBV (~130 mCi (4.8 GBq)) in an isotonic solution that was submitted for quality control (QC) testing. After synthesis was complete, the semipreparative HPLC column was flushed with 70% ethanol.

3 QC PROCEDURES

CAUTION: All radiochemicals produced for clinical use must have local regulatory approval (e.g., FDA, EMEA, MHRA, PFSB, etc.) prior to human use. QC procedures must be carried out by trained personnel, and each dose must meet all established QC criteria before release to the clinic.

QC procedures for [^{18}F]FEOBV, based upon the current requirements for research use radiopharmaceuticals outlined in the US Pharmacopeia [25], are summarized in the following text. Complete QC data for three repeat batches of [^{18}F]FEOBV produced using the method disclosed herein are summarized in Table 1. Each of the three doses met all of the established QC criteria.

3.1 Visual Inspection

The [^{18}F]FEOBV dose is examined behind a PET L-block and must be clear, colorless, and free of particulate matter.

TABLE 1 QC Data for Three Repeat Runs of [^{18}F]FEOBV

QC Test	Release Criteria	Run 1	Run 2	Run 3
Yield/mCi (GBq)	N/A	84.5 (3.1)	126 (4.66)	180 (6.66)
Visual inspection	Clear, colorless	Clear, colorless	Clear, colorless	Clear, colorless
Radiochemical identity	RRT=0.9–1.1	1.01	1.01	1.01
Radiochemical purity	≥95%	99	99	98
Specific activity	No limit established	2410 Ci/μmol	2905 Ci/μmol	6505 Ci/μmol
Residual solvent analysis	DMSO <5000 ppm	Pass	Pass	Pass
	Acetone <5000 ppm	Pass	Pass	Pass
	Acetonitrile <410 ppm	Pass	Pass	Pass
Dose pH	4.5–7.5	5	5	5
Residual Kryptofix [2.2.2]	≤50 μg/ml	≤10 μg/ml	≤10 μg/ml	≤10 μg/ml
Sterile filter integrity test	>50 psi	Pass	Pass	Pass
Radionuclidic identity ($t_{1/2}$)	105–115 min	110	110	112
Endotoxin analysis	≤17.5 EU/ml	≤2 EU/ml	≤2 EU/ml	≤2 EU/ml
Sterility testing	No colony growth out to 14 days	Pass	Pass	Pass

3.2 Radiochemical Identity

HPLC analysis for radiochemical identity was conducted using a Shimadzu LC-2010A$_{HT}$ Liquid Chromatograph [26] fitted with UV detectors and Bioscan γ-detectors [27], a Phenomenex Gemini 5 μ C18 (2 × 50 mm) column [28] at a flow rate of 0.8 ml/min; mobile phase, 40% acetonitrile, 60% 0.2% aqueous diethylamine [29], $\lambda = 260$ nm. The retention time of [^{18}F]FEOBV is compared to that of the [^{19}F]FEOBV reference standard [18] and must be ±10% (relative retention time (RRT) must be 0.9–1.1).

3.3 Radiochemical Purity

HPLC analysis of radiochemical purity was conducted using the HPLC system described in Section 3.2. Radiochemical purity must be >90%. Representative analytical HPLC traces are displayed in Fig. 3.

3.4 Specific Activity

A sample of the product is analyzed using the chromatographic conditions detailed in Section 3.2. The area of the product UV peak is measured to calculate the total amount of FEOBV in the drug product. The amount of radioactivity (Ci) from [^{18}F]FEOBV is then divided by the total amount (mmol) of FEOBV to determine the specific activity of the drug product.

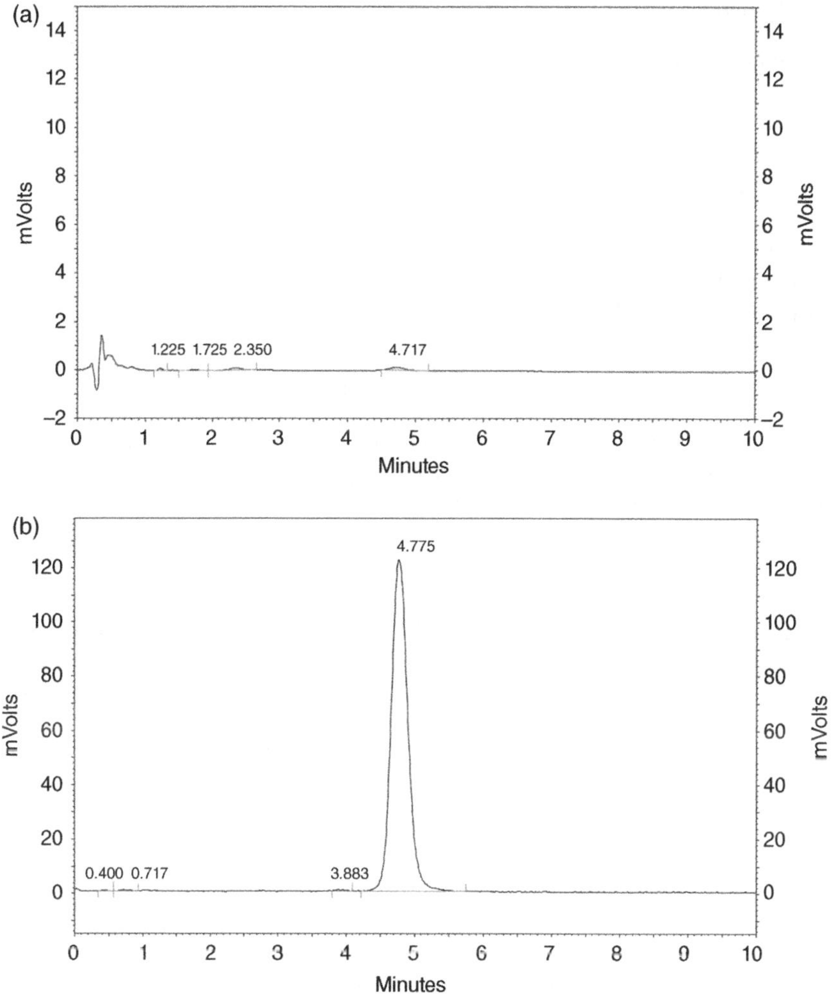

FIGURE 3 Analytical UV (a) and radioactive (b) HPLC trace for [^{18}F]FEOBV.

3.5 Residual Solvent Analysis

Analysis of residual solvent levels (acetone (from the TRACERlab FX_{FN} drying cycle), DMSO, and acetonitrile) in [^{18}F]FEOBV doses was performed using a Shimadzu GC-2010 gas chromatograph (GC) equipped with a split/splitless inlet and flame ionization detector and a Restek Stabilwax GC column (35 m × 0.25 mm) using conditions outlined in the following text. Limits imposed by the ICH Harmonised Tripartite Guidelines are <410 ppm/day for acetonitrile and <5000 ppm/day for class 3 solvents (DMSO, acetone).

GC Conditions

Oven temperature profile:

1. Hold at 70°C for 2 min.
2. Increase 70–180°C at 40°C/min.
3. Hold at 180°C for 5 min.

Inlet temperature: 180°C
Column linear velocity: 30.2 cm/s
FID temperature: 250°C

3.6 Dose pH

The pH of the [^{18}F]FEOBV dose was analyzed by applying a small amount of the dose to colorpHast® pH 2.0–9.0 nonbleeding pH-indicator strips [30] and determined by visual comparison to the scale provided. Dose pH must be 4.5–7.5.

3.7 Residual Kryptofix 222 Analysis

Residual Kryptofix 222 levels in [^{18}F]FEOBV doses were analyzed using the established spot test [31]. Strips of plastic-backed silica gel TLC plates saturated with iodoplatinate reagent were spotted with water (negative control), 50 µg/ml Kryptofix [2.2.2] standard (positive control), and [^{18}F]FEOBV dose. If Kryptofix 222 is present in a sample, a blue-black spot appears. Spots for the three samples were compared, and a visual determination of residual Kryptofix 222 in the [^{18}F]FEOBV dose was made. Less than 50 µg/mL is acceptable.

3.8 Sterile Filter Integrity Test

The sterile filter from the [^{18}F]FEOBV (with needle still attached) was connected to a nitrogen supply via a regulator. The needle was submerged in water and the nitrogen pressure was gradually increased. If the pressure was raised above the filter acceptance pressure (50 psi) without seeing a stream of bubbles, the filter was considered intact. If a stream of bubble occurs <50 psi, the test fails.

3.9 Radionuclidic Identity

Activities were measured using a Capintec CRC®-15R Radioisotope Dose Calibrator [32], and half-life was calculated using Equation 1. Calculated half-life must be 105–115 min:

$$T_{1/2} = -\ln 2\left(\frac{\text{time difference}}{\left(\ln\left(\text{ending activity} / \text{starting activity}\right)\right)}\right) \tag{1}$$

3.10 Endotoxin Analysis

Endotoxin content in doses of [^{18}F]FEOBV was analyzed by a Charles River Laboratories Endosafe® Portable Testing System [33] and according to the US Pharmacopeia (USP) [34]. Doses must contain ≤17.5 endotoxin units (EU)/ml.

3.11 Sterility Testing

Samples of [^{18}F]FEOBV were placed onto fluid thioglycolate media (FTM) tubes and soybean-casein digest agar media (SCDM) tubes. FTM is used to test for anaerobes, aerobes, and microaerophiles, while SCDM is used to test for nonfastidious and fastidious microorganisms. [^{18}F]FEOBV tubes were incubated along with positive and negative controls for 14 days. FTM tubes were incubated at 32°C and SCDM tubes were incubated at 22°C according to current USP ⟨71⟩. Tubes were visually inspected on the 3rd, 8th, and 14th days of the test period and compared to the positive and negative standards. Positive standards must show growth (turbidity) in the tubes, and [^{18}F]FEOBV/negative controls must have no culture growth after 14 days to be indicative of sterility.

WASTE DISPOSAL INFORMATION

All hazardous chemicals and toxic materials were disposed of according to Prudent Practices in the Laboratory (Washington, D.C.: National Academy Press, 1995).

CHEMICAL ABSTRACTS NOMENCLATURE (REGISTRY NUMBER)

Acetonitrile (75-05-8)
Ammonium acetate (631-61-8)
Carbonic acid, potassium salt (1:2) (584-08-7)
DMSO (67-68-5)
Ethanol (64-17-5)
4,7,13,16,21,24-Hexaoxa-1,10-diazabicyclo[8.8.8]hexacosane (23978-09-8)
Methane, 1,1′-sulfinylbis-(67-68-5)
(2*R*,3*R*)-2-Napthalenol-5-[2-fluoro-^{18}F)ethoxy]-1,2,3,4-tetrahydro-3-(4-phenyl-1-piperidinyl) (153215-75-9)

REFERENCES AND NOTES

For detailed supplier information, see Appendix 1.

1. M. R. Kilbourn, B. G. Hockley, L. Lee, P. Sherman, C. Quesada, K. A. Frey, R. A. Koeppe, *Nucl Med Biol*, 2009, 36, 489.
2. M. Parent, M. A. Bedard, A. Aliaga, J. P. Soucy, E. Landry St-Pierre, M. Cyr, A. Kostikov, E. Schirrmacher, G. Massarweh, P. Rosa-Neto, *Neuroimage*, 2012, 62, 555.
3. N. I. Bohnen, J. Klaus, *Mov Disord*, 2013, 28, 1492.
4. M. Parent, M. Cyr, A. Aliaga, A. Kostikov, E. Schirrmacher, J. P. Soucy, N. Mechawar, P. Rosa-Neto, M. A. Bedard, *EJNMMI Res*, 2013, 3, 70.

5. N. Giboureau, P. Emond, R. R. Fulton, D. J. Henderson, S. Chalon, L. Garreau, P. Roselt, S. Eberl, S. Mavel, S. Bodard, M. J. Fulham, D. Guilloteau, M. Kassiou, *Synapse*, 2007, 61, 962.
6. N. Giboureau, I. M. Som, A. Boucher-Arnold, D. Guilloteau, M. Kassiou, *Curr Top Med Chem*, 2010, 10, 1569.
7. G. K. Mulholland, D. M. Wieland, M. R. Kilbourn, K. A. Frey, P. S. Sherman, J. E. Carey, D. E. Kuhl, *Synapse*, 1998, 30, 263.
8. T. R. Degrado, G. K. Mulholland, D. M. Wieland, M. Schwaiger, *Nucl Med Biol*, 1994, 21, 189.
9. M. Petrou, K. A. Frey, M. R. Kilbourn, P. J. H. Scott, D. M. Raffel, N. I. Bohnen, M. L. T. M. Müller, R. L. Albin, *R A Koeppe J Nucl Med*, 2014, 55, 396.
10. M. Petrou, R. A. Koeppe, P. J. H. Scott, N. I. Bohnen, M. R. Kilbourn, K. A. Frey, *J Nucl Med*, 2012, 53 (Suppl. 1), 89 p.
11. Virgin [^{18}O]H_2O purchased from ABX, Amic, Rotem, or Medical Isotopes and used as received.
12. GE Silver High Yield [^{18}F]fluoride target.
13. GE Healthcare, United States.
14. GE Healthcare, United States.
15. Sep-Pak® QMA-Light cartridges were purchased from Waters (part no. WAT023525) and conditioned with 10 ml ethanol, 10 ml water, 10 ml 0.5 M sodium bicarbonate, and a further 10 ml water prior to use.
16. Potassium carbonate purchased from Aldrich (part no. 209619) and used as received. Sterile water purchased from Hospira (part no. 0409-4887-50) and used as received.
17. Kryptofix 222 was purchased from Acros (part no. 29195-0010) and used as received. Anhydrous acetonitrile was purchased from Acros (part no. 61096-1000) and used as received.
18. [^{18}F]FEOBV precursor and reference standard were prepared in-house.
19. Anhydrous DMSO was purchased from Acros (part no. 61042-10000) and used as received. We notice a gradual decline in yield throughout the lifetime of a bottle of DMSO and stress the importance of using fresh solvents in this synthesis.
20. Semipreparative HPLC mobile phase: 45% MeCN : 55% aqueous 50 mM ammonium acetate, flow rate = 4 ml/min.
21. Semipreparative HPLC column: 10×250 mm, Phenomenex Synergi Polar-RP (part no. 00G-4336-N0); semipreparative column was equilibrated with 200–300 ml of mobile phase prior to synthesis.
22. Waters, pn: WAT054955.
23. Millex-GV sterile 0.22 μm filters were purchased from Millipore (part no. SLGV013SL) and used as received.
24. 10 ml sterile vials were purchased from Hollister-Stier (part no. 7515ZA) and vented with a sterile Millex-FG vent filter (part no. SLFG025LS).
25. US Pharmacopeia <823>. Radiopharmaceuticals for positron emission tomography-compounding. US Patent 32–NF 27. 2009.
26. Shimadzu Corporation, United States.
27. Bioscan, Inc., United States.
28. Analytical HPLC column: purchased from Phenomenex (part no. 00B-4435-B0); analytical column was equilibrated with mobile phase for 30–60 min prior to use.
29. Fisher Scientific (pn: 373837). Used as received.
30. EMD Chemicals Inc., United States (part no. 9578-3).
31. B. H. Mock, W. Winkle, M. T. Vavrek, *Nucl Med Biol*, 1997, 24, 193.
32. Capintec, Inc., United States.
33. Charles River Laboratories, United States.
34. US Pharmacopeia <85>. Bacterial endotoxins test. US Patent 32–NF 27. 2009.

CHAPTER 3

SYNTHESIS OF [^{18}F]FLUOROMETHYLCHOLINE ([^{18}F]FCH) VIA [^{18}F]FLUOROMETHYL TOSYLATE

Melissa E. Rodnick, Allen F. Brooks, Brian G. Hockley, Bradford D. Henderson, and Peter J. H. Scott

Department of Radiology, University of Michigan School of Medicine, Ann Arbor, Michigan, USA

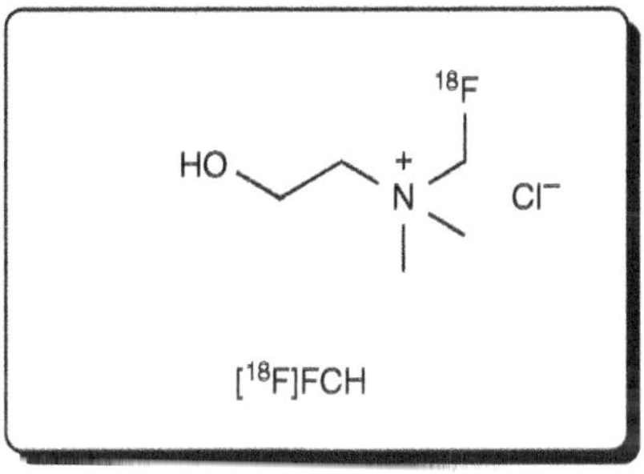

1 INTRODUCTION

Prostate cancer therapy ranges from curative therapy for localized prostate cancer to life-prolonging treatment and palliation for in the case of disseminated prostate cancer [1]. Accurate staging of prostate cancer is essential to in selecting the most appropriate treatment for a given patient. This has created an urgent need for accurate and noninvasive methods for staging of prostate cancer [2], and [^{18}F]fluoromethylcholine-PET is one such method. Primary prostate cancer cells, along with its metastases, are known to upregulate choline kinase, which, in turn, leads to an elevated uptake of choline for the biosynthesis of phospholipids. Intracellular phosphorylation traps choline inside the cell, and therefore, PET imaging with [^{18}F]fluoromethylcholine can readily detect this trapping, and differentiate prostate cancer cells from neighboring nonmalignant tissue [3–5].

Radiochemical Syntheses, Volume 2: Further Radiopharmaceuticals for Positron Emission Tomography and New Strategies for Their Production, First Edition. Edited by Peter J. H. Scott.

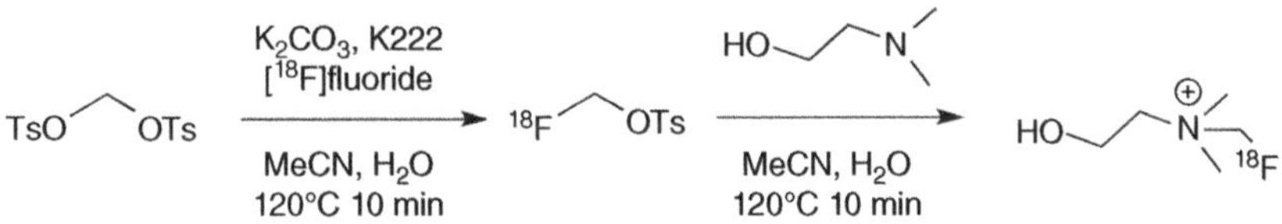

FIGURE 1 Synthesis of [18F]fluoromethylcholine.

Historically, [18F]fluoromethylcholine has been prepared by fluoromethylation of dimethylaminoethanol (DMAE) with [18F]fluorobromomethane [3–10]. However, we have only obtained low yields of product with this method in our laboratory because of compatibility issues between gaseous [18F]fluorobromomethane and the liquid phase synthesis modules we employ for most routine radiosyntheses of fluorine-18-labeled radiotracers. We therefore desired a liquid-phase method of preparing [18F]fluoromethylcholine that is complementary to the gas-phase method reported in Volume 1 of this series [7], so that radiochemists can select the method that is most compatible with their own synthesis equipment. We recently developed a fully automated one-pot liquid-phase synthesis of [18F]fluoromethylcholine (Fig. 1) via *in situ* formation of [18F]fluoromethyl tosylate [11] and present an expanded discussion of the synthesis herein.

2 SYNTHESIS PROCEDURES

CAUTION: All radiochemical syntheses must be carried out using appropriate equipment in a facility authorized for the use of radioactive materials. Personal protective equipment must be worn, and all local radiation safety laws followed.

2.1 Production of [18F]Fluoride

[18O]H_2O (2 ml) [12] was loaded into the [18F]fluoride target [13] of a General Electric Medical Systems (GEMS) PETtrace cyclotron [14]. The target was bombarded (60 μA beam for 30 min) to generate approximately 1.5 Ci (55.5 GBq) of [18F]fluoride by the $^{18}O(p,n)^{18}F$ nuclear reaction.

2.2 Azeotropic Drying of [18F]Fluoride

The [18F]fluoride was delivered to a GEMS TRACERlab FX_{FN} synthesis module [14] as a solution in [18O]H_2O (2 ml). This solution was passed through a Sep-Pak® QMA-Light cartridge [15] to trap the [18F]fluoride and recycle the [18O]H_2O. The [18F]fluoride was then eluted into the TRACERlab FX_{FN} glassy carbon reaction vessel using a solution of aqueous potassium carbonate (3.5 mg in 0.5 ml H_2O) [16]. A solution of Kryptofix 222 (5 mg in 1 ml MeCN) [17] was added and the reaction mixture was azeotropically dried, initially at 80°C under vacuum for 4 min and subsequently at 60°C with both vacuum and argon flow for an additional 4 min.

2.3 Synthesis of [^{18}F]Fluoromethylcholine

A solution of methylene ditosylate (7–8 mg) [18] in anhydrous MeCN (750 µl) and sterile water (10 µl) was added to the dried [^{18}F]fluoride, and the reaction was heated to 120°C with stirring for 10 min. Subsequently, the reaction mixture was cooled to 50°C, followed by the addition of DMAE (40 µl in 350 µl MeCN), which was heated to 120°C with stirring for an additional 10 min. The reaction mixture was then cooled to 60°C and underwent evaporation of the reaction solvent by maintaining 60°C and subjecting the reaction to both a continuous argon stream and vacuum draw for 5 min.

2.4 Purification and Formulation of [^{18}F]Fluoromethylcholine

Sterile water (5.5 ml) was added to the dried reaction mixture and passed through the Sep-Pak C18 Plus [19] into the round bottom flask containing ethanol (10 ml) to trap unreacted ditosylmethane and [^{18}F]fluoromethyl tosylate, as well as any tosylmethylcholine generated as a by-product. This was repeated with an additional aliquot of sterile water (5.5 ml). The water/ethanol mixture was transferred through the Sep-Pak CM-Light [20] to trap the desired [^{18}F]fluoromethylcholine. The Sep-Pak CM-Light was washed with ethanol (15 ml) to remove unreacted DMAE and water (20 ml) to remove residual ethanol to waste. Subsequently, [^{18}F]fluoromethylcholine was eluted off into a collection vial containing 0.9% sodium chloride for injection, USP (7 ml) with 0.9% sodium chloride for injection, USP (3 ml). The final formulation (10 ml) was then passed through a 0.22 µm sterile filter [21] into a sterile dose vial [22] to provide doses of [^{18}F]fluoromethylcholine as an isotonic solution submitted for quality control (QC) testing.

3 QUALITY CONTROL

CAUTION: All radiopharmaceuticals produced for clinical use must have local regulatory approval before use in humans. Trained personnel must carry out QC procedures, and each dose must meet all established QC criteria before release to the clinic.

The QC of radiopharmaceuticals prepared at the University of Michigan is carried out in accordance with the US Pharmacopeia (USP) [23], which is summarized in the following text. QC data for three repeat batches of [^{18}F]fluoromethylcholine produced using the method disclosed herein are summarized in Table 1. Each of the three doses met all of the established QC criteria.

3.1 Visual Inspection

Doses were examined visually and had to be clear, colorless, and free of particulate matter.

TABLE 1 Quality Control Data for [^{18}F]FCH

QC Test	Release Criteria	Run 1	Run 2	Run 3
Yield/mCi (GBq)	N/A	102 (3.8)	81 (3.0)	86 (3.2)
Visual inspection	Clear, colorless	Clear, colorless	Clear, colorless	Clear, colorless
Radiochemical identity	RRT=0.9–1.1	1.06	1.03	1.03
FCH concentration	No limit established	0.10 µg/ml	0.10 µg/ml	0.07 µg/ml
Radiochemical purity	≥95%	100	99.9	99.2
Specific activity	No limit established	15.5 Ci/µmol	13.2 Ci/µmol	19.7 Ci/µmol
Residual solvent analysis	DMAE <20 ppm	Pass	Pass	Pass
	Ethanol <5000 ppm	Pass	Pass	Pass
	Acetonitrile <410 ppm	Pass	Pass	Pass
	Acetone <5000 ppm	Pass	Pass	Pass
Dose pH	4.5–7.5	5.0	5.0	5.0
Residual Kryptofix 222	≤50 µg/ml	≤50 µg/ml	≤50 µg/ml	≤50 µg/ml
Sterile filter integrity test	>50 psi	>50 psi	>50 psi	>50 psi
Radionuclidic identity ($t_{1/2}$)	105–115 min	115	106	110
Endotoxin analysis	≤17.5 EU/ml	≤2 EU/ml	≤2 EU/ml	≤2 EU/ml
Sterility testing	No colony growth out to 14 days	Pass	Pass	Pass

3.2 Dose pH

The pH of the doses was analyzed by applying a small amount of the dose to colorpHast pH 2.0–9.0 nonbleeding pH-indicator strips [24] and determined by visual comparison with the scale provided.

3.3 Radionuclidic Identity

Activities were measured using a Capintec CRC®-15R Radioisotope Dose Calibrator [25], and half-life was calculated using Equation 1. Calculated half-life must be 105–115 min:

$$T_{1/2} = -\ln 2\left(\frac{\text{time difference}}{\ln\,(\text{ending activity/starting activity})}\right) \quad (1)$$

3.4 Chemical and Radiochemical Purity

The radiochemical purity (RCP) and concentration of fluoromethylcholine and DMAE in each batch were determined using a Shimadzu HPLC system with the following components: SCL-10Avp system controller, DGU-14A in-line degassing unit, LC-10ADvp pump, CDD-10Avp conductivity detector with temperature-controlled cell, CTO-20A oven [26], and Bioscan FC3300 flow count radioactivity detector [27]. The system was not equipped with an ion suppressor: Column, Waters SCX column, IC-Pak™ Cation M/D, 3.9 × 150 mm, pn WAT036570; mobile phase, 5 mM HCl (Fisher Scientific); flow rate, 1.25 ml/min; RT ~ 6.5 min; a representative HPLC trace is shown in Fig. 2. RCP must be >95%, and residual DMAE must be <20 μg/ml.

3.5 Radiochemical Identity

Using the HPLC system described in Section 3.4, the retention time of [^{18}F]fluorocholine is compared to that of the [^{19}F]fluorocholine-unlabeled reference standard [28] and must be ±10% (relative retention time (RRT) must be 0.9–1.1):

$$\mathrm{RRT} = \frac{t_{\mathrm{r}}\ ([^{18}\mathrm{F}]\mathrm{FCH})}{t_{\mathrm{r}}\ ([^{19}\mathrm{F}]\mathrm{FCH})}$$

3.6 Residual Kryptofix 222

Residual Kryptofix 222 (K222) levels in [^{18}F]FCH doses were analyzed using the established spot test [29]. Strips of plastic-backed silica gel TLC plates saturated with iodoplatinate reagent were spotted with water (negative control), 50 mg/ml Kryptofix 222 standard (positive control), and [^{18}F]FCH dose. If K222 was present in a sample, a blue-black spot appeared. Spots for the three samples were compared and a visual determination of residual K222 in the dose was made. Less than 50 μg/ml is acceptable, and all doses of fluorocholine prepared in this study were found to contain residual K222 below this level.

3.7 Sterile Filter Integrity Test

Sterile filters from doses (with needle still attached) were connected to a nitrogen supply via a regulator. The needle was then submerged in water and the nitrogen pressure gradually increased. If the pressure was raised above the filter acceptance pressure without seeing a stream of bubbles, the filter was considered intact.

3.8 Bacterial Endotoxins

Endotoxin content in radiopharmaceutical doses was analyzed using a Charles River Laboratories Endosafe® Portable Testing System [30] and according to the USP [31]. Doses must contain <175 endotoxin units.

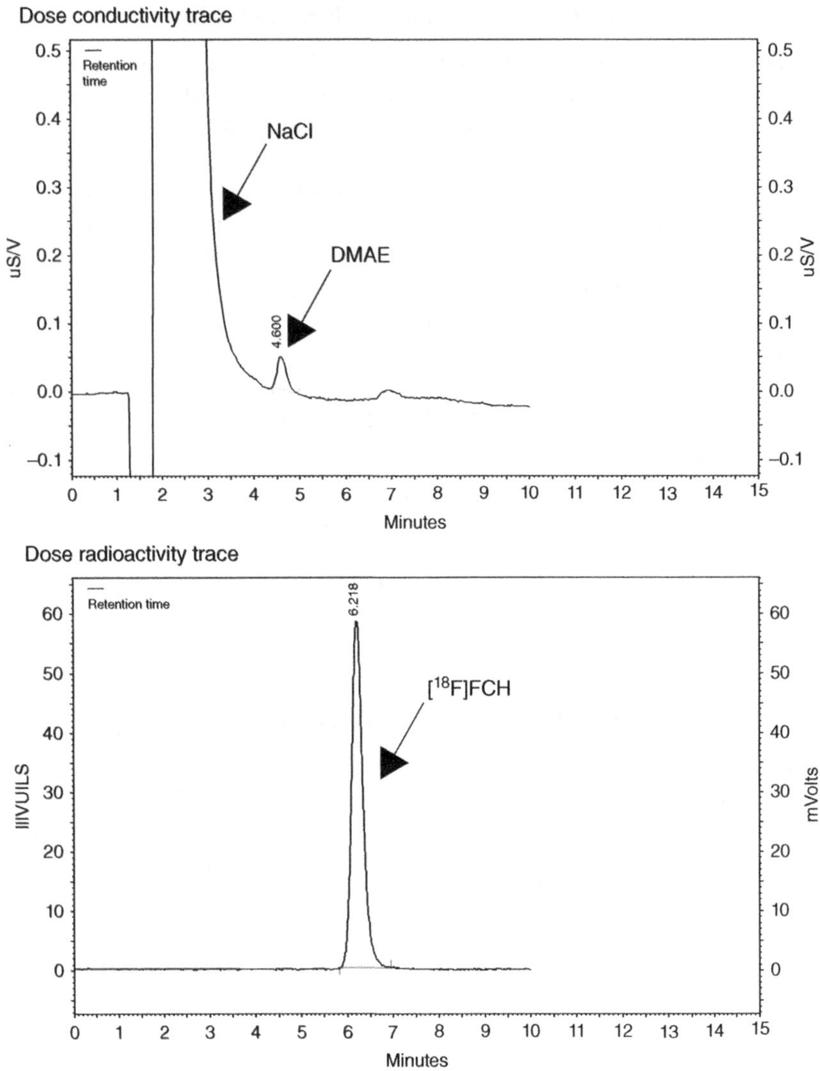

FIGURE 2 Analytical HPLC tracers for [^{18}F]fluoromethylcholine.

3.9 Sterility

Culture tubes of fluid thioglycolate media (FTM) and soybean-casein digest agar media (SCDM) [32] were inoculated with samples of [^{18}F]FCH doses and incubated (along with positive and negative controls) for 14 days. FTM tubes were incubated at 32°C and SCDM tubes were incubated at 22°C according to current USP guidelines [33]. FTM is used to test for anaerobes, aerobes, and microaerophiles, while SCDM is used to test for nonfastidious and fastidious microorganisms. Culture tubes were visually inspected on the 3rd, 8th, and

14th days of the test period and compared with the positive and negative standards. Positive standards must show growth (turbidity) in the tubes, and dose/negative controls must have no culture growth after 14 days to be indicative of sterility.

WASTE DISPOSAL INFORMATION

All hazardous chemicals and toxic materials were disposed of according to Prudent Practices in the Laboratory (Washington, D.C.: National Academy Press, 1995).

CHEMICAL ABSTRACTS NOMENCLATURE (REGISTRY NUMBER)

Acetonitrile (75-05-8)
Carbonic acid, potassium salt (1:2) (584-08-7)
2-Dimethylaminoethanol (108-01-0)
Ethanaminium, *N*-(fluoromethyl)-2-hydroxy-*N,N*-dimethyl-, chloride (459424-38-5)
Ethanol (64-17-5)
4,7,13,16,21,24-Hexaoxa-1,10-diazabicyclo[8.8.8]hexacosane (23978-09-8)
Hydrochloric acid (7647-01-0)
Methanol, 1-(fluoro-^{18}F)-,1-(4-methylbenzenesulfonate) (113426-16-7)
Methylene ditosylate (24124-59-2)

REFERENCES AND NOTES

For detailed supplier information, see Appendix 1.

1. M. H. Poulsen, K. Bouchelouche, P. F. Høilund-Carlsen, H. Petersen, O. Gerke, S. I. Steffansen, N. Marcussen, N. Svolgaard, W. Vach, U. Geertsen, S. Walter, *BJU Int*, 2012, 110, 1666.
2. A. Heidenreich, J. Bellmunt, M. Bolla, S. Joniau, M. Mason, V. Matveev, N. Mottet, H. P. Schmid, T. van der Kwast, T. Wiegel, F. Zattoni, *Eur Urol*, 2011, 59, 61.
3. T. R. DeGrado, S. W. Baldwin, S. Wang, M. D. Orr, R. P. Liao, H. S. Friedman, R. Reiman, D. T. Price, R. E. Coleman, *J Nucl Med*, 2001, 42, 1805–1814.
4. T. R. DeGrado, R. E. Coleman, S. W. Baldwin, D. T. Price, M. D. Orr, S. Wang. ^{18}F-Labeled Choline Analogs, WO 2001/82864 A2, 2001. 74 pp.
5. T. R. DeGrado, R. E. Coleman, S. Wang, S. W. Baldwin, M. D. Orr, C. N. Robertson, T. J. Polascik, D. T. Price, *Cancer Res*, 2001, 61, 110.
6. D. Kryza, V. Tadino, M. A. Filannino, G. Villeret, L. Lemoucheux, *Nucl Med Biol*, 2008, 35, 255.
7. D. Kryza, Synthesis of [^{18}F]fluorocholine ([^{18}F]FCH), in *Radiochemical Syntheses Volume 1: Radiopharmaceuticals for Positron Emission Tomography* P. J. H. Scott and B. G. Hockley, editors, John Wiley and Sons, Hoboken, NJ, 2012.
8. M. Nader, A. Hoepping, *Nuklearmedizin*, 2005, 44 (A192), 7.
9. A. Sperandeo, A. Cistaro, N. Paligoric, D. Busetta, R. Mangiapane, U. Ficola, *J Nucl Med*, 2009, 50 (Suppl. 2), 28.
10. X. Shao, B. G. Hockley, R. Hoareau, P. L. Schnau, P. J. H. Scott, *Appl Rad Isotop* 2011, 69, 403.
11. M. E. Rodnick, A. F. Brooks, B. G. Hockley, B. D. Henderson, P. J. H. Scott, *Appl Rad Isotop*, 2013, 78, 26.
12. Virgin [^{18}O]H_2O purchased from ABX or Rotem and used as received.
13. GEMS silver high-yield [^{18}F]fluoride target.

14. GE Healthcare, United States.
15. Sep-Pak® QMA-Light cartridges were purchased from Waters (part no. WAT023525) and conditioned with 10 ml ethanol, 10 ml water, 10 ml 0.5 M sodium bicarbonate, and a further 10 ml water prior to use.
16. Potassium carbonate purchased from Aldrich (part no. 209619) and used as received. Sterile water purchased from Hospira (part no. 0409-4887-50) and used as received.
17. Kryptofix 222 was purchased from Acros (part no. 29195-0010) and used as received. Anhydrous acetonitrile was purchased from Acros (part no. 61096-1000) and used as received.
18. Ditosylmethane can either be purchased from ABX advanced biochemicals (part no. 6177.0010) or prepared in-house as follows: silver *p*-toluenesulfonate (0.52 g, 1.87 mmol) was suspended in MeCN (4 ml) in an oven-dried flask. To this suspension, diiodomethane (0.06 ml, 0.75 mmol) was added dropwise, and the reaction was stirred at reflux for 16 h. After this time, the reaction was cooled to room temperature and passed through a sintered glass frit to remove the silver salt side product. The filtrate was collected and concentrated under vacuum, and the resulting residue was purified by flash chromatography (eluting with hexane/ethyl acetate, 4:1) to give 0.27 g (83% yield) of ditosylmethane as a crystalline white solid: R_f 0.3 (4:1, hexane/ethyl acetate); ^{1}H NMR (400 MHz, DMSO-d_6) δ 7.58 (d, J ¼ 8, 4H), 7.38 (d, J ¼ 8, 4H), 5.90 (s, 2H), 2.40 (s, 6H); MS (ESI): m/z 379 $(M+Na)^+$.
19. Sep-Pak C18 Plus was purchased from Waters (part no. WAT020515).
20. Sep-Pak CM-Light was purchased from Waters (part no. WAT023531).
21. Millex-GV sterile 0.22 μm filters were purchased from Millipore (part no. SLGV013SL) and used as received.
22. 10 ml sterile vials were purchased from Hollister-Stier (part no. 7515ZA) and vented with a sterile Millex-FG vent filter (part no. SLFG025LS).
23. US Pharmacopeia <823>. Radiopharmaceuticals for positron emission tomography-compounding. US Patent 32–NF 27. 2009.
24. EMD Chemicals, Inc., United States (part no. 9578-3).
25. Capintec, Inc., United States.
26. Shimadzu Corporation, United States.
27. Bioscan, Inc., United States.
28. Unlabeled reference standard was purchased from ABX advanced biochemicals (part no. 6130).
29. B. H. Mock, W. Winkle, M. T. Vavrek, *Nucl Med Biol*, 1997, 24, 193.
30. Charles River Laboratories, USA.
31. US Pharmacopeia <85>. Bacterial endotoxins test. US Patent 32–NF 27. 2009.
32. Becton, Dickinson and Company, USA.
33. US Pharmacopeia <71>. Sterility tests. US Patent 32–NF 27. 2009.

CHAPTER 4

RADIOSYNTHESIS OF [^{18}F]FLOTEGATIDE ([^{18}F]RGD-K5)

Hartmuth C. Kolb, Fanrong Mu, Umesh Gangadharmath,
Vani P. Mocharla, Zhihong Zhu, Ashok Chaudhary,
and Joseph C. Walsh

Siemens Molecular Imaging, Culver City, California, USA

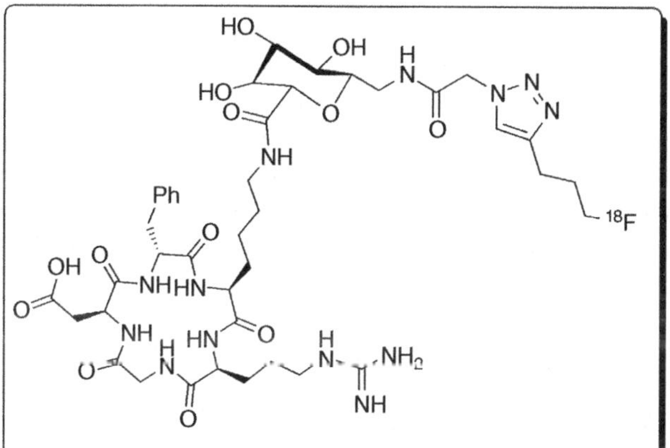

^{18}F-flotegatide (^{18}F-RGD-K5)

1 INTRODUCTION

Angiogenesis is the sprouting of neovasculature in response to the controlled growth of tissues [1]. The process of initiating and maintaining normal angiogenesis is orchestrated through different peptides and proteins, including growth factors (e.g., vascular endothelial growth factor, VEGF) and cell-surface-expressed cell adhesion proteins known as integrins. The integrins are a family of transmembrane proteins comprised of two noncovalently linked α and β subunits that promote both cell survival and cell migration through cellular adhesion and signal

Radiochemical Syntheses, Volume 2: Further Radiopharmaceuticals for Positron Emission Tomography and New Strategies for Their Production, First Edition. Edited by Peter J. H. Scott.

transduction processes [2]. In tumorigenesis, the *angiogenic switch* is believed to cause a tumor's transition from dormancy to malignancy, in part to support the metabolic demands of the tumor [3]. During tumorigenesis, the increased expression and activation of specific integrin subtypes (e.g., integrin $\alpha_v\beta_3$) facilitates both tumor invasion and metastasis [4]. Because integrin $\alpha_v\beta_3$ is not widely expressed in normal tissues yet is overexpressed on the surface of angiogenic endothelial cells in malignant tumors [5, 6], it is an attractive imaging target for both single-photon emission computed tomography (SPECT) and positron emission tomography (PET) imaging applications. The unique role of integrin $\alpha_v\beta_3$ in angiogenesis may support noninvasive diagnostic imaging of candidate patients who might respond favorably to anti-integrin or antiangiogenesis treatment while enabling antiangiogenesis therapy monitoring in patients with integrin-positive tumors [7].

Most PET tracers targeting overexpression of integrin $\alpha_v\beta_3$ *in vivo* utilize the tripeptide binding motif of arginine, glycine, and aspartate (*R–G–D*) contained within a cyclic pentapeptide core. Several integrin $\alpha_v\beta_3$ PET ligands whose binding is both specific and high have been successfully radiolabeled and imaged [8]. Examples of current integrin $\alpha_v\beta_3$ targeting tracers used to detect integrin overexpression in humans include [^{18}F]-galactosyl-RGD, a cyclic pentapeptide [7], and [^{18}F]-fluciclatide, a disulfide-linked R–G–D cyclic peptide [9]. The [^{18}F]-radiolabeling of these molecules utilizes a *traditional* approach requiring protecting groups and multiple purification steps, which can impede the efficiency of production.

As an alternate to traditional radiolabeling techniques, copper-catalyzed triazole formation (*click chemistry*) facilitates easy radiolabeling because it is rapid, high yielding, and highly selective and does not require the use of protecting groups [10–14]. Click chemistry is especially compatible for labeling highly functionalized molecules, such as peptides. Early reports have disclosed the Cu(I) catalyzed radiolabeling of thymidine analogs [15], with additional reports disclosing the click radiolabeling of various RGD peptides analogs [16–19].

We have developed an RGD-based integrin PET imaging agent, [^{18}F]-flotegatide ([^{18}F]-RGD-K5), which is an analog of [^{18}F]-galactosyl-RGD. The incorporation of a 1,2,3-triazole moiety within [^{18}F]-RGD-K5 is a by-product of click chemistry; however, the 1,2,3-triazole imparts a unique pharmacodynamic profile compared to other RGD-based imaging agents [20]. We report herein the radiosynthesis of [^{18}F]-RGD-K5 for use as a PET imaging agent for detecting integrin $\alpha_v\beta_3$ expression both *in vitro* and *in vivo*.

2 SYNTHESIS PROCEDURES

CAUTION: All radiochemical syntheses must be carried out using appropriate equipment in a facility authorized for the use of radioactive materials. Personal protective equipment must be worn and all local radiation safety laws followed.

The synthetic scheme for the production of [^{18}F]RGD-K5 is shown in Fig. 1.

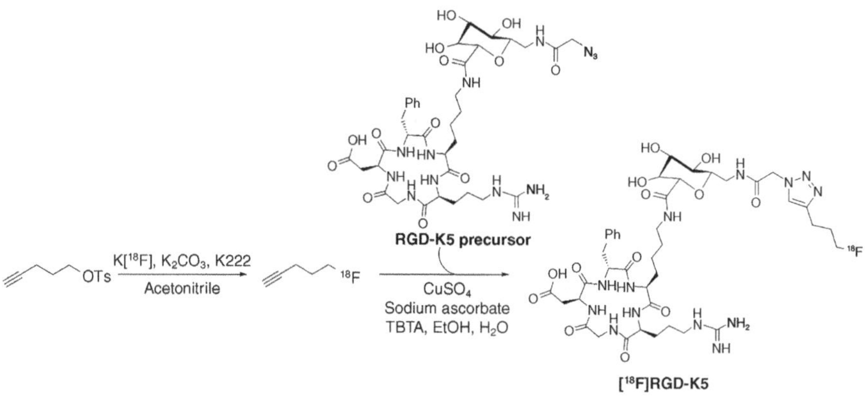

FIGURE 1 Synthesis of [^{18}F]RGD-K5.

2.1 Production of [^{18}F]Fluoride

[^{18}F]-labeling was performed using an automated synthesis module (*Explora® RN Series*) with hardware modifications (see Fig. 2) to accommodate the volatility of ^{18}F-fluoropentyne [21, 22]. The *main reaction vessel* was connected to a secondary vial using valve V11 for the purposes of performing the click chemistry reaction. Both lines connected to V11 were pneumatically controlled to raise and lower the transfer lines during the distillation process (denoted by *X* in Fig. 2). The HPLC transfer line, which connects the click chemistry vial to the HPLC load vial, was also pneumatically controlled to facilitate the transfer process. The transfer line was equipped with a pneumatic device designed to remotely raise/lower the transfer lines. Both the click and HPLC load vials were fitted with charcoal vents to capture volatile [^{18}F]-fluoropentyne [23].

[^{18}O]H_2O (2 ml) [24] is loaded into the [^{18}F]fluoride tantalum target of a Siemens RDS-111 cyclotron [25]. The target is bombarded (60 μA beam for 45–60 min) to generate approximately 1.5–2.0 Ci (56–74 GBq) of [^{18}F]fluoride by the ^{18}O(p,n)^{18}F nuclear reaction. The [^{18}F]fluoride is delivered through polyethylene, polypropylene, or PEEK tubing into a v-vial within a shielded area prior to delivery into the automated synthesis module. If high specific activity [^{18}F]RGD-K5 is required, the target and delivery lines are prerinsed with [^{18}O]H_2O (2 ml) through the target and delivery lines into a separate collection vial for recycling.

2.2 Azeotropic Drying of [^{18}F]Fluoride

Aqueous [^{18}F]fluoride ion is trapped on an anion-exchange resin cartridge [26] located on a Siemens Explora RN automated synthesis module, and the [^{18}O]H_2O is collected for recycling. The fluoride ion is then eluted into a glassy carbon reaction vessel using a solution of aqueous potassium carbonate (3 mg in 0.4 ml of water) [27, 28]. A solution of Kryptofix® 222 (20 mg in 1 ml of acetonitrile) [29, 30] is added directly into the glassy carbon reaction vessel after the elution of the [^{18}F]fluoride ion.

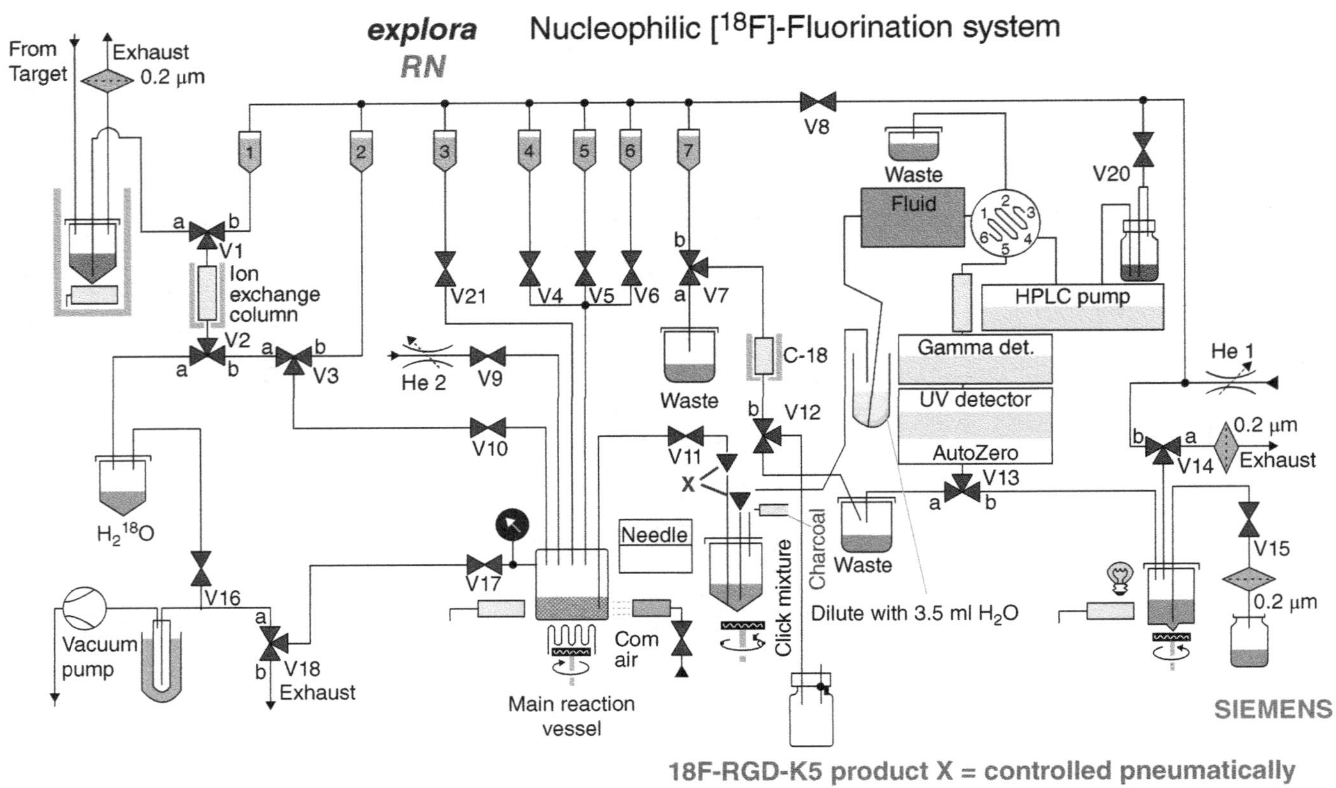

FIGURE 2 Explora® RN automated synthesis module.

The solution is concentrated to dryness to remove both the acetonitrile and water. Typically, the drying begins at 68°C for 5 min followed by 95°C for 30 s under reduced pressure [31] (250 mbar) and a stream of argon [32]. After removal of the solvents, a hard vacuum is pulled for 30 s to 2 min at 70°C until a low, steady-state pressure is obtained [33].

2.3 Synthesis of [^{18}F]RGD-K5

A solution of pent-4-ynyl 4-methylbenzenesulfonate (pentyne tosylate [34], 20 mg, 8.4 μmol) in anhydrous acetonitrile (0.8–1.0 ml) is added to the activated [^{18}F]fluoride while the reaction vessel remains at 70°C. The mixture is then heated at 110°C up to 3 min. A separate vial is prepared containing the click chemistry mixture of RGD-K5 azide precursor (4 mg) [35], $CuSO_4$ (0.1 M, 0.25 ml) [36], tris((1-benzyl-1*H*-1,2,3-triazol-4-yl)methyl)amine (TBTA) (15 mg) [37] [38], sodium ascorbate (40 mg) [39], anhydrous acetonitrile (0.25 ml), and ethanol/H_2O (2:1 ratio, 0.25 ml), which is connected to V11.

After fluorination, the transfer line in the reaction vessel was kept above the surface of the reaction mixture. Valve V11 was opened for approximately 20 s to allow the transfer of [^{18}F]-pentyne from the reaction pot into the click vial (via bubbling of the labeled pentyne into the click solution) with the HPLC transfer line held above the reaction mixture. After the transfer was complete, V11 was closed and the click reaction was allowed to proceed for 10–35 min.

2.4 Purification and Formulation of [^{18}F]RGD-K5

The crude [^{18}F]RGD-K5 solution is manually transferred to the HPLC load loop (5.0 ml) and purified by semipreparative HPLC [40] with a flow of 5.0 ml/min using the step-gradient elution profile below:

0–5 min: A, 0%; B, 100%
5–10 min: A, 5%; B, 95%
10–15 min: A, 10%; B, 90%
15 min: A, 18%; B, 82%
A: acetonitrile (0.08% 12N HCl)
B: water (0.08% 12N HCl)

The product is collected as monitored by flow-through radioactivity and UV (206 or 254 nm) detectors. Representative UV and radioactivity chromatograms from a semi-preparative HPLC purification of a crude [^{18}F]-RGD-K5 solution are shown in Fig. 3.

The desired product has a retention time of approximately 33 min. Purified [^{18}F]-RGD-K5 was collected from the column, diluted with sterile water (40 ml), and captured onto a previously activated [41] C18 Sep-Pak cartridge [42]. The C18 Sep-Pak cartridge was then washed with sterile water (10 ml) followed by elution of the product with 0.5–1.0 ml of ethanol. The sample was then diluted with sterile water (9.0–4.5 ml of water) to afford a final formulation of [^{18}F]-RGD-K5 in a maximum of 10% ethanol/water. The solution was then sterile filtered [43] into in sterile vials [44].

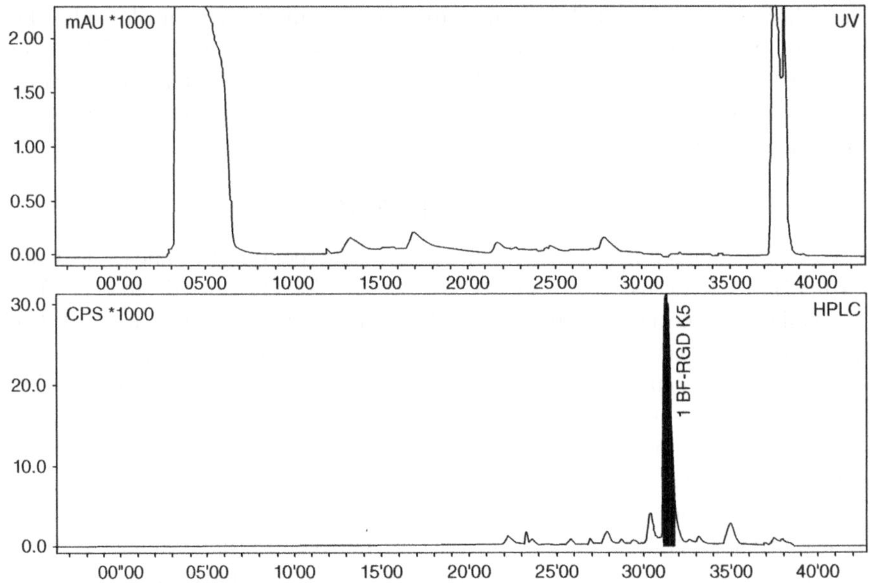

FIGURE 3 Semipreparative HPLC UV (upper) and radioactive gamma (lower) traces from the purification of a crude [^{18}F]RGD-K5 solution.

3 QUALITY CONTROL PROCEDURES

CAUTION: All radiochemicals produced for human use must comply with local regulatory requirements as required (e.g., FDA, EMEA, MHRA, PFSB, RDRC, IRB, etc.). Quality control procedures must be performed by trained personnel, and each batch must meet preestablished quality control criteria before release.

Three batches of [^{18}F]RGD-K5 were produced according to the above conditions and analyzed according to the quality control (QC) tests summarized in Table 1. Each of the three batches met all of the established QC criteria.

3.1 Radioactivity Concentration

The radioactivity concentration is determined by measuring the radioactivity of a sample of the [^{18}F]RGD-K5 final product contained in a low dead volume syringe. Our specification for radioactivity concentration requires that the product must be within the range of 5–50 mCi/ml at end of synthesis (EOS) [45]. The radioactivity concentration for all three batches was within the range of 5–50 mCi/ml at EOS.

3.2 Visual Inspection

The [^{18}F]RGD-K5 final product is visually examined through leaded glass. The solution must be clear, colorless, and free of particulate matter. All three batches met this specification.

TABLE 1 QC Specifications and Data for Three Batches of [^{18}F]RGD-K5

QC Test	Specification	Batch 1	Batch 2	Batch 3
Radioactivity concentration	5–50 mCi/ml	45.1 mCi/ml	48.3 mCi/ml	48.1 mCi/ml
Visual inspection	Clear, colorless solution, free of particulate matter	Pass	Pass	Pass
Radiochemical identity	Retention time must correspond to standard peak	Pass	Pass	Pass
Radiochemical purity	≥95%	99.9%	99.9%	99.9%
Specific activity	≥0.4 Ci/μmol	5.70 Ci/μmol	2.17 Ci/μmol	8.18 Ci/μmol
Acetonitrile concentration	≤0.04% wt/wt	Pass	Pass	Pass
Ethanol concentration	4.0–8.8% wt/wt	Pass	Pass	Pass
pH	5.0–7.5	7	5.5	6
Residual Kryptofix 222	<50 μg/ml	Pass	Pass	Pass
Membrane filter integrity test	Meets or exceeds manufacturer's specifications	Pass	Pass	Pass
Radionuclidic identity ($t_{1/2}$)	Half-life ($t_{1/2}$) of 105–115 min	Pass	Pass	Pass
Bacterial endotoxins	<175 EU/injected volume	Pass	Pass	Pass
Sterility testing[a]	Absence of microbial growth after 14-day incubation in two media	Pass	Pass	Pass
Radionuclidic purity[a]	99.5% of the gamma emissions are due to ^{18}F	Pass	Pass	Pass

[a] [^{18}F]RGD-K5 final product is released for administration before these tests are completed.

3.3 Radiochemical Identity and Purity

The radiochemical identity and purity of the [^{18}F]RGD-K5 final product is determined by analytical reverse-phase high performance liquid chromatography (RP-HPLC) [46]. The system is equipped with a gamma detector (NaI), a UV detector (267 nm), and a C18 analytical column [47] operated in a linear gradient mode (5% acetonitrile/water to 40% acetonitrile/water over 10 min at 2 ml/min). The radiochemical identity of the tracer is confirmed by use of [^{19}F]RGD-K5 standard and by comparing the retention times between the UV peak for the [^{19}F]RGD-K5 standard and the radioactive peak for [^{18}F]RGD-K5 from the final product solution. The average offset between the UV and gamma detectors is 0.10 min. Our specification for radiochemical purity is greater than or equal to 95%. All three batches met this specification; each batch had a radiochemical purity of 99.9%. In addition, the doses were stable (i.e., ≥95% radiochemical purity) up to 8 h after the EOS. The maximal activity concentrations ranged from 45.1 to 48.3 mCi/ml.

Representative UV and radioactivity chromatograms from the analysis of an [^{18}F]-RGD-K5 final product solution are shown in Fig. 4.

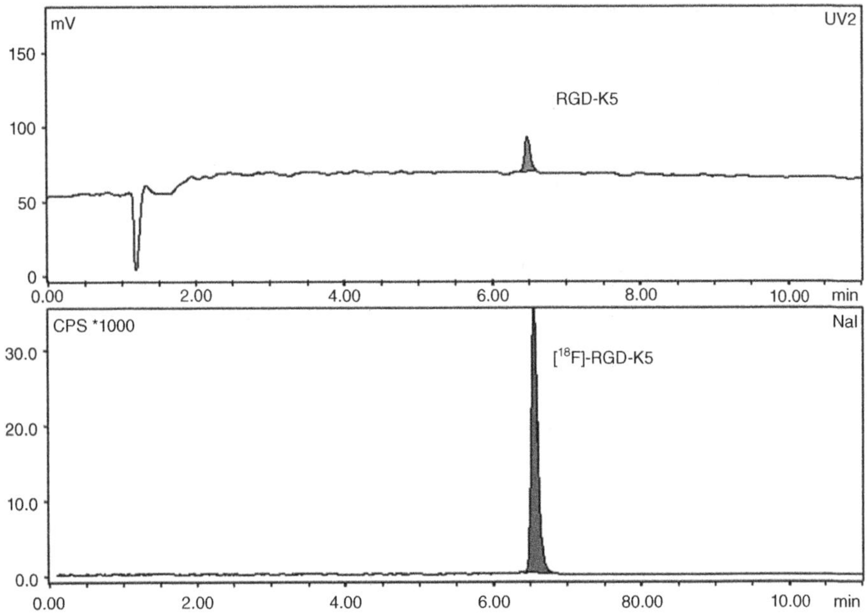

FIGURE 4 Analytical RP-HPLC traces for both the UV (upper) and radioactivity (lower) traces from the analysis of an undiluted [^{18}F]RGD-K5 final product sample.

3.4 Specific Activity

The mass of [^{19}F]RGD-K5 in the [^{18}F]RGD-K5 final product solution is determined by RP-HPLC (see Section 3.3). Our specification for specific activity is ≥0.4 Ci/μmol at EOS. The specific activity of the three [^{18}F]RGD-K5 batches ranged from 2.17 to 8.18 Ci/μmol at EOS, with [^{19}F]RGD-K5 mass concentrations ranging from 6.32 to 23.92 μg/ml.

3.5 Residual Solvents and Ethanol Concentration

The presence of residual solvents (acetonitrile) and the concentration of ethanol in the [^{18}F]RGD-K5 final product is determined by gas chromatography (GC) analysis. The GC system is equipped with a flame ionization detector (FID) and a polyethylene glycol column (helium carrier gas at 10 ml/min). Two injections are made into the GC [48] for analysis: first, an undiluted sample is injected to determine the acetonitrile concentration; second, a diluted (60:1) sample to determine the ethanol concentration. Our specification for acetonitrile is ≤0.04% (w/w), which is also the USP limit concentration for acetonitrile, and the allowed concentration range of ethanol is 4.0–8.8% (w/w). The acetonitrile concentration in all three [^{18}F]RGD-K5 batches was less than 0.04% (w/w). The ethanol concentration in the three batches ranges from 5.0 to 6.4% (w/w). These ethanol concentrations are considered within the tolerances for IV injection [49].

3.6 pH

The pH of the [^{18}F]RGD-K5 final product is determined using pH strips. Our specification for pH requires that the pH must be within the range of 5.5–7.5. The pH values for all three [^{18}F]RGD-K5 batches fell within this range.

3.7 Residual Kryptofix 222

The testing for residual Kryptofix 222 in the [^{18}F]RGD-K5 final product uses an iodine spot test. Aluminum oxide plates are spotted with a Kryptofix 222 standard (50 μg/ml) and the final product solution, followed by drying and visualization of the plate using iodine vapor. The product contains less than 50 μg/ml of Kryptofix 222 if the spot for the standard is darker than that for the product. Our specification for Kryptofix 222 requires that the concentration of Kryptofix 222 must be <50 μg/ml. All three [^{18}F]RGD-K5 batches met this specification.

3.8 Membrane Filter Integrity Test

The integrity of the membrane filter used for sterilization of the [^{18}F]RGD-K5 is determined by a bubble point test [50]. Before performing this test, the membrane filter is carefully rinsed with 10 ml of water to eliminate ethanol present in the membrane. A gas supply is then attached to the inlet of the membrane filter and the pressure slowly increased until a continuous stream of bubbles appears at the outlet. Our specification requires that the bubble point must meet or exceed the manufacturer's specification. The filters for all three [^{18}F]RGD-K5 batches met this specification.

3.9 Sterility Testing

The sterility testing of the [^{18}F]RGD-K5 final product is performed via inoculation of the product into fluid thioglycolate media (FTM) and tryptic soy broth (TSB). In each case, the volume of the inoculum is 0.25 ml into a media volume of approximately 10 ml [50]. The inoculated samples are incubated for 14 days at either 30–35°C (FTM) or 20–25°C (TSB) and visually inspected daily during the test period. Due to the length of the sterility test, the [^{18}F]RGD-K5 final product is released for administration before completing the test. All three [^{18}F]RGD-K5 batches were shown to have an absence of microbial growth at the completion of the 14-day incubation period.

3.10 Radionuclidic Identity

The radionuclidic identity of the [^{18}F]RGD-K5 final product is determined by half-life measurement over a time period of 10 min. The calculated half-life must be within a range of 105–115 min. All three [^{18}F]RGD-K5 batches met this specification.

3.11 Bacterial Endotoxin Test

The testing for the presence of bacterial endotoxins in the [^{18}F]RGD-K5 final product is performed using the kinetic chromogenic method according to USP requirements and specifications [51]. All three [^{18}F]RGD-K5 batches met the USP specification of <175 EU/dose.

3.12 Radionuclidic Purity

The radionuclidic purity of the [^{18}F]RGD-K5 final product is determined on a decayed sample using a multichannel analyzer (MCA) that detects the presence of long-lived radionuclides by gamma spectrometry [52]. Our specification for radionuclidic purity requires that ≥99.5% of the gamma emissions are due to ^{18}F. All three [^{18}F]RGD-K5 batches met this specification. The [^{18}F]RGD-K5 final product is released for administration before completing this test.

WASTE DISPOSAL INFORMATION

All hazardous chemicals and toxic materials were disposed of according to Prudent Practices in the Laboratory (Washington, DC: National Academy Press, 1995).

CHEMICAL ABSTRACTS NOMENCLATURE (REGISTRY NUMBER)

Full compound name (CAS No.)
Acetonitrile (75-05-8)
Carbonic acid and potassium salt (1:2) (584-08-7)
Copper (II) sulfate (7758-98-7)
Cyclo[L-arginylglycyl-L-α-aspartyl-D-phenylalanyl-N6-[2,6-anhydro-7-[(2-azidoacetyl)amino]-7-deoxy-L-glycero-L-galacto-heptonoyl]-L-lysyl] (RGD-K5 azide precursor) (1010702-98-3)
Cyclo[L-arginylglycyl-L-α-aspartyl-D-phenylalanyl-N6-[2,6-anhydro-7-deoxy-7-[[2-[4-[3-(fluoro-18F)propyl]-1*H*-1,2,3-triazol-1-yl]acetyl]amino]-L-glycero-L-galacto-heptonoyl]-L-lysyl]([^{18}F]-Flotegatide, [^{18}F]-RGD-K5)(1010702-75-6)
Ethanol (64-17-5)
4,7,13,16,21,24-Hexaoxa-1,10-diazabicyclo[8.8.8]hexacosane (23978-09-8)
Hydrochloric acid (7647-01-0)
Pentyne tosylate (77758-50-0)
Sodium ascorbate (134-03-2)
Tris((1-benzyl-1*H*-1,2,3-triazol-4-yl)methyl)amine (TBTA) (510758-28-8)

REFERENCES AND NOTES

For detailed supplier information, see Appendix 1.

1. E. M. Conway, D. Collen, P. Carmeliet, *Cardiovasc. Res.* 2001, 49, 507.
2. H. Jin, J. Varner, *Br. J. Cancer.* 2004, 90, 561.

3. E. Y. Lin, J. W. Pollard, *Cancer. Res.* 2007, 67, 5064.
4. J. S. Kerr, A. M. Slee, S. A. Mousa, *Expert Opin Investig Drugs* 2002, 11, 1765.
5. E. Garanger, D. Boturyn, P. Dumy, *Anti. Cancer. Agents Med. Chem.* 2007, 7, 552.
6. M. Schottelius, H. J. Wester, *Methods* 2009, 48, 161.
7. A. J. Beer, R. Haubner, M. Sarbia, M. Goebel, S. Luderschmidt, A. L. Grosu, O. Schnell, M. Niemeyer, H. Kessler, H. J. Wester, W. A. Weber, M. Schwaiger, *Clin. Cancer. Res.* 2006, 12, 3942.
8. W. Cai, G. Niu, X. Chen, *Curr. Pharm. Des.* 2008, 14, 2943.
9. L. M. Kenny, R. C. Coombes, I. Oulie, K. B. Contractor, M. Miller, T. J. Spinks, B. McParland, P. S. Cohen, A. M. Hui, C. Palmieri, S. Osman, M. Glaser, D. Turton, A. Al-Nahhas, E. O. Aboagye, *J. Nucl. Med.* 2008, 49, 879.
10. H. C. Kolb, M. G. Finn, K. B. Sharpless, *Angew. Chem. Int. Ed. Engl.* 2001, 40, 2004.
11. H. C. Kolb, K. B. Sharpless, *Drug. Discov. Today* 2003, 8, 1128.
12. C. W. Tornoe, C. Christensen, M. Meldal, *J. Org. Chem.* 2002, 67, 3057.
13. M. Meldal, C. W. Tornoe, *Chem. Rev.* 2008, 108, 2952.
14. V. V. Rostovtsev, L. G. Green, V. V. Fokin, K. B. Sharpless, *Angew. Chem. Int. Ed. Engl.* 2002, 41, 2596.
15. H. C. Kolb, J. C. Walsh, K. Chen. The preparation of molecular imaging probes using click chemistry. I. Siemens Medical Solutions USA. WO2006116629. 2006, p. 53.
16. Z. B. Li, Z. Wu, K. Chen, F. T. Chin, X. Chen, *Bioconjug. Chem.* 2007, 18, 1987.
17. M. Glaser, M. Solbakken, D. R. Turton, R. Pettitt, J. Barnett, J. Arukwe, H. Karlsen, A. Cuthbertson, S. K. Luthra, E. Arstad, *Amino Acids* 2009, 37, 717.
18. S. Maschauer, J. Einsiedel, R. Haubner, C. Hocke, M. Ocker, H. Hubner, T. Kuwert, P. Gmeiner, O. Prante, *Angew. Chem. Int. Ed. Engl.* 2010, 49, 976.
19. M. Colombo, A. Bianchi, *Molecules* 2010, 15, 178.
20. J. C. Walsh, H. C. Kolb, *Chimia (Aarau)* 2010, 64, 29.
21. D. H. Kim, Y. S. Choe, B. T. Kim, *Appl. Radiat. Isot.* 2010, 68, 329.
22. S. H. Hausner, J. Marik, M. K. Gagnon, J. L. Sutcliffe, *J. Med. Chem.* 2008, 51, 5901.
23. Supelclean Coconut Charcoal SPE Tube.
24. Virgin ^{18}O-water with an enrichment of greater than 98% was used for these runs.
25. Siemens Medical Solutions, USA, Inc., Molecular Imaging, 2501 N. Barrington Road, Hoffman Estates, IL, 60192-5203, USA.
26. Chromafix cartridges were purchased from ABX (Macherey Nagel PS-HCO3 Chromafix). The cartridges were washed with deionized water and dried under a flow of Argon prior to use.
27. Potassium carbonate was purchased from Sigma (ACS reagent grade, 99+%) and used directly.
28. Sterile water was purchased from Hospira, Inc.
29. Kryptofix®-[2.2.2] was purchased from Sigma (98%) and used directly.
30. Acetonitrile was purchased from Sigma (anhydrous, ≥99%).
31. A vacuum pump from ILMVAC GmbH (type: 400861) was used for these runs.
32. Due to the delay in temperature ramps, the drying times for this step may be longer than 5.5 min.
33. This low and constant pressure reading provides an indication that the reaction vial is sufficiently free of moisture to begin the fluorination process.
34. T. R. Hoye, J. R. Vyvyan, *J. Org. Chem.* 1995, 60, 4184.
35. Prepared in-house by Siemens MI Biomarker Research.
36. Sourced from Sigma-Aldrich.
37. T. R. Chan, R. Hilgraf, K. B. Sharpless, V. V. Fokin, *Org. Lett.* 2004, 6, 2853.
38. Sourced from Sigma-Aldrich.
39. Sourced from Sigma-Aldrich.
40. Phenomenex Luna, C18, 5 μ, 10×250 mm.
41. The cartridge was washed with ethanol (5 ml) followed by sterile water (10 ml) prior to use.
42. Sep-Pak Vac 3 cc tC18, waters part number: WAT054925.
43. Care must be taken to ensure that the sterile filters contain low amounts of leachable materials, which may give rise to UV impurities in the final dose.
44. Sterile vials used were sourced from either Greer or Hospira.

45. This specification is based on stability studies and the ease of handling doses. The radioactivity concentration could be increased to values beyond this specification, but this would require additional stability studies to demonstrate the shelf life of the product.
46. An Agilent 1200 series HPLC was used for the analytical analysis.
47. Phenomenex Luna C18, 5 μ, 10 × 250 mm.
48. Varian model 3800 gas chromatograph using either a DB-Wax or CP-Wax 52CB column (0.53 mm × 30 m, 1.0–1.2 mm film thickness).
49. R. G. Strickley, *Pharm. Res.* 2004, 21, 201.
50. S. Yu, *Biomed. Imaging Intervention J.* 2006, 2, e57.
51. The kits were purchased from Charles River Endosafe, Inc. Prior to performing the pyrogen test, the EtOH/water formulation was found to neither enhance nor inhibit the test for pyrogens.
52. Canberra multichannel analyzer with germanium detector and PROcount-2000 software.

CHAPTER 5

SYNTHESIS OF 3-[^{18}F]FLUORO-5-(2-PYRIDINYLETHYNYL) BENZONITRILE ([^{18}F]FPEB)

STEVEN H. LIANG[1,2], DANIEL L. YOKELL[1], RAUL N. JACKSON[1], PETER A. RICE[1], ELI LIVNI[1,2], DAVID ALAGILLE[3,4], GILLES TAMAGNAN[3,4], THOMAS LEE COLLIER[1,2,5], AND NEIL VASDEV[1,2]

[1] Division of Nuclear Medicine and Molecular Imaging, Massachusetts General Hospital, Boston, Massachusetts, USA
[2] Department of Radiology, Harvard Medical School, Boston, Massachusetts, USA
[3] Molecular NeuroImaging, LLC, New Haven, Connecticut, USA
[4] Institute for Neurodegenerative Disorders, New Haven, Connecticut, USA
[5] Advion Inc., Ithaca, New York, USA

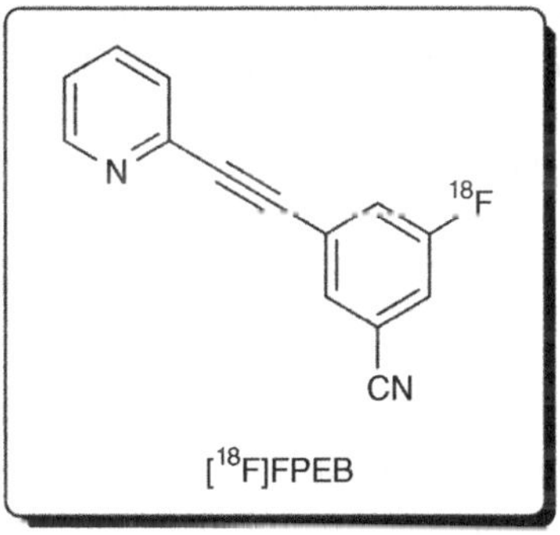

1 INTRODUCTION

Glutamatergic signaling abnormalities have long been implicated in several neuropsychiatric disorders. Several lines of evidence point to a critical role for metabotropic glutamate receptor subtype 5 (mGluR5) in the pathologic process of Parkinson's disease, Alzheimer's disease, fragile X syndrome, as well as other neurodegenerative diseases. Imaging mGluR5 with positron emission tomography (PET) [1–10] has proven useful for quantitative assessment of this receptor in preclinical models and clinical studies and enables novel therapeutics to be evaluated.

Radiochemical Syntheses, Volume 2: Further Radiopharmaceuticals for Positron Emission Tomography and New Strategies for Their Production, First Edition. Edited by Peter J. H. Scott.

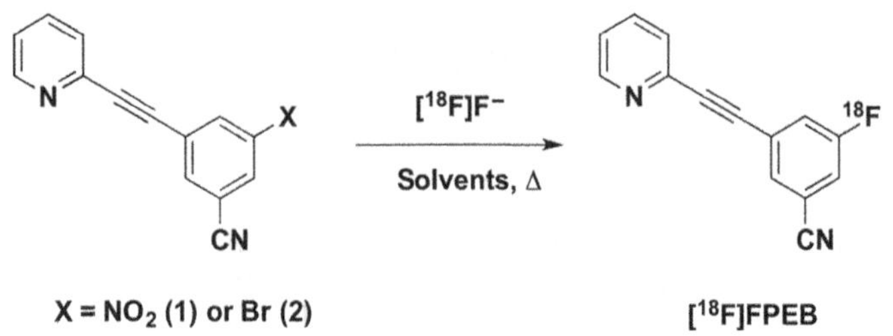

FIGURE 1 Synthesis of [18F]FPEB.

3-Fluoro-5-[(pyridin-3-yl)ethynyl]benzonitrile (FPEB) is a mGluR5 antagonist with high potency, selectivity, and brain penetrance. Recent work with [18F]FPEB has demonstrated that this radiopharmaceutical is suitable for quantifying regional brain concentrations of mGluR5 in healthy human subjects [9, 10].

The preparation of clinically useful quantities of [18F]FPEB is presently a challenge as the literature procedures are low yielding (1–5% uncorrected radiochemical yield, relative to [18F]fluoride, at the end of synthesis) [1, 5]. The lack of an efficient radiosynthesis of [18F]FPEB has limited the use of this radiotracer to a few PET centers worldwide. Herein, we describe an improved radiosynthesis with two different precursors (Fig. 1, [11, 12]) and modified purification procedures using a GE medical systems commercial TRACERlab™ FX_{FN} radiosynthesis module, as well as on an Advion NanoTek® continuous-flow microfluidic platform [13]. The procedures described herein are suitable for routine human use.

2 SYNTHESIS PROCEDURES

CAUTION: All radiochemical syntheses must be carried out using appropriate equipment in a facility authorized for the use of radioactive materials. Personal protective equipment must be worn and all local radiation safety regulations followed.

2.1 Production of [18F]Fluoride

A GE PETtrace 16/8.4 MeV cyclotron was used for [18F]fluoride radionuclide production. A GE high-yield niobium target containing >97% enriched O-18 water (Isotec, Taiyo Nippon Sanso, or Rotem) was bombarded with protons at integrated currents up to 65 μA to produce [18F]fluoride. Following completion of bombardment, the [18F]fluoride was transferred to the GE TRACERlab™ FX_{FN} radiosynthesis module or Advion NanoTek platform via helium gas overpressure.

2.2 Azeotropic Drying of [18F]Fluoride

2.2.1 GE TRACERlab™ FX_{FN} Method

A schematic diagram of the GE medical systems commercial TRACERlab™ FX_{FN} radiosynthesis module used for the synthesis of [18F]FPEB is shown in Fig. 2. [18F]

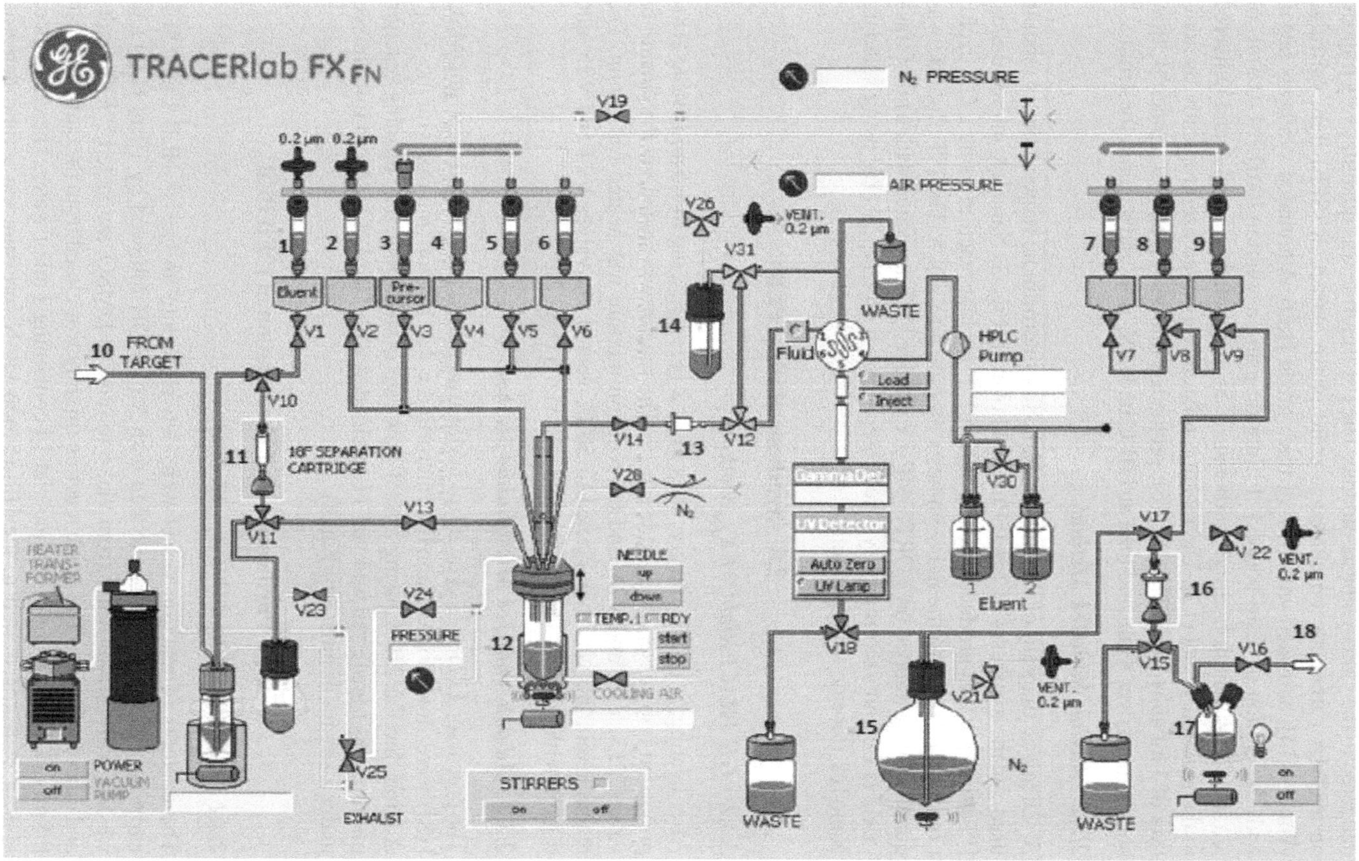

FIGURE 2 TRACERlab™ FX_{FN} radiosynthesis module used for the synthesis of [^{18}F]FPEB.

Fluoride was produced by the $^{18}O(p,n)^{18}F$ nuclear reaction using a GE cyclotron and delivered to the radiosynthesis module. The [^{18}F]fluoride was quantitatively trapped on a QMA carbonate ion exchange solid phase extraction (SPE) light cartridge (Waters [14]; activated with 6 ml of trace grade H_2O). Automated synthesis began with the elution of resin-bound [^{18}F]fluoride using a solution (0.075 M, 0.6 ml) of tetrabutylammonium hydrogencarbonate [15], preloaded into vial 1 and delivered to the reaction vessel (12). The reaction mixture was dried azeotropically by addition of 1 ml anhydrous CH_3CN [16], preloaded into vial 4, at 85°C under N_2 flow and vacuum over 8 min and then at 110°C under N_2 flow and vacuum for 4 min.

2.2.2 *Advion NanoTek Method*

A schematic diagram of the Advion NanoTek microfluidic radiosynthesis module used for the synthesis of [^{18}F]FPEB is shown in Fig. 3, NanoTek microfluidics. [^{18}F]Fluoride was produced by the $^{18}O(p,n)^{18}F$ nuclear reaction using a GE cyclotron and delivered to the radiosynthesis module into vial 8. The [^{18}F]fluoride was quantitatively trapped on a ORTG MP-1 cartridge [17] (activated with 6 ml of trace grade H_2O) and azeotropically dried three times with CH_3CN [16] using standard NanoTek fluoride drying macro.

2.3 Synthesis of [^{18}F]FPEB

2.3.1 *GE TRACERlab™ FX_{FN} Method*

After heating to 150°C, the precursor (**1** or **2** [11, 12]; 2 mg in 1.5 ml DMSO [18]) preloaded into vessel 3 was added to 12. The reactor was sealed via the closure of valves V13, V20, and V24, and the reaction mixture was heated for 15 min. The reaction mixture was then cooled to 50°C, vented via valves V24 and V25, and diluted with 10 ml of H_2O, preloaded into vial 5 (Fig. 2).

2.3.2 *Advion NanoTek Microfluidic Method*

The precursor **1** (2 mg [11] in 200 μl DMSO) and dried [^{18}F]nBu$_4$NF (250 μl DMSO [18]) were flowed into NanoTek microfluidic 4 m reactor (internal volume 32 μl, Advion, Inc.) at 210°C using a total flow rate of 40 μl/min. The ensuing reaction mixture was transferred into vial 11, preloaded with 10 ml of H_2O and fully mixed under a stream of nitrogen (Fig. 3).

2.4 Purification and Formulation of [^{18}F]FPEB

2.4.1 *GE TRACERlab™ FX_{FN} Method*

The contents of reaction vessel were delivered onto an Oasis® HLB Light SPE cartridge (Waters [19]; preactivated with 5 ml EtOH followed by 10 ml H_2O) and washed with 5 ml of water from vial 6 to remove DMSO, unreacted [^{18}F]fluoride, and other impurities. The crude reaction mixture was eluted from the cartridge with 0.8 ml of CH_3CN from vial 2 into vial 14 containing 1 ml of water. The contents of vial 14 were transferred to the HPLC loop via N_2 pressure via a fluid detector, injected onto a semipreparative column (X-Select HSS T3, 250 × 10.00 mm, 5 μm),

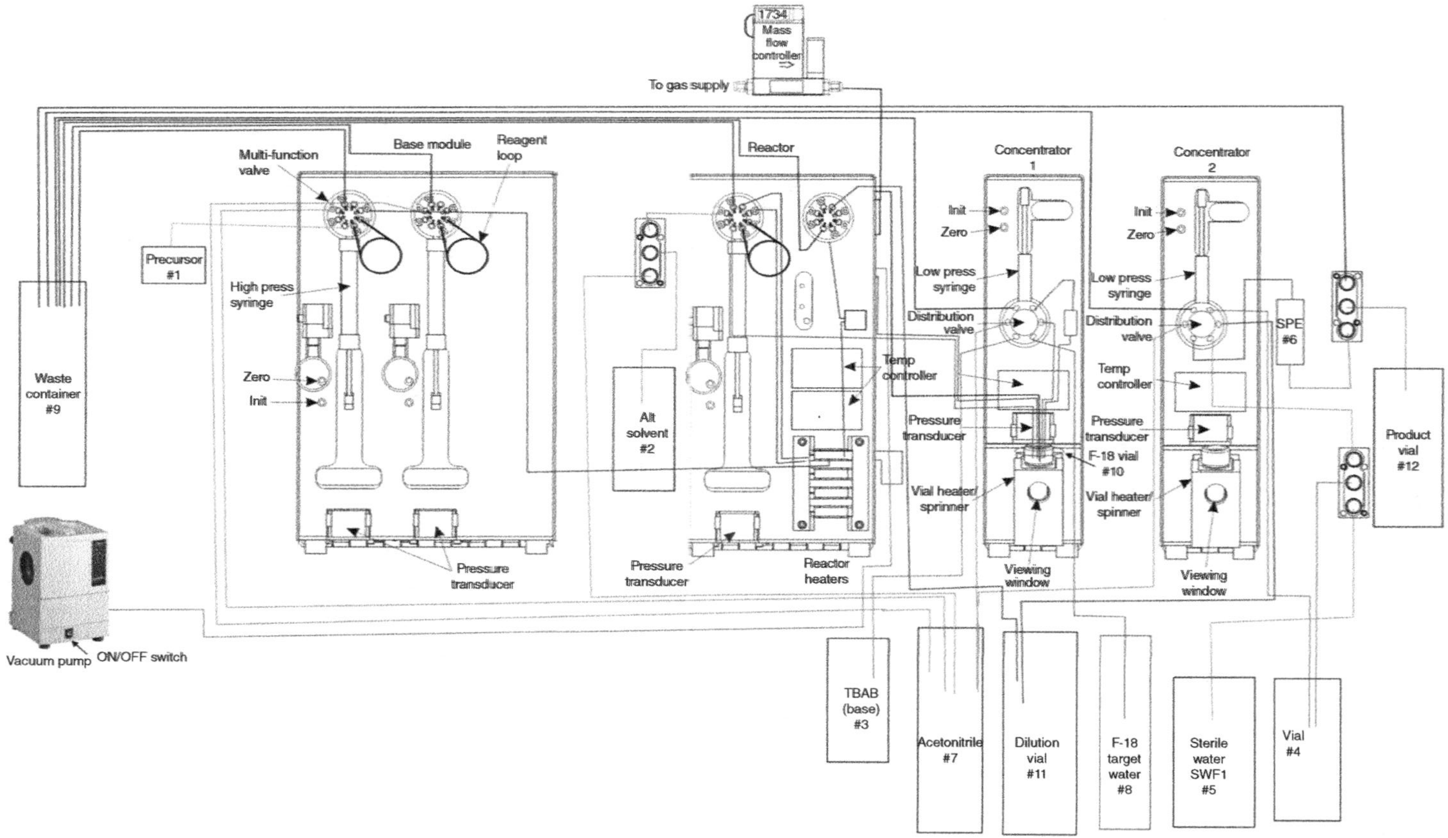

FIGURE 3 NanoTek microfluidics radiosynthesis module used for the synthesis of [^{18}F]FPEB.

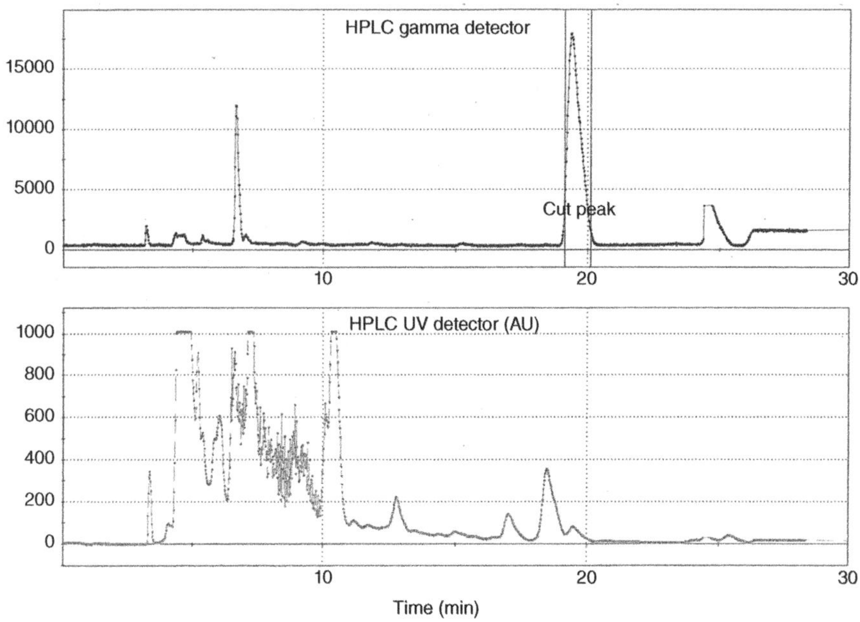

FIGURE 4 Semipreparative UV and radioactive HPLC traces for [^{18}F]FPEB.

and eluted with 45:55 CH_3CN/20 mM ammonium acetate [20] (pH 6) at a flow rate of 4 ml/min. The eluent was monitored by UV ($\lambda = 254$ nm) and radiochemical detectors connected in series. The retention time of [^{18}F]FPEB is 19 min (Fig. 4). The [^{18}F]FPEB product peak was collected into the HPLC fraction collect vessel (15), where it was diluted with 20 ml of sterile water for injection, USP [21]. The diluted HPLC fraction was then loaded on a C18 light SPE cartridge (16) for HPLC solvent removal [22]. The C18 SPE was washed with 10 ml sterile water for injection, USP [21], preloaded into vial 7, and then the [^{18}F]FPEB product was recovered in 1 ml dehydrated alcohol for injection, USP [23] (ethanol; vial 8), into vial 17 and 10 ml of 0.9% sodium chloride for injection, USP [24] (vial 9). Following product formulation, the [^{18}F]FPEB was transferred to the PET production facility clean room via sanitized product transfer line for cold sterile filtration in the ISO Class 5 isolator. The solution was transferred and passed through a 0.22 μm Millipore GV sterilizing filter (EMD Millipore) into a vented sterile 30 ml dose vial (Hospira).

2.4.2 Advion NanoTek Method

The contents of vial 11 were delivered onto an Oasis HLB Light SPE cartridge on line 6 (Waters [19]; preactivated with 5 ml EtOH followed by 10 ml H_2O) and washed with 25 ml of water to remove DMSO, unreacted [^{18}F]fluoride, and other impurities. The crude reaction mixture was eluted from the cartridge with 1 ml of

CH_3CN, followed by 3 ml of sterile water into vial 12. The contents of vessel 12 were transferred to the HPLC loop under vacuum, injected onto a semipreparative column (X-Select HSS T3, 250×10.00 mm, 5 µm), and eluted with 45:55 CH_3CN/20 mM ammonium acetate (pH 6) at a flow rate of 4 ml/min. The eluent was monitored by UV (λ=254 nm) and radiochemical detectors connected in series. The retention time of [^{18}F]FPEB is 19 min (Fig. 4). The formulation procedure for [^{18}F]FPEB is identical to that of GE TRACERlab™ FX_{FN} method. For further details on the radiosynthesis and characterization of [^{18}F]FPEB by continuous-flow microfluidics, refer to reference [25].

3 QUALITY CONTROL PROCEDURES

CAUTION: All radiopharmaceuticals produced for clinical use must have local regulatory approval before use in humans. Trained personnel must carry out QC procedures, and each dose must meet all established QC criteria before release to the clinic.

Quality control procedures and release criteria for [^{18}F]FPEB are summarized in the following and in Table 1.

TABLE 1 Quality Control Data for [^{18}F]FPEB from the GE Tracerlab FX_{FN} Method

Quality Control Test	Release Criteria	Results (n=3)
Isolated product (mCi)	>20 mCi per run	108±26 mCi
Visual inspection	Clear, colorless, particulate free	Pass
Radiochemical identity	±10% of FPEB reference standard retention time	Pass
Radiochemical purity	≥95%	100%
Chemical purity	≥90% corrected to specific activity of 800 mCi/µmol	94.9%±3.1% (corrected to 800Ci/µmol)
Specific activity	≥800 mCi/µmol at time of administration	7.07±1.38 Ci/µmol
Residual solvent analysis	DMSO (Class III) <5 mg/ml Acetone (Class III) <5 mg/ml Acetonitrile (Class II) <0.4 mg/ml Ethanol (Class III) <10% v/v ±10%	Pass
pH assay	4.5–8.5	5–5.5
Sterile filter integrity test	≥50 psi	Pass
Radionuclidic ID—photopeak	>99.5% emission @ 511 KeV, 1.022 MeV	Pass
Radionuclidic ID—half-life	105–115 min	Pass
Endotoxin analysis	Not greater than 17.5 EU/ml	Less than 1 EU/ml
Sterility testing	No evidence of growth at 14 days post inoculation	Pass

3.1 Visual Inspection

The [^{18}F]FPEB dose is examined behind a lead glass L-block (suitable for PET isotopes) and must be clear, colorless, and free of particulate matter.

3.2 Radiochemical Identity, Radiochemical Purity, Chemical Purity, and Specific Activity

To determine the identity of [^{18}F]FPEB, aliquots of the formulated product were injected onto an analytical HPLC system using a Novapak C18 column, 150×4.6 mm, 4 μm, and eluted with 45:55 EtOH/water at a flow rate of 1 ml/min, monitored at $\lambda = 280$ nm. After completion of the chromatograph, peaks on UV and radioactivity detector were integrated, and the radiochemical and chemical purities were determined by the area of integration.

The major radiochemical product was identified as [^{18}F]FPEB ($t_R = \sim 4.7$ min; Fig. 5), confirmed by coinjection with the reference standard FPEB. The retention time of [^{18}F]FPEB is compared to that of the standard FPEB and must be within ±10% error. The radiochemical purity must be >95%. Chemical purity may be recalculated based on a specific activity of 800 mCi/μmol if the ≥90% specification is not met. Specific activity was determined using standard [^{18}F]FPEB specific activity calibration curve. Specific activity must be ≥800 mCi/μmol at time of administration.

3.3 Residual Solvent Analysis

Residual solvent assay was performed to verify that residual solvents from the synthesis and maintenance of the synthesis units are within acceptable limits. Gas chromatography (GC) was used to determine the solvent residue and the results must meet the following specifications:

DMSO (Class III) < 5 mg/ml
Acetone (Class III) < 5 mg/ml
Acetonitrile (Class II) < 0.4 mg/ml
Ethanol (Class III) < 10% v/v ±10% (formulation excipient)

3.4 pH Assay

The pH of [^{18}F]FPEB was determined by applying a few drops of the dose to pH-indicator paper. Match the reference color and the pH value must be 4.5–8.5.

3.5 Sterile Filter Integrity Test

Sterile filter integrity test was performed as per manufacturer specification and the pressure must be ≥50 psi for the Millipore Millex-GV 0.22 μm sterilizing filter.

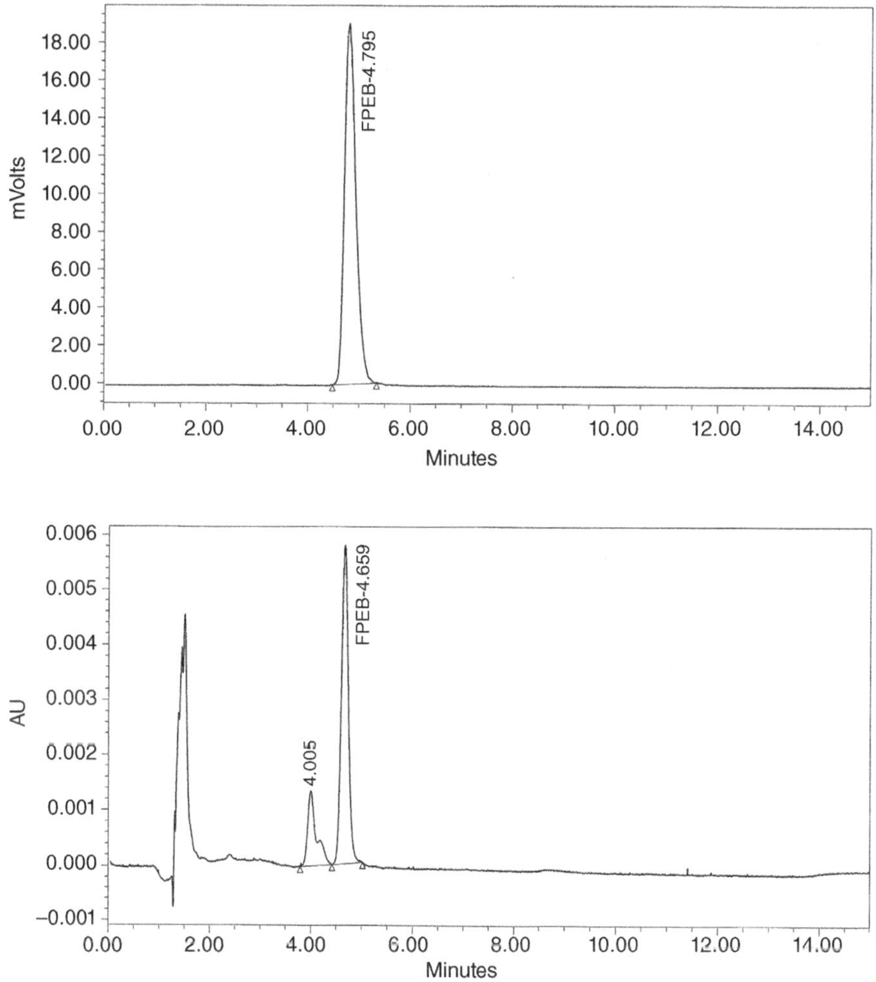

FIGURE 5 Analytical radioactive (top) and UV (bottom) HPLC traces for [^{18}F]FPEB.

3.6 Radionuclidic ID: Photopeak and Half-Life

Half-life was determined based on the following protocol:

Measure the radioactivity of the formulated product at two separated time points. Perform the half-life calculation using the following equation:

$$t_{½} = \frac{-\ln(2)\,(\mathrm{T2}-\mathrm{T1})}{\ln(\mathrm{A2}-\mathrm{A1})}$$

T1: the first time point; T2: the second time point; A1: radioactivity measured at T1; A2: radioactivity measured at T2. The half-life must be 105–115 min.

Photopeak was determined based on the following protocol:

Introduce small amount of radioactivity of formulated product into gamma spectrometer. Record the spectrum and integrate the areas under the signals of the spectrum. The result must be >99.5% emission @511 KeV, 1.022 MeV.

3.7 Endotoxin Analysis

Endotoxin analysis was performed on a Charles River Laboratories Endosafe PTS system. Doses must contain ≤17.5 EU/ml.

3.8 Sterility Testing

Sterility testing was performed postrelease and must be started within 30 h from end of synthesis. [^{18}F]FPEB sample was inoculated into Trypticase soy broth (TSB) and fluid thioglycolate medium (FTM) media tubes. TSB tubes were incubated at 20–25°C and FTM tubes were incubated at 30–35°C for 14 days and must be free of culture growth after 14 days.

WASTE DISPOSAL INFORMATION

All hazardous chemicals and toxic materials were disposed of according to Prudent Practices in the Laboratory (Washington, DC: National Academy Press, 1995).

CHEMICAL ABSTRACT NOMENCLATURE (REGISTRY NUMBER)

Acetonitrile (75-05-8)
Ammonium acetate (631-61-8)
Dimethyl sulfoxide (67-68-5)
Ethanol (64-17-5)
3-[2-(Pyridine-2-yl)ethynyl]-5-nitrobenzene (1031370-96-3)
Sodium acetate (127-09-3)
Tetrabutylammonium hydrogencarbonate (17351-62-1)

REFERENCES AND NOTES

For detailed supplier information, see Appendix 1.

1. T. G. Hamill, S. Krause, C. Ryan, C. Bonnefous, S. Govek, T. J. Seiders, N. D. Cosford, J. Roppe, T. Kamenecka, S. Patel, R. E. Gibson, S. Sanabria, K. Riffel, W. Eng, C. King, X. Yang, M. D. Green, S. S. O'Malley, R. Hargreaves, H. D. Burns, *Synapse*, 2005, 56, 205.
2. S. Patel, O. Ndubizu, T. Hamill, A. Chaudhary, H. D. Burns, R. Hargreaves, R. E. Gibson, *Mol. Imaging Biol.*, 2005, 7, 314.
3. D. Pellegrino, F. Cicchetti, X. Wang, A. Zhu, M. Yu, M. Saint-Pierre, A. L. Brownell, *J. Nucl. Med.*, 2007, 48, 1147.
4. J-Q. Wang, A-L. Brownell, *Curr. Med. Imaging Rev.*, 2007, 3, 186.
5. J-Q. Wang, W. Tueckmantel, A. Zhu, D. Pellegrino, A. L. Brownell, *Synapse* 2007, 61, 951.
6. A. Zhu, X. Wang, M. Yu, J. Q. Wang, A. L. Brownell, *J. Cereb. Blood Flow Metab.*, 2007, 27, 1623.

7. R. Sanchez-Pernaute, J-Q. Wang, D. Kuruppu, L. Cao, W. Tueckmantel, A. Kozikowski, O. Isacson, A. L. Brownell, *NeuroImage*, 2008, 42, 248.
8. D. Alagille, H. DaCosta, Y. Chen, K. Hemstapat, A. Rodriguez, R. M. Baldwin, P. J. Conn, G. D. Tamagnan, *Bioorg. Med. Chem. Lett.*, 2011, 21, 3243.
9. D. F. Wong, R. Waterhouse, H. Kuwabara, J. Kim, J. R. Brašić, W. Chamroonrat, M. Stabins, D. P. Holt, R. F. Dannals, T. G. Hamill, P. D. Mozley, *J. Nucl. Med.*, 2013, 54, 388.
10. J. M. Sullivan, K. Lim, D. Labaree, S. F. Lin, T. J. McCarthy, J. P. Seibyl, G. Tamagnan, Y. Huang, R. E. Carson, Y. S. Ding, E. D. Morris, *J. Cereb. Blood Flow Metab.*, 2013, 33, 532.
11. FPEB nitro precursor **1**: 3-[2-(pyridine-2-yl)ethynyl]-5-nitrobenzene; product# 3571.0002, ABX.
12. (a) FPEB bromo precursor **2** (3-[2-(pyridine-2-yl)ethynyl]-5-bromobenzene): To a solution of 2-(trimethylsilyl)ethynyl)pyridine (Alfa Aesar, 2.85 mmol) in 10 ml of DMF was successively added Et_3N (1.59 ml, 11.42 mmol), 3,5-dibromobenzonitrile (Oakwood Chemicals, 2.86 mmol), CuI (53 mg, 0.28 mmol), and *trans*-dichlorobis(triphenylphosphine)palladium (198 mg, 0.28 mmol). The resulting solution was stirred at room temperature, and TBAF 1 M in THF (3.13 ml, 3.13 mmol) was added dropwise. After 2 h at this temperature, the mixture was hydrolyzed with 50 ml of H_2O and extracted with AcOEt (4×30 ml). The organic layer was washed with saturated NaCl (3×30 ml), dry over anhydrous Na_2SO_4, and concentrated on a rotary evaporator. Purification of the residue by column chromatography (hexanes/AcOEt 80/20) yielded 434 mg (58%) of 4e as an off-white solid. (b) ^1H NMR, 500 MHz ($CDCl_3$), δ=7.30–7.32 (m, 1H, CHAr); 7.54 (d, 1H, J=7.4 Hz, CHAr); 7.72 (td, 1H, J=7.7, 1.6 Hz, CHAr); 7.78 (dd, 2H, J=8.3, 1.3 Hz, CHAr); 7.95 (d, 1H, J=1.3 Hz, CHAr); 8.65 (d, 1H, J=3.6 Hz, CHAr). ^{13}C NMR, 125 MHz ($CDCl_3$), δ=84.7 (1C, C≡C); 91.9 (1C, C≡C); 114.4 (1C, Cq); 116.5 (1C, Cq); 122.8 (1C, Cq); 123.6 (1C, CHAr); 125.6 (1C, Cq); 127.5 (1C, CHAr); 133.6 (1C, CHAr); 134.6 (1C, CHAr); 136.3 (1C, CHAr); 138.8 (1C, CHAr); 142.0 (1C, Cq); 150.3 (1C, CHAr). HRMS (cal: 282.9871; obs: 282.9869) Anal ($C_{14}H_7N_3O_2 \cdot HCl$) cal:C 52.62%, H 2.52%, N 8.77%; obs: C 52.36%, H 2.63%, N 8.69%. Purity HPLC 99.98%.
13. Stephenson, N.A., Holland, J.P., Kassenbrock, A., Yokell, D.L., Livni, E., Liang, S.H. & Vasdev, N. Iodonium Ylide Mediated Radiofluorination of 18F-FPEB and Validation for Human Use. Journal of Nuclear Medicine (2015). DOI. 10.2967/jnumed.114.151332
14. QMA Sep-Pak; part# WAT023525, Waters.
15. Tetrabutylammonium hydrogencarbonate, 0.075 M; product#808.0000.6, ABX.
16. Acetonitrile; product#34998-4L, Sigma-Aldrich; acetonitrile, extra dry; product#448391000, Acros Organics.
17. ORTG (Oakdale, TN).
18. Dimethyl sulfoxide (DMSO); product#TS20684, Thermo-Scientific.
19. HLB light cartridge; part#186005125, Waters.
20. Ammonium acetate (NH_4OAc) >98%; product#A7330, Sigma-Aldrich.
21. Sterile water for injection, USP; product# 0409-4887-99, Hospira
22. C18 Sep-Pak; part# WAT023501, Waters.
23. Dehydrated alcohol (EtOH), USP; product# 0517-8575-10, American Regent 1 ml.
24. 0.9% sodium chloride for injection, USP; product# 0409-4888-10, Hospira.
25. S. H. Liang, D. Yokell, R. N. Jackson, P. A. Rice, R. Callahan, K. A. Johnson, D. Alagille, G. Tamagnan, T. L. Collier, N. Vasdev. *Med. Chem. Commun.*, 2014, 5, 432–435.

CHAPTER 6

RADIOSYNTHESIS OF 2′-DEOXY-2′-[^{18}F]FLUORO-5-METHYL-1-β-D-ARABINOFURANOSYLURACIL ([^{18}F]FMAU)

KAI CHEN AND PETER S. CONTI

Molecular Imaging Center, Department of Radiology, Keck School of Medicine, University of Southern California, Los Angeles, California, USA

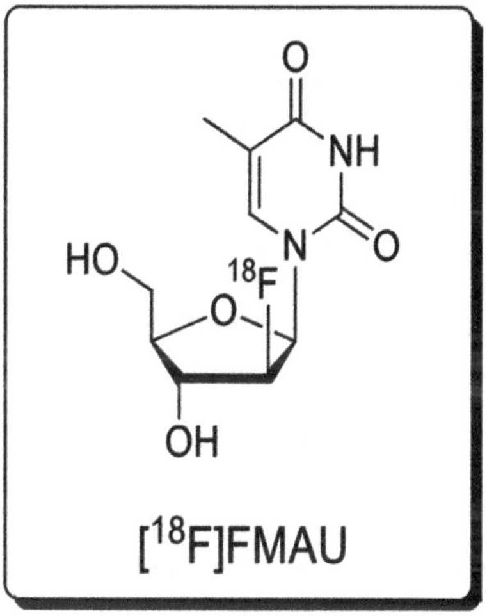

1 INTRODUCTION

Over the last two decades, we and others have been exploring the radiosynthesis and *in vivo* pharmacology of antiviral and antileukemic nucleoside derivatives as potential imaging agents for a variety of applications, including agents such as 2′-fluoro-5-[^{125}I]iodo-1-β-D-arabinofuranosylcytosine (FIAC), 2′-fluoro-5-[^{125}I, ^{131}I, ^{123}I]iodo-1-β-D-arabinofuranosyluracil (FIAU), 2′-deoxy-1-[^{11}C]methyl-pseudouridine, [methyl-^{11}C] 3′-azido-thymidine, and 2′-fluoro-5-[methyl-^{11}C]-1-β-D-arabinofuranosyluracil (FMAU) [1–8]. Of the above agents, FMAU has received considerable attention as a potential *in vivo* cell proliferation marker for use with

Radiochemical Syntheses, Volume 2: Further Radiopharmaceuticals for Positron Emission Tomography and New Strategies for Their Production, First Edition. Edited by Peter J. H. Scott.

positron emission tomography (PET) [9]. FMAU is a 2′-fluoropyrimidine where a hydroxyl group is located on the 3′-position of deoxyribose. *In vitro* studies demonstrated that FMAU is transported into cells and phosphorylated by mammalian cytosolic kinases and displays substantial incorporation into host DNA [10]. Radiolabeling of FMAU with F-18 was achieved at the 2′-fluoro position of the sugar [11–13]. Preclinical studies have shown that [^{18}F]FMAU retention in tumors and non-tumor tissues with rapid cell turnover (e.g., marrow and small intestine) reflects its incorporation into DNA [14–17]. In addition, preliminary clinical studies have shown that (i) FMAU is highly resistant to catabolism in humans and (ii) there is good tumor uptake of radiolabeled FMAU in a variety of tumors comparable to that seen in human studies with [^{18}F]FLT [18–21]. Taken together, [^{18}F]FMAU is a promising PET tracer for imaging cell proliferation.

The radiosynthesis of [^{18}F]FMAU [11, 12] and other 5-substituted thymidine analogs [13] was reported from our group. In that procedure, the radiosynthesis of [^{18}F]FMAU involves four steps, including (1) radiofluorination of 2-trifluoromethanesulfonyl-1,3,5-tri-*O*-benzoyl ribofuranose to 2-[^{18}F]-fluoro-1,3,5-tri-*O*-benzoyl arabinofuranose derivative, (2) conversion of 2-[^{18}F]-fluoro-1,3,5-tri-*O*-benzoyl arabinofuranose derivative to 1-bromo-2-[^{18}F]-fluoro-1,3,5-tri-*O*-benzoyl derivative, (3) coupling of 1-bromo-2-[^{18}F]-fluoro-1,3,5-tri-*O*-benzoyl derivative with 2,4-bis-trimethylsilyluracil derivative, and (4) hydrolysis of the protecting groups to produce the desired product [^{18}F]FMAU. The complexity of this method is often accompanied by frequent production failures. In addition, this radiosynthesis approach requires significant modification of commercial automated synthesis module. Recently, we and others reported new approaches of [^{18}F]FMAU synthesis [22–24], including one with a simplified one-pot reaction condition by using Friedel–Crafts catalysts [22, 23]. The new approach affords an optimized process with shorter synthesis time and higher radiochemical yield as compared to previous method. Our new method is general that can be applied to the radiosynthesis of [^{18}F]FMAU and other 2′-[^{18}F]fluoro-2′-deoxy-arabino-5-substituted pyrimidine nucleoside derivatives and compatible with commercially available radiosynthesis modules used for production of cGMP-compliant radiotracers.

2 SYNTHESIS PROCEDURES

CAUTION: All radiochemical syntheses must be carried out using appropriate equipment in a facility authorized for the use of radioactive materials. Personal protective equipment must be worn and all local radiation safety laws followed.

2.1 Production of [^{18}F]Fluoride

[^{18}O]H_2O (2 ml) [25] was loaded into the [^{18}F]fluoride tantalum target of a Siemens RDS-112 cyclotron [26]. The target was bombarded (60 μA beam for 1.5 h) to generate approximately 900 mCi (33.3 GBq) of [^{18}F]fluoride by the ^{18}O(p, n)^{18}F nuclear reaction.

2.2 Azeotropic Drying of [^{18}F]Fluoride

Aqueous [^{18}F]fluoride ion was trapped on an anion exchange resin (Dowex (Cl^-) 1×8, 200–400 mesh). The [^{18}O]H_2O was collected for recycling, and the [^{18}F]fluoride ion was then eluted into a glassy carbon reaction vessel using a solution of aqueous potassium carbonate (7.5 mg in 0.75 ml of water) [27]. A solution of Kryptofix® [2.2.2] (20 mg in 1 ml of acetonitrile) [28] was added directly into the reaction vessel after the elution of the [^{18}F]fluoride ion. The solution was concentrated to dryness to remove both the acetonitrile and water using a stream of argon at 100°C and a gentle vacuum of about 250 mbar. Azeotropic drying was then repeated twice with 1 ml portions of acetonitrile.

2.3 Synthesis of [^{18}F]FMAU

After the heating block was cooled down to 85°C, a solution of the precursor, 1,3,5-tri-*O*-benzoyl-2-*O*-trifluoromethanesulfonyl-α-furanose [23] (10.0 mg, 17 μmol), in anhydrous acetonitrile (0.8 ml) was added to the activated [^{18}F]fluoride. The mixture was then heated at 85°C for 20 min. After evaporating acetonitrile at 85°C under a steam of argon, *O,O′*-bis(trimethylsilyl)thymine (20 mg, 73.9 μmol) in a solution of 1,2-dichloroethane (300 μl), hexamethyldisilazane (HMDS) (200 μl), and trimethylsilyl trifluoromethanesulfonate (TMSOTf) (150 μl) was added [29]. The mixture was heated at 85°C for 1 h. The solvent was then evaporated, and potassium methoxide [30] (25 wt% in CH_3OH, 400 μl) in 400 μl CH_3OH [31] was added. After 5 min hydrolysis at 85°C, solvent was evaporated and the reaction mixture was cooled and neutralized by addition of aqueous hydrochloric acid [32] (6.0 M, 0.2 ml) (Fig. 1).

2.4 Purification and Formulation of [^{18}F]FMAU

The reaction mixture was diluted with 1 ml of HPLC mobile phase (6% acetonitrile:94% water) and purified by semipreparative HPLC using a Phenomenex Luna C18(2) reversed-phase column (5 μm, 250 × 10 mm) [33] with a flow rate of 3.5 ml/min. The product was collected from the column as monitored by flow-through radioactivity and UV (254 nm) detectors (Fig. 2). The HPLC fraction containing the [^{18}F]FMAU (typically eluting between 19 and 22 min) was collected, diluted with water (10 times the volume of the HPLC fraction), and passed through an

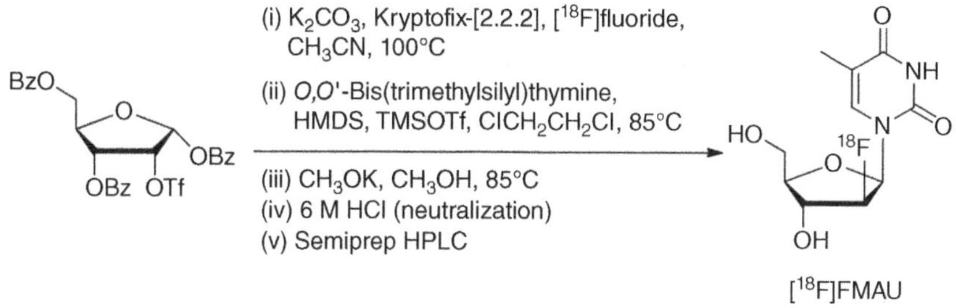

FIGURE 1 Radiosynthesis of [^{18}F]FMAU.

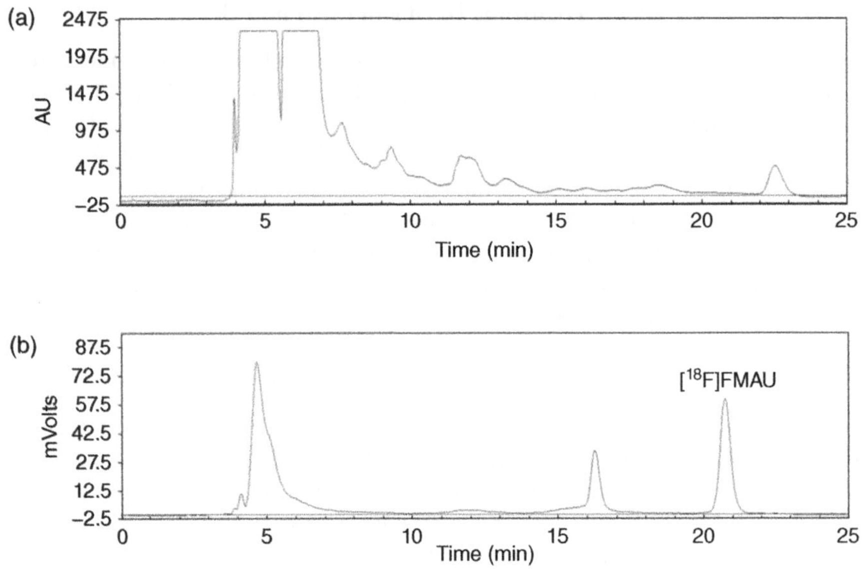

FIGURE 2 Semipreparative HPLC UV (a) and radioactive γ (b) traces for [18F]FMAU.

Oasis® HLB Plus cartridge [34]. The [18F]FMAU was then eluted with 50% EtOH/water solution (1 ml), passed through a Pall® Supor AEF sterile filter (0.22 μm) [35], and collected directly into a sterile empty vial. The resultant solution was adjusted to 8% EtOH/water by the addition of water (5.25 ml) for quality control (QC) testing.

3 QC PROCEDURES

CAUTION: All radiopharmaceuticals produced for clinical use must have local regulatory approval before use in humans. Trained personnel must carry out QC procedures, and each dose must meet all established QC criteria before release to the clinic.

Three batches of [18F]FMAU were produced according to the aforementioned conditions and analyzed according to the QC test summarized in Table 1. Each of the three batches met all of the established QC criteria.

3.1 Radioactivity Concentration

The radioactivity concentration is assessed by measuring the radioactivity of a sample of the [18F]FMAU final product contained in a low dead volume syringe. Our internal specification for radioactivity concentration is within the range of

TABLE 1 QC Specifications and Data for Three Batches of [^{18}F]FMAU

QC	Specification	Batch 1	Batch 2	Batch 3
Radioactivity concentration mCi/ml (GBq/ml)	1–75 (0.037–2.8)	2.96 (0.11)	3.28 (0.12)	2.72 (0.10)
Visual inspection	Clear, colorless solution, free of particulate matter	Pass	Pass	Pass
Radiochemical identity	Retention time must correspond to standard peak	Pass	Pass	Pass
Radiochemical purity (%)	≥95	99.9	99.9	99.9
Specific activity (Ci/μmol)	≥0.4	0.85	0.72	0.93
Residue solvent analysis (in ppm)	Acetonitrile <410	Pass	Pass	Pass
	1,2-Dichloroethane <5	Pass	Pass	Pass
	Methanol <3000	Pass	Pass	Pass
pH	5.5–7.5	5.5	6	6
Residual kryptofix[2.2.2] (μg/ml)	<50	Pass	Pass	Pass
Membrane filter integrity test	Meets or exceeds manufacturer's specifications	Pass	Pass	Pass
Radionuclidic identity ($t_{1/2}$) (min)	105–115	Pass	Pass	Pass
Bacterial endotoxins	<175 EU/injected volume	Pass	Pass	Pass
Sterility testing	Absence of microbials after 14-d incubation in two media	Pass	Pass	Pass

1–75 mCi/ml (0.037–2.8 GBq/ml) [36]. The radioactivity concentration for all three batches was within the range of 1–75 mCi/ml (0.037–2.8 GBq/ml).

3.2 Visual Inspection

The [^{18}F]FMAU final product is visually examined through leaded glass. The solution must be clear, colorless, and free of particulate matter. All three batches met this specification.

3.3 Radiochemical Identity and Purity

The radiochemical identity and purity of the [^{18}F]FMAU final product are assessed by analytical RP-HPLC [37]. The system is equipped with a γ-detector (NaI), a UV detector (254 nm), and a Phenomenex Luna C18(2) reversed-phase analytic column (5 μm, 250 × 4.6 mm) [38], operated in isocratic mode (8% acetonitrile/92% water) with a flow rate of 1 ml/min. Representative analytical HPLC traces are displayed in Fig. 3. The average offset between the UV and γ detectors is 0.35 min.

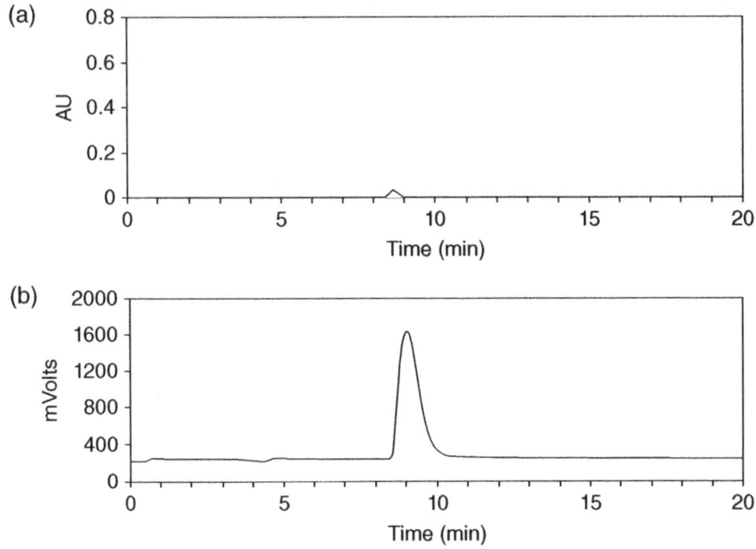

FIGURE 3 Analytical UV (a) and radioactive γ (b) HPLC traces for [^{18}F]FMAU.

Our internal specification for radiochemical purity is ≥95%. All three batches met this specification.

3.4 Specific Activity

The mass of [^{19}F]FMAU in the [^{18}F]FMAU final product solution is determined by RP-HPLC (see preceding text). Our internal specification for specific activity is ≥0.4 Ci/µmol at the end of synthesis. The specific activity of the three [^{18}F]FMAU batches ranged from 0.72 to 0.93 Ci/µmol.

3.5 Residual Solvents and Ethanol Concentration

Gas chromatography (GC) is used to determine the residual solvents levels in the dose. The GC system used was a Thermo Scientific TRACE™ 1310 with an AI/AS 1310 autosampler, split/splitless inlet, a flame ionization detector (FID), and a Zebron column (ZB-WAX plus 30 m × 0.53 mm, 1.0 µm film thickness) [39]. Limits imposed by the United States Pharmacopeia (USP) are <410 ppm/day for acetonitrile, <5 ppm/day for 1,2-dichloroethane, and <3000 ppm/day for methanol [40]. The residue solvent analysis from all three batches met the specifications.

3.6 pH

The pH of the [^{18}F]FMAU final product is assessed with pH strips. Our internal specification for pH is within the range of 5.5–7.5. The pH values for all three [^{18}F]FMAU batches fell within this range.

3.7 Residual Kryptofix[2.2.2]

The concentration of Kryptofix[2.2.2] in the [^{18}F]FMAU final product is assessed with an iodine spot test. Aluminum oxide plates are spotted with a Kryptofix[2.2.2] standard (50 μg/ml) and the final product solution, followed by drying and visualization of the plate using iodine vapor. The product contains <50 μg/ml of Kryptofix[2.2.2] if the spot for the standard is darker than that for the product. Our internal specification for Kryptofix[2.2.2] is that the concentration must be <50 μg/ml. All three [^{18}F]FMAU batches met this specification.

3.8 Membrane Filter Integrity Test

The integrity of the membrane filter used for sterilization of the [^{18}F]FMAU is assessed by the bubble point test [41]. Prior to the test, the membrane filter is carefully rinsed with 10 ml of water to eliminate ethanol present in the membrane. A gas supply is then attached to the inlet of the membrane filter, and the pressure slowly increased until a continuous stream of bubbles appears at the outlet. Out internal specification is that the bubble point must meet or exceed the manufacturer's specifications. The filters for all three [^{18}F]FMAU batches met this specification.

3.9 Radionuclidic Identity

The radionuclidic identity for the [^{18}F]FMAU final product is assessed by half-life determination over a period of 10 min. The calculated half-life must be within a range of 105–115 min. All three [^{18}F]FMAU batches met this specification.

3.10 Radionuclidic Purity

The radionuclidic purity of the [^{18}F]FMAU final product is assessed on a decayed sample for the presence of long-lived radionuclides by a gamma-ray spectroscopy system [42]. Our specification for radionuclidic purity is that 99.5% of the γ-emissions are due to F-18. All three [^{18}F]FMAU batches met this specification.

3.11 Bacterial Endotoxin Test

The presence of bacterial endotoxins in the [^{18}F]FMAU final product is assessed with the limulus amebocyte lysate (LAL) reagent. The test is performed by the gel-clot method according to USP requirements and specifications [43]. All three [^{18}F]FMAU batches met USP specifications of <175 EU/dose.

3.12 Sterility Test

The sterility of the [^{18}F]FMAU final product is assessed by inoculation of the product into fluid thioglycolate media (FTM) and tryptic soy broth (TSB). For each run, 0.25 ml of the inoculum is added into a media volume of ~10 ml [41]. The inoculated samples are incubated for 14 days at either 30–35°C (FTM) or 20–25°C (TSB) and visually inspected daily during the test period [44]. Positive standards must show microbial growth (turbidity), and [^{18}F]FMAU/negative controls must demonstrate the absence of microbial growth at the completion of the 14-day incubation period to be indicative of sterility.

WASTE DISPOSAL INFORMATION

All hazardous chemicals and toxic materials were disposed of according to Prudent Practices in the Laboratory (Washington, DC: National Academy Press, 1995).

CHEMICAL ABSTRACTS NOMENCLATURE (REGISTRY NUMBER)

Acetonitrile (75-05-8)
O,*O*′-Bis(trimethylsilyl)thymine (7288-28-0)
1,2-Dichloroethane (107-06-2)
Ethanol (64-17-5)
HMDS (999-97-3)
4,7,13,16,21,24-Hexaoxa-1,10-diazabicyclo[8.8.8]hexacosane (23978-09-8)
Hydrochloric acid (7647-01-0)
Methanol (67-56-1)
Potassium carbonate (584-08-7)
Potassium methoxide (865-33-8)
1,3,5-Tri-*O*-benzoyl-2-*O*-trifluoromethanesulfonyl-α-furanose (97614-41-0)
TMSOTf (27607-77-8)

REFERENCES AND NOTES

For detailed supplier information, see Appendix 1.

1. M. E. Perlman, P. S. Conti, B. Schmall, K. A. Watanabe, *Int. J. Nucl. Med. Biol.,* 1984, 11, 215.
2. H. K. Misra, E. E. Knaus, L. I. Wiebe, D. L. Tyrrell, *Appl. Radiat. Isot.,* 1986, 37, 901.
3. J. R. Grierson, A. F. Shields, J. H. Courter, *J. Label. Compd. Radiopharm.,* 1994, 35, 290.
4. J. R. Grierson, A. F. Shields, J. M. Link, *J. Label. Compd. Radiopharm.,* 1995, 37, 608.
5. P. S. Conti, M. M. Alauddin, J. R. Fissekis, B. Schmall, K. A. Watanabe, *Nucl. Med. Biol.,* 1995, 22, 783.
6. P. S. Conti, J. R. Bading, M. M. Alauddin, J. H. Fissekis, B. Berenj, S. Hayles, H. Jadvar, *J. Nucl. Med.,* 2002, 43, 26P.
7. A. F. Shields, J. R. Grierson, S. M. Kozawa, M. Zheng, *J. Nucl. Med.,* 1995, 36, 208P.
8. A. F. Shields, J. R. Grierson, B. M. Dohmen, H. J. Machulla, J. C. Stayanoff, J. M. Lawhorn-Crews, J. E. Obradovich, O. Muzik, T. J. Mangner, *Nat. Med.,* 1998, 4, 1334.
9. J. R. Bading, A. F. Shields, *J. Nucl. Med.,* 2008, 49 (Suppl. 2), 64S.
10. T. C. Chou, X. B. Kong, M. P. Fanucchi, Y. C. Cheng, K. Takahashi, K. A. Watanabe, J. J. Fox, *Antimicrob. Agents Chemother.,* 1987, 31, 1355.
11. M. M. Alauddin, P. S. Conti, J. D. Fissekis, *J. Label. Compd. Radiopharm.,* 2002, 45, 583.
12. M. M. Alauddin, P. S. Conti, J. D. Fissekis, *J. Label. Compd. Radiopharm.,* 2003, 46, 285.

13. T. J. Mangner, R. W. Klecker, L. Anderson, A. F. Shields, *Nucl. Med. Biol.*, 2003, 30, 215.
14. J. R. Bading, A. H. Shahinian, P. Bathija, P. S. Conti, *Nucl. Med. Biol.*, 2000, 27, 361.
15. J. R. Bading, A. H. Shahinian, A. Vail, P. Bathija, G. W. Koszalka, R. T. Koda, M. M. Alauddin, J. D. Fissekis, P. S. Conti, *Nucl. Med. Biol.*, 2004, 31, 407.
16. H. Sun, T. J. Mangner, J. M. Collins, O. Muzik, K. Douglas, A. F. Shields, *J Nucl Med*, 2005, 46, 292.
17. H. Sun, A. Sloan, T. J. Mangner, U. Vaishampayan, O. Muzik, J. M. Collins, K. Douglas, A. F. Shields, *Eur. J. Nucl. Med. Mol. Imaging*, 2005, 32, 15.
18. A. K. Buck, H. Schirrmeister, M. Hetzel, M. Von Der Heide, G. Halter, G. Glatting, T. Mattfeldt, F. Liewald, S. N. Reske, B. Neumaier, *Cancer Res.*, 2002, 62, 3331.
19. L. B. Been, P. H. Elsinga, J. de Vries, D. C. Cobben, P. L. Jager, H. J. Hoekstra, A. J. Suurmeijer, *Eur. J. Surg. Oncol.*, 2006, 32, 39.
20. A. K. Buck, M. Bommer, S. Stilgenbauer, M. Juweid, G. Glatting, H. Schirrmeister, T. Mattfeldt, D. Tepsic, D. Bunjes, F. M. Mottaghy, B. J. Krause, B. Neumaier, H. Dohner, P. Moller, S. N. Reske, *Cancer Res.*, 2006, 66, 11055.
21. M. Muzi, A. M. Spence, F. O'Sulliva, D. A. Mankoff, J. M. Wells, J. R. Grierson, J. M. Link, K. A. Krohn, *J. Nucl. Med.*, 2006, 47, 1612.
22. Z. Li, H. Cai, P. S. Conti, *Nucl. Med. Biol.*, 2011, 38, 201.
23. K. Chen, Z. Li, P. S. Conti, *Nucl. Med. Biol.*, 2012, 39, 1019.
24. V. Paolillo, S. Riese, J. G. Gelovani, M. M. Alauddin, *J. Label. Compd. Radiopharm.*, 2009, 52, 553.
25. [^{18}O]H_2O purchased from ISOFLEX USA, Inc., and used as received.
26. Siemens Medical Solutions USA, Inc. Molecular Imaging, 2501 N. Barrington Road, Hoffman Estates, IL 60192-5203, USA.
27. Potassium carbonate was purchased from Sigma (ACS reagent grade, 99+%) and used directly. Sterile water was purchased from Hospira, Inc.
28. Kryptofix-[2.2.2] was purchased from Sigma (98%) and used directly. Acetonitrile was purchased from Sigma (anhydrous, ≥99%).
29. *O,O'*-Bis(trimethylsilyl)thymine (97%), 1,2-dichloroethane (anhydrous, 99.8%), hexamethyldisilazane (HMDS) (99.9%), and trimethylsilyl trifluoromethanesulfonate (99%) were purchased from Sigma.
30. Potassium methoxide (95%) was purchased from Sigma.
31. Methanol (anhydrous, 99.8%) was purchased from Sigma.
32. Hydrogen chloride (ACS reagent grade, 37%) was purchased from Sigma.
33. Phenomenex Luna C18(2) reversed-phase column (5 μm, 250 × 10 mm), part number 00G-4252-N0.
34. The cartridge was washed with EtOH (5 ml) followed by water (10 ml) before use.
35. Care must be taken to ensure that the sterile filters contain low amounts of leachable materials, which may give rise to UV impurities in the final dose.
36. This specification is based on stability studies and the ease of handling doses. The radioactivity concentration could be increased to values beyond this specification, but this would require additional stability studies to demonstrate the shelf life of the product.
37. Analytical reversed-phase HPLC was performed on two Waters 515 HPLC pumps, a Waters 2487 absorbance UV detector, and a Ludlum Model 2200 radioactivity detector, which were operated by Waters Empower 2 Software.
38. Phenomenex Luna C18(2) reversed-phase analytic column (5 μm, 250 × 4.6 mm), part number 00G-4252-E0.
39. GC column: ZB-WAX plus 30 m × 0.53 mm (1.0 μm film thickness), purchased from Phenomenex (part number 7HK-G013-22).
40. International Conference on Harmonisation of Technical Requirements for Registration of Pharmaceuticals for Human Use (ICH).
41. S. Yu, *Biomed. Imaging Intervention J.*, 2006, 2, e57–e67.
42. A gamma-ray spectroscopy system with a Model 727 from Canberra Industries, Inc., and the Genie 2000 program were used.
43. U.S. Pharmacopeia <85> Bacterial endotoxins test. USP 32–NF 27. 2009.
44. U.S. Pharmacopeia <71> Sterility tests. USP 32–NF 27. 2009.

PART II

CARBON-11 LABELED RADIOPHARMACEUTICALS

CHAPTER 7

SYNTHESIS OF *N*-[^{11}C]METHYL-4-PIPERIDINYL PROPIONATE ([^{11}C]PMP)

Xia Shao and Peter J. H. Scott

Department of Radiology, University of Michigan School of Medicine, Ann Arbor, Michigan, USA

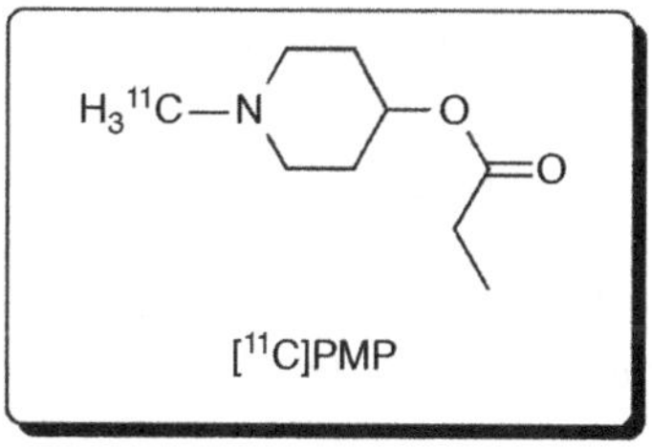

1 INTRODUCTION

Acetylcholinesterase (AChE) catalyzes the hydrolysis of the neurotransmitter acetylcholine to choline and acetic acid, leading to the termination of cholinergic neurotransmission. The activity of AChE is an important marker for degeneration of the central cholinergic system.

Clinical and postmortem studies provide evidence that the biochemical changes in the brains of Alzheimer's disease (AD) patients include decrease in AChE activity [1–3]. ^{11}C-labeled piperidinyl ester, 1-[^{11}C]methylpiperidin-4-yl propionate ([^{11}C]PMP), serves as AChE substrate and is hydrolyzed to a hydrophilic product, *N*-[^{11}C]methyl-4-piperidinol (MPOH), which does not cross the blood–brain barrier, and is irreversibly trapped locally in the brain according to the distribution of AChE enzyme activity. Kinetic analysis of this radioactivity trapping provides a quantitative estimate of the regional AChE activity [4–6]. This compound is currently in routine clinical use for the study of AChE function in patients with AD [7, 8]. In addition, actual measures of central AChE activity in AD patients could also support new drug development and be used to optimize the drug dosage schedule.

Radiochemical Syntheses, Volume 2: Further Radiopharmaceuticals for Positron Emission Tomography and New Strategies for Their Production, First Edition. Edited by Peter J. H. Scott.

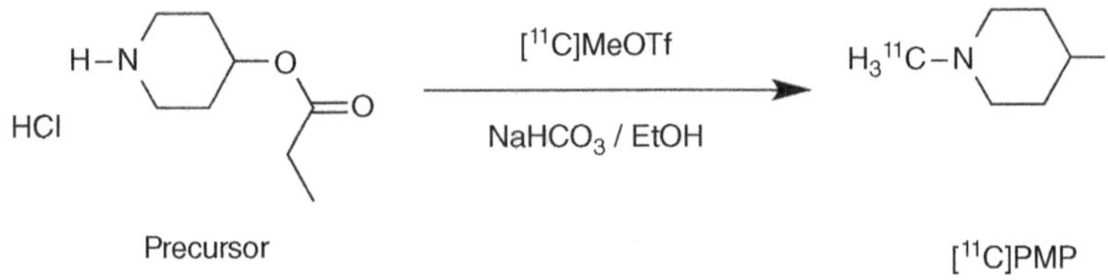

FIGURE 1 Radiosynthesis of [^{11}C]PMP.

In the past, when [^{11}C]PMP was synthesized, its free base precursor was prepared and dried prior to every synthesis [4, 9]. Recently, after we discovered that [^{11}C] methylation can effectively be conducted in a ethanol/water mixture, we developed a new method using ethanol as the only solvent [10]. This method eliminates all other organic solvents and also simplifies the procedure by adding aqueous carbonate solution directly to the precursor (Fig. 1).

2 SYNTHESIS PROCEDURES

CAUTION: All radiochemical syntheses must be carried out using appropriate equipment in a facility authorized for the use of radioactive materials. Personal protective equipment must be worn and all local radiation safety laws followed.

2.1 Production of [^{11}C]CO_2

[^{11}C]CO_2 was produced by General Electric (GE) PETtrace cyclotron. The target [11] was loaded with [^{14}N]N_2 gas [12] and bombarded with a proton beam (40 μA beam for 30 min) to generate ~3 Ci of [^{11}C]CO_2 by the $^{14}N(p,\alpha)^{11}C$ nuclear reaction.

2.2 Preparation of [^{11}C]Methyl Triflate

[^{11}C]CO_2 was delivered from the target by nitrogen gas pressure to a column filled with a mixture of molecular sieve (0.3 g) [13] and Shimalite-Ni catalyst (0.2 g) [14] at room temperature. The column was sealed under H_2 and then heated to 350°C for 20 s to reduce the [^{11}C]CO_2 to [^{11}C]CH_4. [^{11}C]CH_4 was passed through a phosphorus pentoxide desiccant [15] column and trapped in a carbosphere [16] column at −78°C. Gaseous [^{11}C]CH_4 was entered into a circulation loop system including a gas pump, an iodine [17] column at 100°C, a reactor tube at 720°C, two adjacent Ascarite II [18] columns, and a Porapak [19] column at room temperature. The gaseous mixture was then circulated for 5 min, while [^{11}C]CH_3I accumulated on the Porapak column. [^{11}C]CH_3I was converted to [^{11}C]CH_3OTf by passing it through a silver triflate-Graphpac [20] column heated to 190°C.

2.3 Synthesis of [^{11}C]PMP

To prepare [^{11}C]PMP, the TRACERlab synthesis module was configured as illustrated in Fig. 2 and loaded as follows: reaction vessel, [^{11}C]PMP precursor (0.5 mg in 100 μl ethanol plus 30 μl of 0.1 M $NaHCO_3$ aqueous solution); vial 1, semipreparative

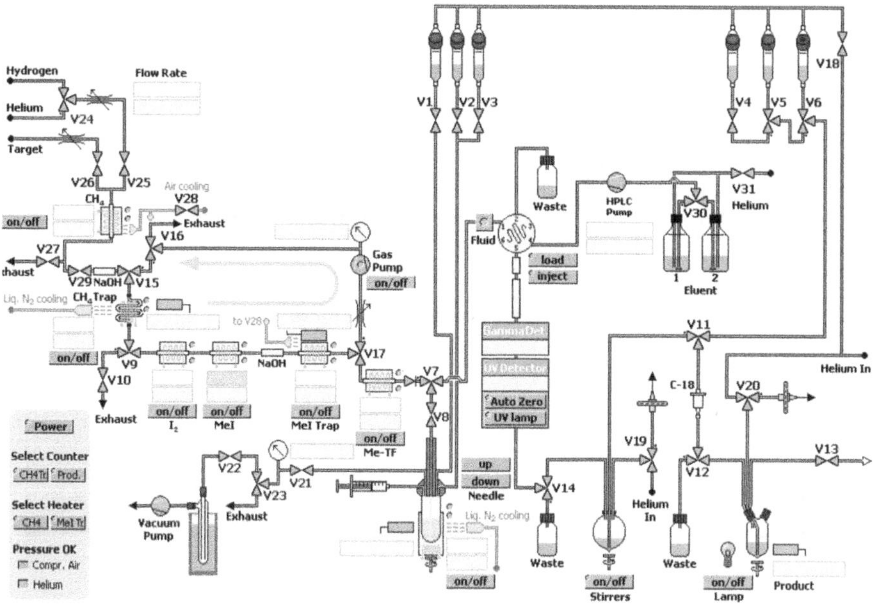

FIGURE 2 TRACERlab $FX_{C\text{-}Pro}$ for the preparation of [^{11}C]PMP.

HPLC mobile phase (20 mM NH_4OAc in 5% EtOH, 1.0 ml); and product collection vial, 0.9% NaCl for injection, USP (2.0 ml). [^{11}C]Methyl triflate was bubbled through the precursor solution at 15 ml/min for 3 min. The reaction mixture was then diluted with 1 ml of HPLC mobile phase and purified using semipreparative HPLC (column, Phenomenex Luna C18, 250×10 mm; mobile phase, 20 mM NH_4OAc in 5% EtOH; flow rate, 4 ml/min). The product peak was collected (RT ~ 12–14 min; see Fig. 3 for a typical HPLC trace) for 2 min (8 ml), diluted with USP saline (2 ml) to provide a final ethanol concentration <5%, and passed through a 0.22 μm sterilizing filter [21] into a sterile dose vial.

3 QUALITY CONTROL PROCEDURES

CAUTION: All radiochemicals produced for clinical use must have local regulatory approval (e.g., FDA, EMEA, MHRA, PFSB, etc.) prior to human use. Quality control (QC) procedures must be carried out by trained personnel, and each dose must meet all established QC criteria before release to the clinic.

QC procedures for [^{11}C]PMP, based upon the current requirements for radiopharmaceuticals laid out in the US Pharmacopeia [22], are summarized in the following text. Complete QC data for 3 repeat batches of [^{11}C]PMP produced using the methods disclosed herein are summarized in Table 1. Each of the three doses met all of the established QC criteria.

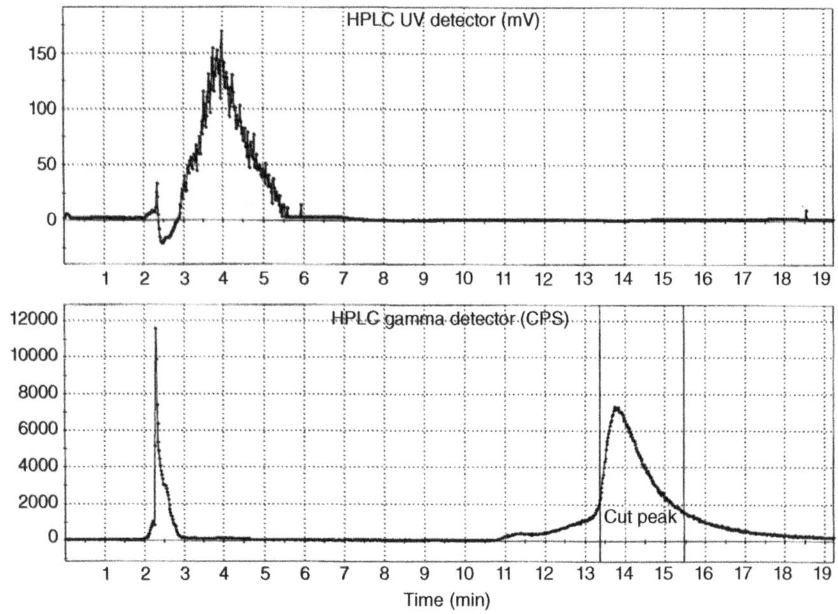

FIGURE 3 Typical semipreparative HPLC trace for [^{11}C]PMP.

TABLE 1 QC Data for 3 Repeat Runs of [^{11}C]PMP

QC Test	Release Criteria	Run 1	Run 2	Run 3
Yield/mCi (GBq)	N/A	213 (7.9)	165 (6.1)	148 (5.5)
Visual inspection	Clear, colorless	Pass	Pass	Pass
Radiochemical identity	RRT = 0.9–1.1	1.0	1.0	1.0
Radiochemical purity	≥95%	97.7%	99.8%	99.0%
Specific activity (mCi/μmol)	N/A	Mass < LOD[a]	Mass < LOD[a]	Mass < LOD[a]
Dose pH	4.5–7.5	6.0	6.0	6.0
Radionuclidic identity	18.5–22.4 min	20 min	20 min	20 min
Sterile filter integrity test	>45 psi	Pass	Pass	Pass
Endotoxin analysis	≤17.5 EU/ml	≤2 EU/ml	≤2 EU/ml	≤2 EU/ml
Sterility testing	No colony growth out of 14 days	Pass	Pass	Pass

[a] LOD = 25 μg/ml, limit of detection for [^{11}C]PMP.

3.1 Visual Inspection

The [^{11}C]PMP QC sample was examined behind shielding lead glass in a clear vial and had to be clear, colorless, and free of visible particulate matter.

3.2 Dose pH

The pH of the [^{11}C]PMP dose was determined by applying a small amount of the QC sample to a colorpHast® pH 2.0–9.0 nonbleeding pH-indicator strip [23] and determined by visual comparison to the scale provided. Dose pH had to be 4.5–7.5.

3.3 Radiochemical Identity

Radiochemical identity was determined using a Shimadzu LC 2010 AHT Liquid Chromatograph fitted with UV detectors and Bioscan gamma detectors. Conditions: column was Luna C8(2) 5 µ 150×2 mm; solvent was 5% MeOH/20 mM NH_4OAc (pH = 4.5); flow rate was 0.35 ml/min; oven was set to 30°C; injection volume was 20 µl; and UV detector operated at 220 nM. The retention time of [^{11}C]PMP (~4.5 min) was compared to that of a known reference standard and had to be within 10% (relative retention time (RRT) had to be 0.9–1.1):

$$\mathrm{RRT} = \frac{t_{\mathrm{r(C11-PMP)}}}{t_{\mathrm{r(C12-PMP)}}}$$

3.4 Radiochemical Purity

Radiochemical purity of each batch was analyzed using the HPLC system described in Section 3.3. Radiochemical purity had to be greater than 95%. Representative analytical HPLC tracers are displayed in Fig. 4.

3.5 Specific Activity

The concentration of [^{11}C]PMP can be determined using the HPLC system described in Section 3.3 by integrating the area under the UV peak of [^{11}C]PMP and comparing with the area of known reference standard peak. The specific activity (As) can be calculated as the division of radioactivity (mCi) by the amount of PMP (µmol):

$$\mathrm{As(mCi/\mu mol)} = A\mathrm{(mCi)}/[C_{\mathrm{PMP}}(\mathrm{\mu g/ml}) \times V\mathrm{(ml)}/171(\mathrm{\mu g/\mu mol})]$$

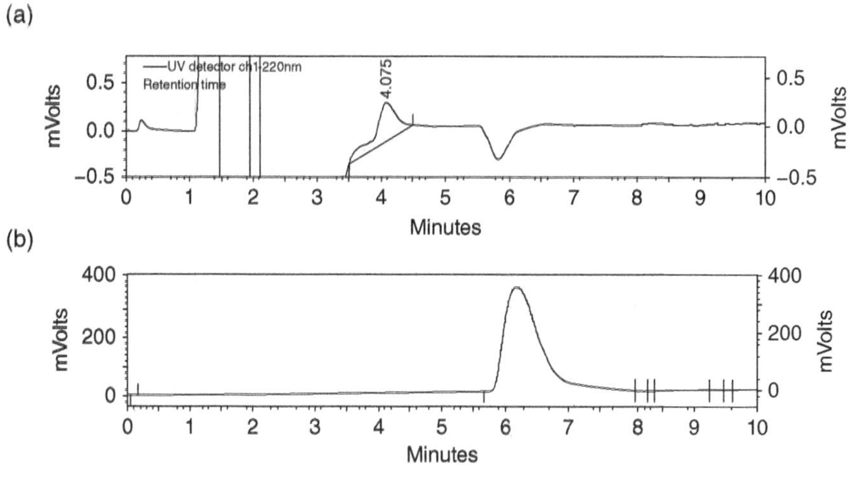

FIGURE 4 Analytical UV (a) and radioactive (b) HPLC trace of [^{11}C]PMP.

A is the measured radioactivity; *V* is the volume of sample measured; *C* is the calculated concentration of PMP.

There is no specific activity release limit for [^{11}C]PMP in place at our institution. The injection volume was based on the injectable mass limit for PMP and was calculated on a dose-by-dose basis by the administering study team.

3.6 Radionuclidic Identity

Activities were measured using a Capintec CRC®-15R Radioisotope Dose Calibrator, and half-life was calculated using the equation below. Calculated half-life had to be 18.5–22.4 min:

$$T_{1/2} = -\ln 2 \left\{ \frac{\text{time difference}}{[\ln(\text{ending activity}/\text{starting activity})]} \right\}$$

3.7 Sterile Filter Integrity Test

The sterile filter with needle that was used to convey the dose into the dose vial was connected to a nitrogen supply via a regulator. The filter was immersed in water and the pressure was gradually increased while observing for a steady stream of bubbles emanating from the needle. If this happened at a pressure greater than the filter acceptance level, then the filter was considered to have been intact and thus passed.

3.8 Endotoxin Analysis

Endotoxin content in doses of [^{11}C]PMP was analyzed by a Charles River Laboratories EndoSafe™ Portable Testing System [24] as per the US Pharmacopeia [25]. Doses had to contain ≤175 endotoxin units (EU) that was adjusted to 17.5 EU/ml as our formulation volume was 10 ml.

3.9 Sterility Testing

Samples of the [^{11}C]PMP dose was added to tubed fluid thioglycolate media (FTM) and to tubed tryptic soy broth (TSB) media. FTM was used to test for anaerobes, aerobes, and microaerophiles, TSB was used to test for fastidious and nonfastidious organisms. These media tubes as well as positive and negative controls were incubated at 30–35°C (FTM) and 22°C (TSB) for 14 days according to current USP guidelines [26]. Tubes were visually inspected on the 3rd, 8th, and 14th day after inoculation and are compared to positive and negative controls. Positive controls had to show growth (turbidity) and the test article samples ([^{11}C]PMP) and negative controls had to have no growth after 14 days in order to pass the test for sterility. Positive controls could be established separately as growth promotion tests and microbial enumeration study [27] according to USP guidelines.

WASTE DISPOSAL INFORMATION

All hazardous chemicals and toxic materials were disposed of according to Prudent Practices in the Laboratory (Washington, DC: National Academy Press, 1995).

REFERENCES AND NOTES

For detailed supplier information, see Appendix 1.

1. D.E. Kuhl, R.A. Koeppe, S. Minoshima, S.E. Snyder, E.P. Ficaro, N.L. Foster, F.A. Frey, M.R. Kilbourn, *Neurology*, 1999, 52, 691.
2. H. Shinotoh, H. Namba, K. Fukushi, S.-L. Nagatsuka, N. Tanaka, A. Aotsuka, S. Tanada, T. Irie, *Alzheimer Dis. Assoc. Disord.*, 2000, 14, S114.
3. L.M. Bierer, V. Haroutunian, S. Gabriel, P.J. Knoott, L.S. Carlin, D.P. Purohit, D.P. Perl, J. Schmeidler, P. Kanof, K.L. Davis, *J. Neurochem.*, 1995, 6, 749.
4. S.E. Snyder, L. Tluczek, D.M. Jewett, T.B. Nguyen, M.R. Kilbourn, D.E. Kuhl, *Nucl. Med. Biol.*, 1998, 25 751.
5. R.A. Koeppe, K.A. Frey, S.E. Snyder, M. Phillipp, M.R. Kilbourn, D.E. Kuhl, *J. Cereb. Blood Flow Metab.*, 1999, 1, 1150.
6. X. Shao, J.M. Lisi, E.R. Butch, M.R. Kilbourn, S.E. Snyder, *Nucl. Med. Biol.*, 2003, 30, 293.
7. R.A. Koeppe, K.A. Frey, S.E. Snyder, M.R. Kilbourn, D.E. Kuhl, *J. Cereb. Blood Flow Metab.*, 1997, 17, S329.
8. S.E. Snyder, N. Gunupudi, P.S. Sherman, E.R. Butch, M.B. Skaddan, M.R. Kilbourn, R.A. Koeppe, D.E. Kuhl, *J. Cereb. Blood Flow Metab.*, 2001, 21, 132.
9. X. Shao, R. Hoareau, A. Runkle, L.J.M. Tluczek, B.G. Hockley, B.D. Henderson, P.J.H. Scott, *J. Label. Compd. Radiopharm.*, 2011, 54, 319.
10. X. Shao, M.V. Fawaz, K.S. Jang, P.J.H. Scott, *Appl. Radiat. Isot.*, 2014, 89, 125.
11. GE standard carbon-11 target with a 75 μm Havar target foil and standard vacuum foil.
12. Ultrahigh-purity nitrogen gas with 1% of oxygen was purchased from Purity Plus.
13. Molecular sieve (4A, 80/100 mesh) was purchased from Alltech, pn: 5624.
14. Shimalite-Ni reduced (80/100 mesh) was purchased from Shimadzu (Kyoto, Japan), pn: 22127719.
15. Phosphorus pentoxide (with moisture indicator) was purchased from Fluka, pn: 79610.
16. Carbosphere (60/80 mesh) was purchased from Alltech (Grace), pn: 5680.
17. Iodine (crystals) was purchased from EMD (Germany), pn: IX01203.
18. Ascarite II (8–20 mesh, 3 in. in TRACERlab standard column) was purchased from Thomas Scientific, pn: C049H40.
19. Porapak type Q (0.3 g, 50–80 mesh) was purchased from Alltech, pn: 2700.
20. Silver triflate-Graphpac was prepared by dissolving 5 g of silver trifluoromethanesulfonate (Sigma-Aldrich) in 100 ml of acetonitrile (HPLC grade, Fisher) and then mixed with 10 g of Graphpac (80/100 mesh, GC grade, Alltech). The solvent was evaporated by rotary vapor under vacuum, and 0.35 g of dried powder was used to pack a TRACERlab standard column.
21. Millex-GS 0.22 μm filter was purchased from Millipore (United States), pn: SLGSV 255F.
22. U.S. Pharmacopeia <823>, Radiopharmaceuticals for positron emission tomography compounding. USP 32-NF 27. 2009.
23. pH-indicator strips were purchased from EMD Chemicals, Inc., pn: 9578-3.
24. Charles River Laboratories, United States.
25. U.S. Pharmacopeia <85>, Bacterial endotoxins test. USP 32-NF 27. 2009.
26. U.S. Pharmacopeia <71>, Sterility tests. USP 32-NF 27. 2009.
27. U.S. Pharmacopeia <61>, Microbial enumeration tests. USP 32-NF 27. 2009.

WASTE DISPOSAL INFORMATION

CHAPTER 8

SYNTHESIS OF 1-(2,4-DICHLOROPHENYL)-4-CYANO-5-(4-[^{11}C] METHOXYPHENYL)-*N*-(PIPERIDIN-1-YL)-1*H*-PYRAZOLE-3-CARBOXAMIDE ([^{11}C]OMAR)

XIA SHAO, KEUNSAM JANG, AND PETER J. H. SCOTT

Department of Radiology, University of Michigan School of Medicine, Ann Arbor, Michigan

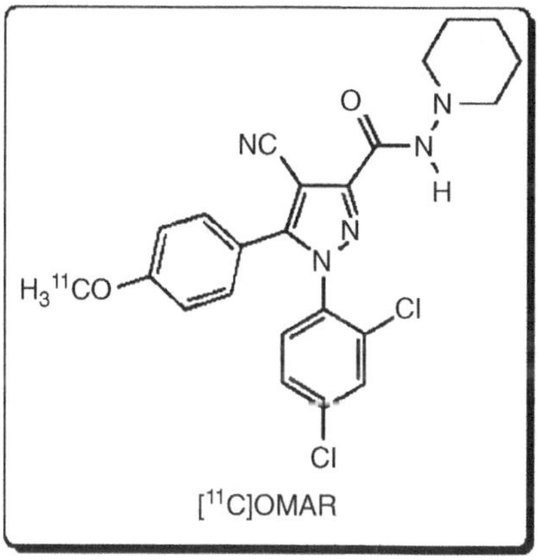

1 INTRODUCTION

The endogenous cannabinoid system consists of two types of G protein-coupled receptors: cannabinoid receptor 1 (CB1) and cannabinoid receptor 2 (CB2). The former is predominantly found in the brain, while the latter tends to be expressed by cancer cells. There is significant interest in evaluating CB1 receptors using PET imaging because their involvement has been implicated in neuropsychiatric and neurodegenerative disorders (e.g., schizophrenia, depression, obesity, as well as addiction to alcohol and other drugs). Of the available PET imaging agents for CB1 receptors, 1-(2,4-dichlorophenyl)-4-cyano-5-(4-[^{11}C]methoxyphenyl)-*N*-(piperidin-1-yl)-1*H*-pyrazole-3-carboxamide ([^{11}C]OMAR or [^{11}C]JHU75528,

Radiochemical Syntheses, Volume 2: Further Radiopharmaceuticals for Positron Emission Tomography and New Strategies for Their Production, First Edition. Edited by Peter J. H. Scott.

originally developed by Horti and colleagues at Johns Hopkins [1]), has emerged as one of the most promising and is currently being evaluated in clinical trials for conditions such as schizophrenia [1b].

2 SYNTHESIS PROCEDURES

CAUTION: All radiochemical synthesis must be carried out using appropriate equipment in a facility authorized for the use of radioactive materials. Personal protective equipment must be worn and all local radiation safety laws followed.

2.1 Production of [^{11}C]CO_2

[^{11}C]CO_2 was produced by General Electric Medical Systems (GEMS) PETtrace cyclotron. The target [2] was loaded with [^{14}N]N_2 gas [3] and bombarded with a proton beam (40 μA beam for 30 min) to generate ~3 Ci (111 GBq) of [^{11}C]CO_2 by the ^{14}N(p,α)^{11}C nuclear reaction.

2.2 Production of [^{11}C]OMAR

[^{11}C]OMAR (**2**) was produced by [^{11}C]methylation of hydroxyl-OMAR tetrabutylammonium salt with [^{11}C]methyl triflate (Fig. 1), in the HPLC loop of a modified TRACERlab $FX_{C\text{-}Pro}$. Hydroxyl-OMAR tetrabutylammonium salt (**1**) was dissolved in 100 μl of ethanol and slowly injected into the HPLC loop. [^{11}C]CO_2 was delivered from the target by nitrogen gas pressure to a column filled with a mixture of molecular sieve (0.3 g) [4] and Shimalite-Ni catalyst (0.2 g) [5] at room temperature. The column was sealed under H_2 and then heated to 350°C for 20 s to reduce the [^{11}C]CO_2 to [^{11}C]CH_4. [^{11}C]CH_4 was passed through a phosphorus pentoxide desiccant [6] column and trapped in a Carbosphere [7] column at −78°C. Gaseous [^{11}C]CH_4 was entered into a circulation loop system including a gas pump, an iodine [8] column at 100°C, a reactor tube at 720°C, two adjacent Ascarite II [9] columns, and a Porapak [10] column at room temperature. The gaseous mixture was then circulated

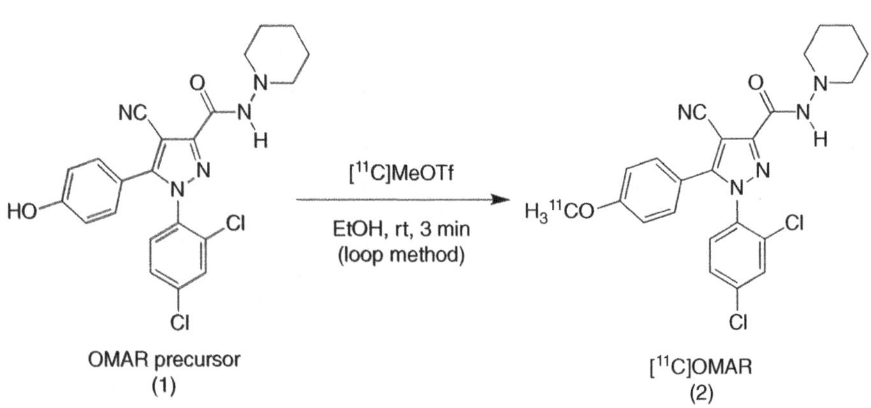

FIGURE 1 Synthesis of [^{11}C]OMAR.

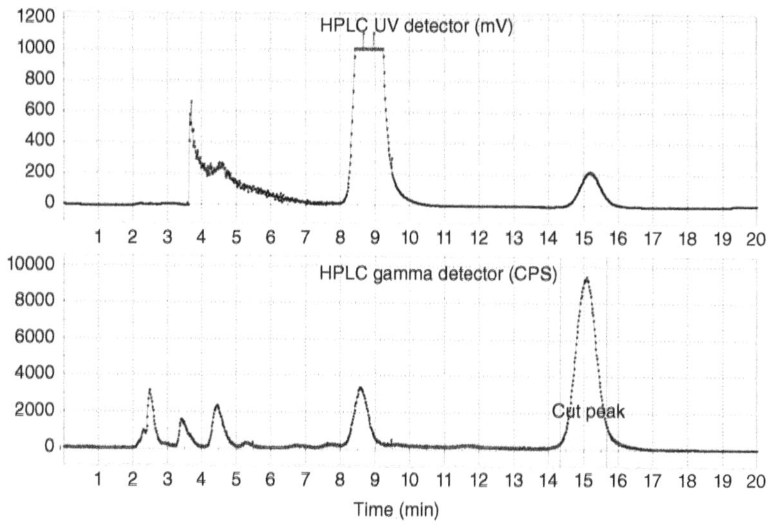

FIGURE 2 Semi-preparative HPLC trace.

for 5 min, while [^{11}C]CH_3I accumulated on the Porapak column. [^{11}C]CH_3I was converted to [^{11}C]CH_3OTf by passing it through a silver triflate-Graphpac [11] column heated to 190°C. [^{11}C]OMAR was then produced by passing [^{11}C]CH_3OTf through the HPLC loop containing OMAR precursor (**1**) at flow rate 15 ml/min.

2.3 Purification and Formulation of [^{11}C]OMAR

After passing [^{11}C]MeOTf through the HPLC loop for 5 min, the reaction mixture was injected onto a HPLC column (Luna C5, 100 × 10 mm) and eluted using 20 mM NH_4OAc in 45% ethanol at 3 ml/min. The product peak (retention time ~15 min; see Fig. 2 for a typical semipreparative HPLC trace) was collected for 60 s into 50 ml water. This mixture was passed through a C18 extraction disk [12] where the [^{11}C]OMAR was trapped. Seven milliliters of Milli-Q water was then passed through to remove impurities from the disk. The disk was dried by passing helium gas through for 1 min, and the [^{11}C]OMAR was eluted off using 0.5 ml ethanol [13] followed by 9.5 ml saline [14]. The formulation was passed through a Millipore Millex-GV filter [15] into a sterile dose vial [16] to produce the final formulated [^{11}C]OMAR product, which was submitted for quality control (QC) testing.

3 QC PROCEDURES

CAUTION: All radiochemicals produced for clinical use must have local regulatory approval (e.g., FDA, EMEA, MHRA, PFSB, etc.) prior to human use. QC procedures must be carried out by trained personnel, and each dose must meet all established QC criteria before release to the clinic.

TABLE 1 QC Data for 3 Repeat Runs of [^{11}C]OMAR

QC Test	Release Criteria	Run 1	Run 2	Run 3
Yield (mCi)	N/A	41	55	41
Visual inspection	Clear, colorless	Pass	Pass	Pass
Radiochemical identity	RRT=0.9–1.1	1.0	1.0	1.0
Radiochemical purity	≥95%	100%	100%	98.0%
Specific activity (mCi/μmol)	N/A	4396	4593	4490
Dose pH	4.5–7.5	5.0	5.0	5.0
Radionuclidic identity	18.5–22.4 min	20 min	20 min	20 min
Sterile filter integrity test	>45 psi	Pass	Pass	Pass
Endotoxin analysis	≤17.5 EU/ml	≤2 EU/ml	≤2 EU/ml	≤2 EU/ml
Sterility testing	No colony growth out of 14 days	Pass	Pass	Pass

QC procedures for [^{11}C]OMAR, based upon the current requirements for radiopharmaceuticals laid out in the US Pharmacopeia [17], are summarized in the following text. Complete QC data for three repeat batches of [^{11}C]OMAR produced using the methods disclosed herein are summarized in Table 1. Each of the three doses met all of the established QC criteria.

3.1 Visual Inspection

The [^{11}C]OMAR QC samples were examined behind a PET L-block and had to be clear, colorless, and free of any particulate matter.

3.2 Dose pH

The pH of the [^{11}C]OMAR doses were determined by applying small amounts of the QC sample to colorpHast® pH 2.0–9.0 nonbleeding pH indicator strips [18] and determined by visual comparison to the scale provided. Dose pH had to be 4.5–7.5.

3.3 Radiochemical Identity

Radiochemical identity was determined using a Shimadzu LC 2010 AHT Liquid Chromatograph [19] fitted with UV detectors and Bioscan gamma detectors [20]. Conditions: column was Jupiter C18(2) 5 μ 250×4.6 mm; solvent was 60% MeCN/20% MeOH/20% 20 mM NaH_2PO_4; flow rate was 1.5 ml/min; oven was set to 30°C; injection volume was 10 μl; and UV detector operated at 254 nm. The retention time of [^{11}C]OMAR (~4.0–4.5 min; Fig. 3) was compared to that of a known reference standard and had to be within 10% (relative retention time (RRT) had to be 0.9–1.1):

$$\mathrm{RRT} = \frac{t_r\,(\text{C11-OMAR})}{t_r\,(\text{C12-OMAR})}$$

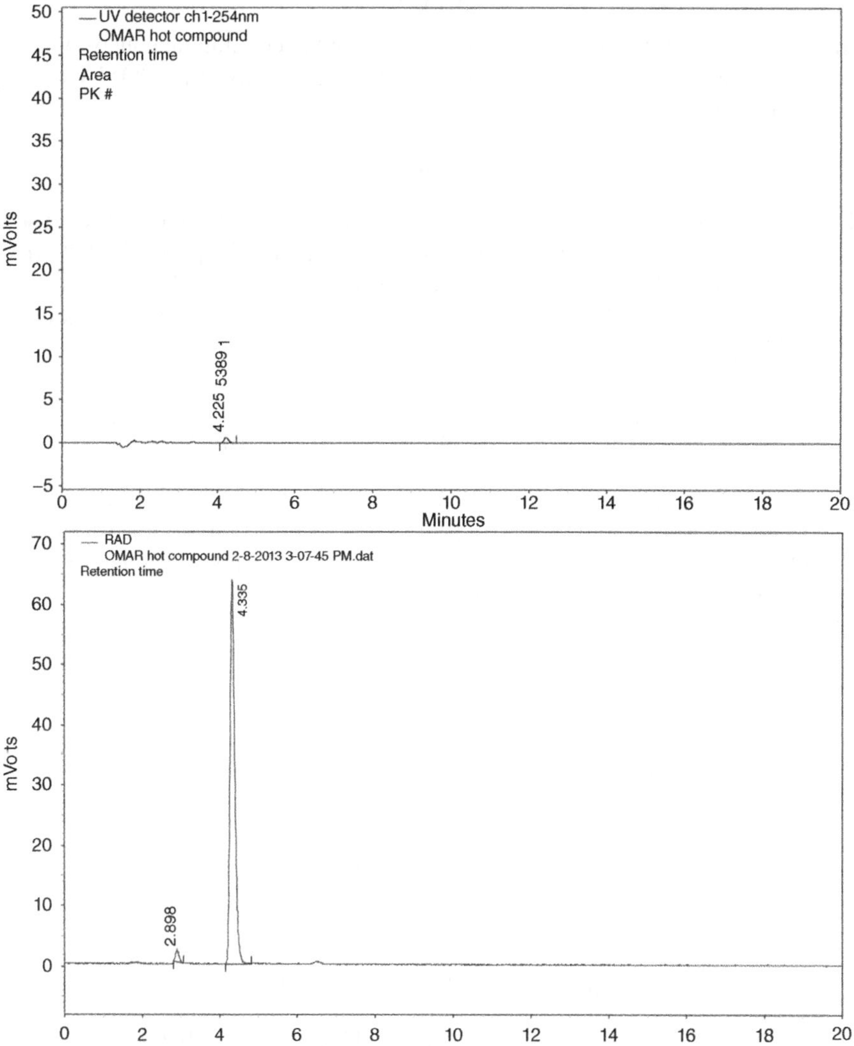

UV detector Ch1-254nm results

retention time	Area	Area %	Height	Width	S/N (ASTM)
4.225	5389	100	661	0.433	3.94

RAD results

retention time	Area	Area %	Width
2.898	12,919	2	0.25
4.335	556,393	98	0.67

FIGURE 3 Analytical UV and radioactive HPLC trace of [^{11}C]OMAR.

3.4 Radiochemical Purity

HPLC analysis of radiochemical purity was conducted simultaneously with the radiochemical identity determination. Radiochemical purity of [^{11}C]OMAR had to be greater than 93.0%.

3.5 Specific Activity

The concentration of [^{11}C]OMAR can be determined using the HPLC system described in Section 3.3 by integrating the area under the UV peak of [^{11}C]OMAR and comparing with the area of known reference standard peak. The specific activity (As) was calculated as the division of radioactivity (mCi) by the amount of raclopride (μmol):

$$\mathrm{As}(\mathrm{mCi}/\mu\mathrm{mol}) = \frac{A(\mathrm{mCi})}{\left[C_{\mathrm{OMAR}}\,(\mu\mathrm{g/ml}) \times V(\mathrm{ml})/469\,(\mu\mathrm{g}/\mu\mathrm{mol})\right]}$$

A is the measured radioactivity; *V* is the volume of sample measured; *C* is the calculated concentration of OMAR. There is no specific activity release limit for [^{11}C]OMAR in place at our institution. The injection volume was based on the injectable mass limit for OMAR and was calculated on a dose-by-dose basis by the administering study team.

3.6 Radionuclidic Identity

Activities were measured using a Capintec CRC®-15R Radioisotope Dose Calibrator [21], and half-life was calculated using the equation below. Calculated half-life must be 18.5–22.4 min:

$$t_{1/2} = -\ln 2\left\{\frac{\text{time difference}}{\left[\ln(\text{ending activity/starting activity})\right]}\right\}$$

3.7 Sterile Filter Integrity Test

The sterile filter with needle that was used to convey the dose into the dose vial was connected to a nitrogen supply via a regulator. The filter was immersed in water and the pressure was gradually increased while observing for a steady stream of bubbles emanating from the needle. If this happened at a pressure greater than the filter acceptance level, then the filter was considered to have been intact and thus passed.

3.8 Endotoxin Analysis

Endotoxin content in doses of [^{11}C]OMAR was analyzed by a Charles River Laboratories Endosafe™ Portable Testing System [22] as per the US Pharmacopeia [23]. Doses had to contain ≤175 endotoxin units (EU) that was adjusted to 17.5 EU/ml as our formulation volume was 10 ml.

3.9 Sterility Testing

Samples of the [^{11}C]OMAR dose were added to tubed fluid thioglycolate media (FTM) and to tubed tryptic soy broth (TSB) media. FTM was used to test for anaerobes, aerobes, and microaerophiles; TSB was used to test for fastidious and non-fastidious organisms [24]. These media tubes as well as positive and negative controls were incubated at 30–35°C (FTM) and 22°C (TSB) for 14 days according to current USP guidelines [25]. Tubes were visually inspected on the 3rd, 8th, and 14th day after inoculation and were compared to positive and negative controls. Positive controls had to show growth (turbidity) and the test article samples ([^{11}C] OMAR) and negative controls had to have no growth after 14 days in order to pass the test for sterility. Positive controls could be established separately as growth promotion tests according to USP guidelines [26].

WASTE DISPOSAL INFORMATION

All hazardous chemicals and toxic materials were disposed of according to Prudent Practices in the Laboratory (Washington, DC: National Academy Press, 1995).

CHEMICAL ABSTRACTS NOMENCLATURE (REGISTRY NUMBER)

Acetonitrile (75-05-8)
Ammonium acetate (631-61-8)
Carbon dioxide (124-38-9)
Ethanol (64-17-5)
Iodine (7553-56-2)
Iodomethane (74-88-4)
Methane (74-82-8)
Methanol (67-56-1)
Methyl triflate (333-27-7)
Monosodium phosphate (7558-80-7)
Phosphorus pentoxide (1314-56-3)
H-Pyrazole-3-carboxamide, 4-cyano-1-(2,4-dichlorophenyl)-5-[4-(methoxy-^{11}*C*)phenyl]-*N*-1-piperidinyl (942063-86-7)
H-Pyrazole-3-carboxamide, 4-cyano-1-(2,4-dichlorophenyl)-5-(4-hydroxyphenyl)-*N*-1-piperidinyl (921590-81-0)

REFERENCES AND NOTES

For detailed supplier information, see Appendix 1.

1. (a) H. Fan, H. T. Ravert, D. P. Holt, R. F. Dannals, A. G. Horti, J. Labelled, *Comp. Radiopharm.*, 2006, 49, 1021; (b) D. F. Wong, H. Kuwabara, A. G. Horti, V. Raymont, J. Brasic, M. Guevara, W. Ye, R. F. Dannals, H. T. Ravert, A. Nandi, A. Rahmim, J. E. Ming, I. Grachev, C. Roy, N. Cascella, *Neuroimage*, 2010, 52, 1505.

2. GE standard carbon-11 target with a 75 μm Havar target foil and standard vacuum foil.
3. Ultrahigh-purity nitrogen gas with 0.5% of oxygen was purchased from Purity Plus.
4. Molecular sieve (4A, 80/100 mesh) was purchased from Alltech, pn: 5624.
5. Shimalite-Ni reduced (80/100 mesh) was purchased from Shimadzu (Kyoto, Japan), pn: 22127719.
6. Phosphorus pentoxide (with moisture indicator) was purchased from Fluka, pn: 79610.
7. Carbosphere (60/80 mesh) was purchased from Alltech (Grace), pn: 5680.
8. Iodine (crystals) was purchased from EMD (Germany), pn: IX01203.
9. Ascarite II (8–20 mesh, 3 in. in TRACERlab standard column) from Thomas Scientific, pn: C049H40.
10. Porapak type Q (0.3 g, 50–80 mesh) was purchased from Alltech, pn: 2700.
11. Silver triflate-Graphpac was prepared by dissolving 5 g of silver trifluoromethanesulfonate (Sigma-Aldrich) in 100 ml of acetonitrile (HPLC grade, Fisher) and then mixed with 10 g of Graphpac (80/100 mesh, GC grade, Alltech). The solvent was evaporated by rotary vapor under vacuum, and 0.35 g of dried powder was used to pack a TRACERlab standard column.
12. Sep-Pak C18 1 cc vac cartridges were purchased from Waters (part no. WAT054955).
13. Dehydrated Alcohol Injection, USP, was purchased from Akorn (NDC: 17478-503-05) and used as received.
14. 0.9% sodium chloride, USP, was purchased from Hospira (part no. 0409-4888-50) and used as received.
15. Millex-GV sterile 0.22 μm filters were purchased from Millipore (part no. SLGV033RS) and used as received.
16. 10 ml sterile vials were purchased from Hollister-Stier (part no. 7515ZA) and vented with a sterile Millex-FG vent filter (part no. SLFG025LS).
17. US Pharmacopeia <823>, Radiopharmaceuticals for positron emission tomography compounding. USP 32-NF 27. 2009.
18. EMD Chemicals, Inc., United States (part no. 9578-3).
19. Shimadzu Corporation, United States.
20. Bioscan, Inc., United States.
21. Capintec, Inc., United States.
22. Charles River Laboratories, United States.
23. US Pharmacopeia <85>, Bacterial endotoxins test. USP 32-NF 27. 2009.
24. Becton, Dickinson and Company, United States.
25. US Pharmacopeia <71> Sterility test. USP 32-NF 27. 2009.
26. US Pharmacopeia <61>, Microbial enumeration tests. USP 32-NF 27. 2009.

CHAPTER 9

SYNTHESIS OF CARBON-11 LABELED (+)-4-PROPYL-3,4,4A,5,6,10B-HEXAHYDRO-2*H*-NAPHTHO[1,2-*B*][1,4]OXAZIN-9-OL ([^{11}C]-(+)-PHNO)

CHRISTOPHE PLISSON, JOAQUIM RAMADA-MAGALHAES, AND JAN PASSCHIER

Imanova Centre for Imaging Sciences, Imperial College London, Hammersmith Hospital, London, UK

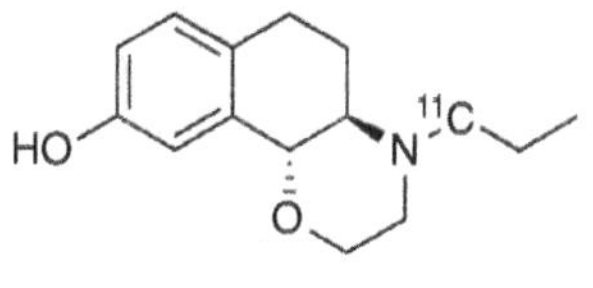

[^{11}C]-(+)-PHNO

1 INTRODUCTION

(+)-PHNO was first reported to be a potent D_2 agonist intended for the treatment of Parkinson's disease in 1984 [1] and reached phase II clinical trials under various names (Dopazinol, Naxagolide, Nazagolide, L647339, MK 458). [^{11}C]-(+)-PHNO was first evaluated as a D_2 agonist radioligand in preclinical species as well as in humans by Wilson and colleagues [2–4].

PET studies in non-human primates demonstrated that the enriched [^{11}C]-(+)-PHNO signal in ventral striatum and globus pallidus compared with [^{11}C]raclopride was selectively decreased by pretreatment with a selective D_3 antagonist or partial agonist BP-897, indicating that the difference in regional distributions between [^{11}C]raclopride and [^{11}C]-(+)-PHNO was driven by a relatively higher affinity of this tracer for D_3 compared with D_{2HIGH} receptor sites [5, 6]. More recently, the pharmacologic nature of [^{11}C]-(+)-PHNO signal in healthy subjects was elucidated and provided quantification of D_3 receptor availability in the living human brain [7]. There is a substantial body of literature to support the role of D_3 receptors in the pathophysiology of several conditions such as addiction, schizophrenia, and

Radiochemical Syntheses, Volume 2: Further Radiopharmaceuticals for Positron Emission Tomography and New Strategies for Their Production, First Edition. Edited by Peter J. H. Scott.

Parkinson's disease. The ability to measure D_3 receptors in healthy controls and patient populations, as well as the level of engagement of this target by therapeutic agents, is valuable to understand the role of D_3 receptors in these conditions.

[^{11}C]-(+)-PHNO is an agonist, and its *in vivo* binding is also expected to be more affected by acute fluctuations in synaptic dopamine than that of antagonist radiotracers such as [^{11}C]raclopride; Shotbolt et al. [8] demonstrated that [^{11}C]-(+)-PHNO is more sensitive than [^{11}C]raclopride in measuring the fluctuations synaptic dopamine in the human striatum.

2 SYNTHETIC PROCEDURES

CAUTION: All radiochemical syntheses must be carried out using appropriate equipment in a facility authorized for the use of radioactive materials. Personal protective equipment must be worn and all local radiation safety laws followed.

2.1 Production of [^{11}C]

[^{11}C]carbon dioxide was produced by the $^{14}N(p,\alpha)^{11}C$ reaction using a Siemens RDS-111 Eclipse cyclotron equipped with a [^{11}C]carbon dioxide target loaded with nitrogen gas containing 1.0% oxygen. Typical irradiation with a proton beam (55 µA) for 50 min provided an estimated 70 GBq of [^{11}C]carbon dioxide, which was transferred with a sweep of helium gas to the automated synthesis module. For a detailed description of the automated synthesis see Sections 2.2, 2.3, and 2.4. All reagents were obtained from commercially available sources [9]. A procedure using the Modular Lab™ synthesis module from Eckert & Ziegler for [^{11}C]-(+)-PHNO production has been described elsewhere [10].

2.2 Production of [^{11}C]propionyl Chloride

In a first step, [^{11}C]carbon dioxide was concentrated by trapping it on a column containing carbon molecular sieves (Alltech molecular sieves 4Å, 60/80 mesh) at room temperature. The progress of the radiosynthesis was continually monitored by way of the control screen (Fig. 1), which indicated radioactivity levels in the two reactors and in the [^{11}C]carbon dioxide trapping module as well as the valves status and helium flow. The [^{11}C]carbon dioxide was released by heating the column to up to 400°C and flushing with helium at an approximately 15 ml/min flow. The concentrated [^{11}C] carbon dioxide was directed to a reaction vessel containing the Grignard solution made of a 1 M solution of ethylmagnesium in tetrahydrofuran (THF) (0.05–0.10 ml) diluted with diethyl ether (0.3 ml). Subsequently, phthaloyl dichloride (0.5 ml) and 2,6-di-*tert*-butylpyridine (0.3 ml) were added, and the resulting mixture was heated up to 170°C. The formed [^{11}C]propionyl chloride was swept by a helium flow at 150 ml/min to a receiving vial placed in the second reactor, cooled at around −5°C (Fig. 2).

2.3 Production of [^{11}C]-(+)-PHNO

The distillation of [^{11}C]propionyl chloride described in Section 2.2 continued until the activity trapped in the receiving vial plateaued. The collection vial containing 1–2 mg of (+)-PHNO hydrochloric salt precursor in THF (0.7 ml) with 50 µl of

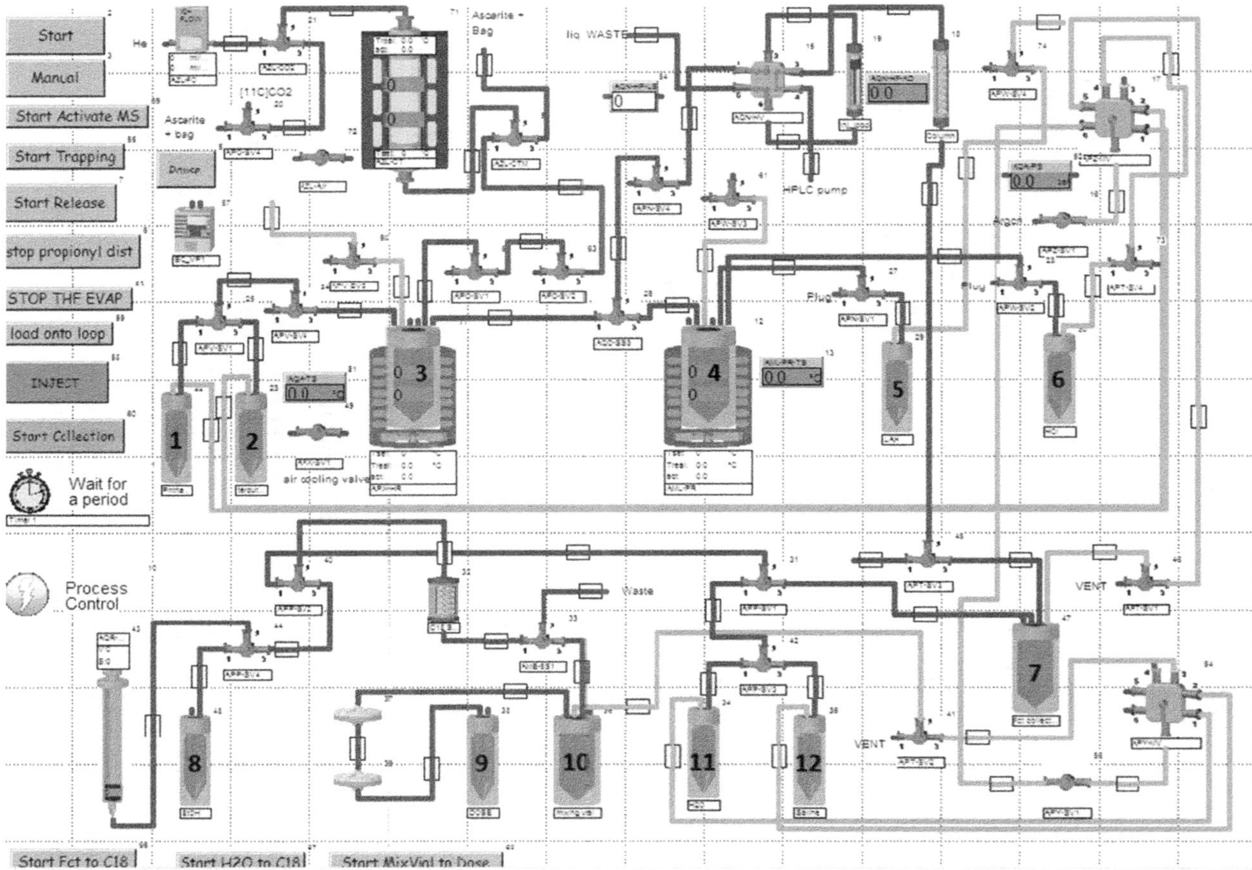

FIGURE 1 Diagram of the Eckert & Ziegler setup for the production of [^{11}C]-(+)-PHNO. Vial 1, phthaloyl dichloride; vial 2, 2,6-ditertbutylpyridine; vial 3, Grignard reaction vessel; vial 4, second reaction vessel; vial 5, $LiAlH_4$; vial 6, 1M HCl; vial 7, HPLC collection; vial 8, EtOH; vial 9, dose; vial 10, mixing vial; vial 11, H_2O; vial 12, saline.

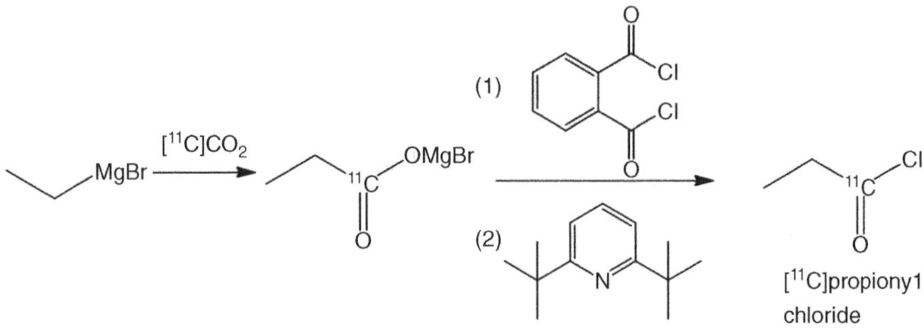

FIGURE 2 Synthesis of [^{11}C]propionyl chloride.

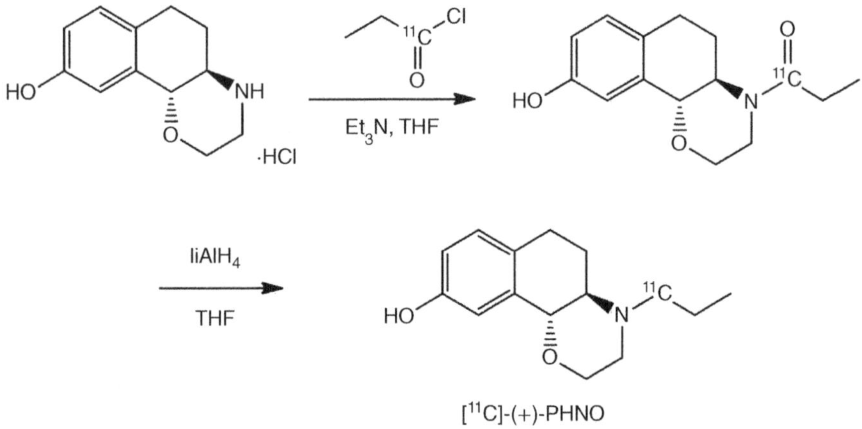

FIGURE 3 Synthesis of [^{11}C]-(+)-PHNO.

triethylamine (TEA) was then heated to 80°C for 3.5 min. The vial was cooled to around 0°C at which point $LiAlH_4$ in THF/diethyl ether was added. The reaction vial was heated to 80°C for 1 min to reduce the amide (Fig. 3).

2.4 Purification and Formulation of [^{11}C]-(+)-PHNO

After reduction of the amide with lithium aluminum hydride, the remaining THF was evaporated by sweeping helium through the solution. Upon dryness, 1N HCl (0.9 ml) was added, and the reaction mixture was stirred and loaded onto the HPLC injection loop for purification. [^{11}C]-(+)-PHNO was purified on a semi-preparative HPLC (Agilent XDB-C18, 9.4×250 mm, 5 µm) using the following multistep gradient conditions:

Mobile phase A: ammonium formate 0.1 M pH 4
Mobile phase B: 50/50 acetonitrile/ammonium formate 0.1 M pH 4

The flow rate was 8 ml/min and the composition was as follows: 10% B from 0 to 1 min, 10 to 28% B from 1 to 1.5 min and 28% B until the end of the HPLC purification. Representative chromatograms are displayed in Fig. 4.

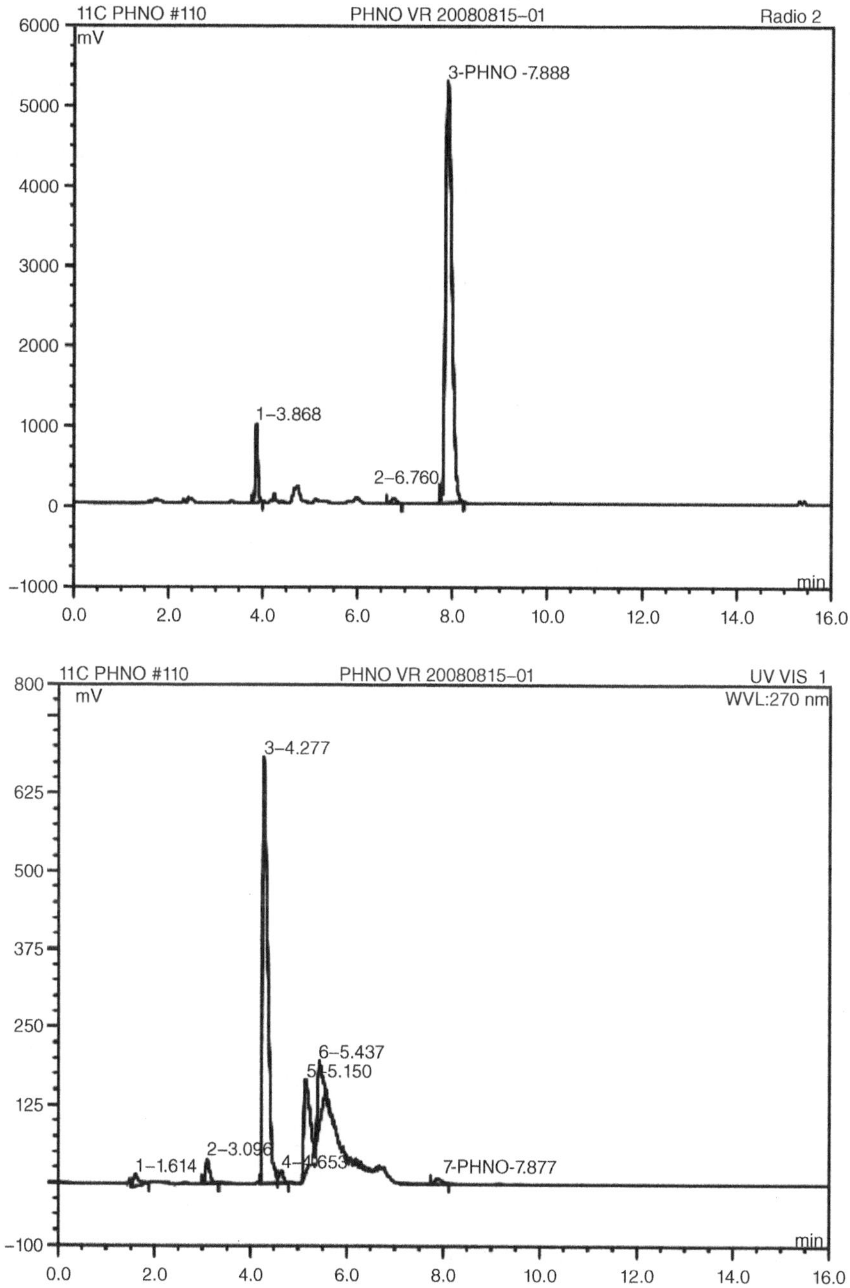

FIGURE 4 Semipreparative UV and radioactive chromatograms for [^{11}C]-(+)-PHNO.

[^{11}C]-(+)-PHNO eluted at approximately 7.9 min and was collected in a vial containing 80–90 mg of sodium carbonate decahydrate in 30 ml of water. The resulting solution was loaded onto a Sep-Pak® C18 classic (Waters) and washed with 10 ml of water. [^{11}C]-(+)-PHNO was eluted with 1.1 ml of ethanol and directed to a 20 ml mixing vial. 10 ml of 0.9% saline were also transferred via the Sep-Pak® C18 to the mixing vial. After mixing of the excipients, the final formulation solution was filtered aseptically through a 0.2 μm sterile filter (Pall sterile Acrodisc® filter with Supor® membrane, PN 4612) into a sterile product vial.

3 QUALITY CONTROL PROCEDURES

CAUTION: All radiopharmaceuticals produced for clinical use must have local regulatory approval before use in humans. Trained personnel must carry out quality control (QC) procedures, and each dose must meet all established QC criteria before release to the clinic.

QC procedures for [^{11}C]-(+)-PHNO, based on the current requirements for compounding PET radiopharmaceuticals described in the Ph. Eur. Pharmacopoeia [11], are summarized in the following text. Complete QC data for three consecutive batches of [^{11}C]-(+)-PHNO produced using the radiosynthetic procedure disclosed herein are summarized in Table 1.

Each of the three batches met all of the established QC criteria. The specific activities are reported at time in QC for these qualifying samples since there was no patient data. The values given for dose mass limit are the total mass in micrograms in a sterile vial (10 ml formulation buffer).

3.1 Visual Inspection

The [^{11}C]-(+)-PHNO dose is examined behind a lead glass window and must be clear, colorless, and free from particles.

3.2 Radionuclidic Identity

The radionuclidic identity was determined by measuring the radioactive decay (using a dose calibrator [12]) of a sample of the [^{11}C]-(+)-PHNO dose at three separate timepoints. The half-life was calculated using Equation 1:

$$t_{\frac{1}{2}} = -\ln 2 \frac{\text{time difference}}{\ln(\text{end activity}/\text{starting activity})} \quad (1)$$

3.3 Radiochemical Identity

Radiochemical identity was determined on an Agilent 1200 Series HPLC [13] equipped with a UV diode array detector (wavelength set at 280 nm, lambda maximum for (+)-PHNO) and a Berthold sodium iodide radiation scintillation detector.

TABLE 1 Quality Control Data for Three Repeat [^{11}C]-(+)-PHNO Runs

QC Test	Release Criteria	Run 1	Run 2	Run 3
Yield (MBq)	n/a	4072	3535	3827
Visual inspection	Clear, colorless, and free from particles	Pass	Pass	Pass
Radionuclidic identity ($t_{1/2}$) (min)	Measured half-life	20.7	20.4	19.9
	Gamma energy: 511 KeV, sum peak at 1022 KeV	Pass	Pass	Pass
Radiochemical identity	The retention time of [^{11}C]-(+)-PHNO for injection (±4%) to that of PHNO reference standard	Pass	Pass	Pass
Radiochemical purity (%)	≥95% [^{11}C]-(+)-PHNO for injection, based on the spike sample	100	100	100
Dose mass	≤5 µg in administered dose	15.5 (limited)	15.3 (limited)	11.9 (limited)
Specificity activity	n/a	64.8	57.1	83.6
Chemical impurity mass limit (UV)	Unknown impurities ≤1.5 µg in administered dose	<LOQ*	<LOQ*	<LOQ*
Residual solvent (in ppm)	Diethyl ether ≤50 mg	<LOQ[a]	<LOQ[a]	<LOQ[a]
	Acetonitrile ≤4.1 mg	1.89	<LOQ[a]	<LOQ[a]
	Tetrahydrofuran ≤7.2 mg	<LOQ[a]	<LOQ[a]	<LOQ[a]
	Triethylamine ≤3.2 mg	<LOQ[a]	<LOQ[a]	<LOQ[a]
Dose pH	Between 4.5 and 8.5	6.55	7.31	7.6
Sterile filter integrity test (psi)	Sterile 0.22 µm filters are singly used. Minimal bubble point >31.2 psig	Pass	Pass	Pass
Endotoxin analysis (EU/ml)	≤175 EU in total administrated dose	Pass	Pass	Pass
Sterility testing	No growth observed in 14 days	Pass	Pass	Pass

[a] LOQ, limit of quantification.

The data were captured and processed using Agilent ChemStation for liquid chromatography systems acquisition software.

The [^{11}C]-(+)-PHNO analysis was performed using an Agilent Zorbax SB-Phenyl column (1.8 µm, 4.6 × 100 mm) [14]. The analytical column is maintained at 35°C and acetonitrile/phosphate buffer 25 mM pH 6.7 [15] (40/60 v/v) at 1 ml/min as mobile phase with a total run time of 6.5 min.

The radiopharmaceutical identity was confirmed by spiking the dose with (+)-PHNO reference standard, and the retention time of the radiolabeled compound was compared with that of nonradioactive PHNO. Retention time difference between the UV and radioactive signal must be within ±4% of the retention time established during the prerelease system suitability test. Representative analytical UV and radioactive chromatograms of a spiked [^{11}C]-(+)-PHNO dose are displayed in Fig. 5.

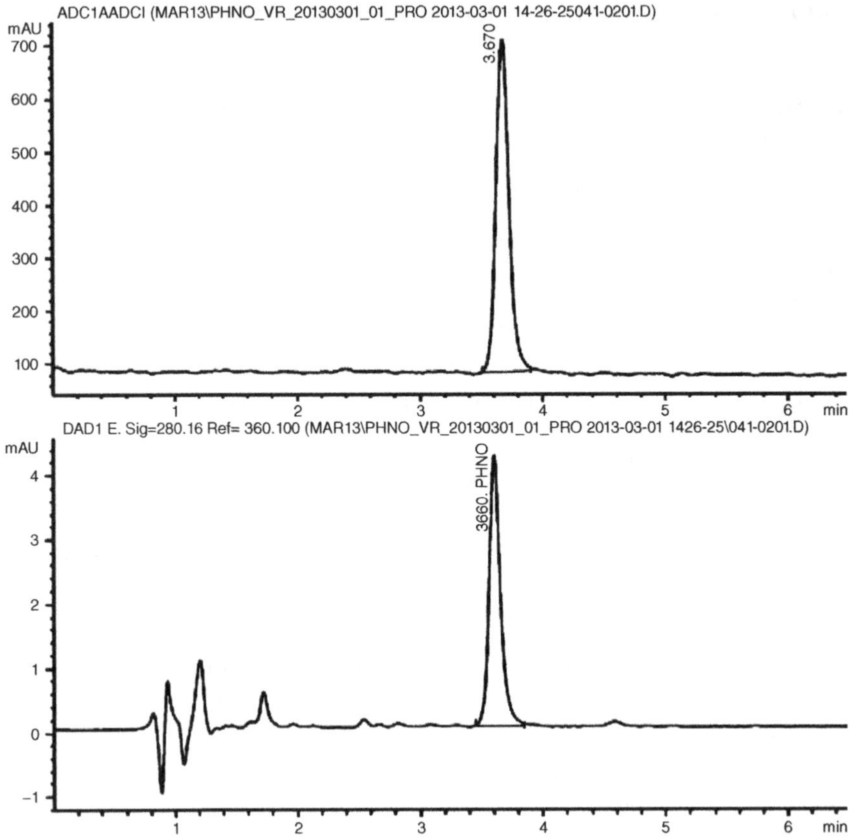

FIGURE 5 Quality control analysis of a [^{11}C]-(+)-PHNO spiked with unlabeled (+)-PHNO at 5 µg/ml (volume ratio: 1/1). Top: radioactivity Chromatogram. Bottom: UV chromatogram (λ=280 nm). Difference in retention time between UV and radioactive peak corresponds to the path difference between the UV and the radiodetector.

3.4 Radiochemical Purity

Radiochemical purity was determined as part of the radiochemical identity test described in Section 3. The radiochemical purity was determined using the radiometric detector and was based on the percentage of all peak areas decay corrected to the start of the sample injection. The radiochemical purity must be ≥95%.

3.5 Dose Mass Limit and Specific Activity

The mass concentration of unlabeled (+)-PHNO (µg/ml) was determined using the same condition as described in Section 3. A UV calibration curve was prepared by injecting 5 PHNO reference samples of known concentration. PHNO concentration of each [^{11}C]-(+)-PHNO dose was determined by integrating the PHNO UV peak as depicted in Fig. 6. The dose injection volume is based on the injectable mass limit for (+)-PHNO and is calculated on a dose-by-dose basis.

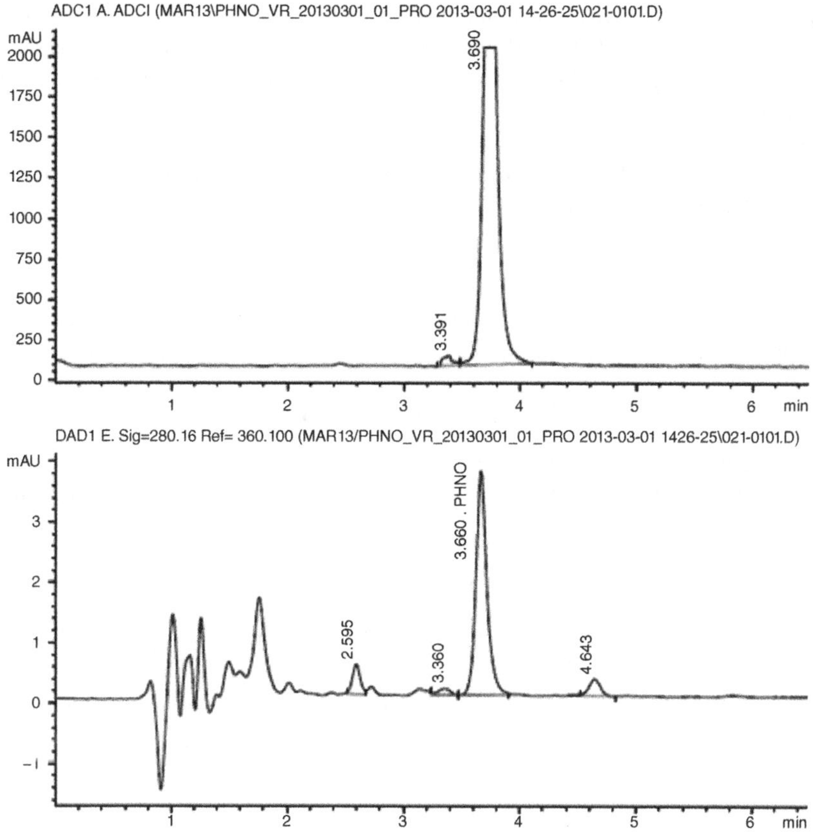

FIGURE 6 Quality control analysis of a [^{11}C]-(+)-PHNO. Top: radioactivity chromatogram. Bottom: UV chromatogram ($\lambda = 280$ nm).

3.6 Chemical Impurity Mass Limit (UV)

The mass concentration of chemical UV impurities was determined on an HPLC using the same condition as described in Section 3. The sum of all impurities (DAD/UV) must be ≤1.5 µg in the administered dose, and the limit of unlabeled (+)-PHNO and precursor to be administered was set at ≤5 µg.

3.7 Residual Solvent Analysis

Gas chromatography (GC) is used to determine the residual solvent levels in the dose. The GC system used was an Agilent 6890N GC [16] equipped with Gerstel automated headspace sampler (RS_HS) [17], split/splitless inlet, a flame ionization detector (FID), and an GC capillary column (Agilent J&W GC Column DB-624, 30 m, 0.32 mm, 1.80 µm, part no: 123-1334) [18]. Limits were set following the ICH Harmonized Guidelines [19].

The method used was based on a headspace GC injection: the dose was diluted with DMSO and incubated for 10 min at 80°C, and 1000 μl of the headspace gas sample was injected using the GC's Gerstel injector. The system operated at constant helium pressure of 15 psi.

Standard calibrations curves were prepared for all solvents entering the manufacture of [^{11}C]-(+)-PHNO: diethyl ether, acetonitrile, THF, and TEA were prepared in formulation buffer [20].

3.8 pH Testing

The pH of the [^{11}C]-(+)-PHNO doses was analyzed using a Jenway pH meter equipped with a micro-pH probe [21]. pH must be between 4.5 and 8.5.

3.9 Sterile Filter Integrity Test

The integrity of the sterilizing membrane filter and filter system was assessed using a Millipore Integritest IT4N Filter Integrity Test Instrument. The filter acceptance pressure was previously determined based on the tracer formulation.

3.10 Endotoxin Analysis

Endotoxin content in doses of [^{11}C]-(+)-PHNO was analyzed by a Charles River Laboratories Endosafe® Portable Testing System (PTS) [22]. The [^{11}C]-(+)-PHNO dose is diluted to 1 in 200 with Charles River Laboratories LAL reagent water <0.005 EU/ml and analyzed using a Endosafe-PTS cartridge of 5–0.05 EU/ml sensitivity. The administrated dose must contain less than 175 EU.

3.11 Sterility Testing

The standard sterility test is designed to determine the presence of mesophilic bacteria and fungi in solutions and is based on the Ph. Eur. Method 2.6.1 Sterility [23]. Aliquots of the [^{11}C]-(+)-PHNO doses were incubated in soybean-casein digest (SCD) broth at 20–25°C and fluid thioglycolate media (FTM) at 30–35°C. Plates were incubated along with positive and negative controls for 14 days. Positive standards must show growth (turbidity) on the plates, and [^{11}C]-(+)-PHNO/negative controls must have no culture growth after 14 days to be indicative of sterility.

WASTE DISPOSAL INFORMATION

All hazardous chemicals and toxic materials were disposed of according to Prudent Practices in the Laboratory (Washington, DC: National Academy Press, 1995).

CHEMICAL ABSTRACTS NOMENCLATURE (REGISTRY NUMBER)

(+)-PHNO precursor (858517-21-2)
Acetonitrile (75-05-8)

Ethanol (64-17-5)
Ammonium formate (540-69-2)
Formic acid (64-18-6)
Diethyl ether (60-29-7)
THF (109-99-9)
TEA (121-44-8)
Phthaloyl chloride (88-95-9)
2,6-Di-tert-butylpyridine (585-48-8)
Sodium carbonate (497-19-8)
Lithium aluminum hydride (16853-85-3)
Ethyl magnesium bromide (925-90-6)

REFERENCES AND NOTES

For detailed supplier information, see Appendix 1.

1. J. H. Jones, P. S. Anderson, J. J. Baldwin, B. V. Clineschmidt, D. E. McClure, G. F. Lundell, W. C. Randall, G. E. Martin, M. Williams, J. M. Hirshfield, G. Smith, P. K. Lumma, *J. Med. Chem.* 1984, 27, 1607.
2. M. Willeit, N. Ginovart, S. Kapur, S. Houle, D. Hussey, P. Seeman, A. A. Wilson, *Biol. Psychiatry* 2006, 59, 389.
3. N. Ginovart, L. Galineau, M. Willeit, R. Mizrahi, P. M. Bloomfield, P. Seeman, S. Houle, S. Kapur, A. A. Wilson, *J. Neurochem.* 2006, 97, 1089.
4. A. A. Wilson, P. McCormick, S. Kapur, M. Willeit, A. Garcia, D. Hussey, S. Houle, P. Seeman, N. Ginovart, *J. Med. Chem.* 2005, 48, 4153.
5. R. Narendran, M. Slifstein, O. Guillin, Y. Hwang, D.-R. Hwang, E. Scher, S. Reeder, E. Rabiner, M. Laruelle, *Synapse* 2006, 60, 485.
6. E. A. Rabiner, M. Slifstein, J. Nobrega, C. Plisson, M. Huiban, R. Raymond, M. Diwan, A. A. Wilson, P. McCormick, G. Gentile, R. N. Gunn, M. A. Laruelle, *Synapse* 2009, 63, 782.
7. G. Searle, J. D. Beaver, R. A. Comley, M. Bani, A. Tziortzi, M. Slifstein, M. Mugnaini, C. Griffante, A. A. Wilson, E. Merlo-Pich, S. Houle, R. Gunn, E. A. Rabiner, M. Laruelle, *Biol. Psychiatry* 2010, 68, 392.
8. P. Shotbolt, A. C. Tziortzi, G. E. Searle, A. Colasanti, J. Van der Aart, S. Abanades, C. Plisson, S. R. Miller, M. Huiban, J. D. Beaver, R. N. Gunn, M. Laruelle, E. A. Rabiner, *J. Cereb. Blood Flow Metab.* 2012, 32, 127.
9. (+)-PHNO precursor and reference were purchased from Toronto Research Chemicals, Inc. (Toronto, Canada); tetrahydrofuran [Hi-Dry anhydrous], triethylamine [Hi-Dry anhydrous], diethyl ether [Hi-Dry stabilized with BHT], water [UPS ultragradient], and acetonitrile [UPS ultragradient] from Romil Ltd. (Cambridge, United Kingdom); sodium chloride [0.9% BP Steriflex® saline for intravenous injection] from Fresenius Kabi Ltd. (Runcorn, United Kingdom); lithium aluminum hydride [1.0 M in tetrahydrofuran], ethylmagnesium bromide [1.0 M in tetrahydrofuran], and 2,6-ditertbutylpyridine [≥97%] from Sigma-Aldrich (Gillingham, United Kingdom); hydrochloric acid [Reag. Ph. Eur., 1 M solution (volumetric)] and sodium carbonate decahydrate [puriss., Ph. Eur., BP, FCC, E500, FG, 99.5–100.5% (ex dried subst.)]; and formic acid [puriss. p.a., ACS reagent, reag. Ph. Eur., ≥98%], ammonium formate [puriss. p.a., for HPLC, ≥99.0% (dried material, NT)], phthaloyl dichloride, anhydrous ethanol [Ph Eur.], and molecular sieves [4Å, 60/80 mesh] from Fisher Scientific (Loughborough, UK). Sep-Pak® C18 classic cartridges from Waters (Elstree, United Kingdom).
10. C. Plisson, M. Huiban, S. Pampols-Maso, G. Singleton, S. P. Hill, J. Passchier, *Appl. Radiat. Isot.* 2012, 70, 380.
11. Ph. Eur monograph, Radiopharmaceutical preparations, 01/2008:0125.

12. ISOMED 2000 Dose Calibrators.
13. High-performance liquid chromatography (HPLC), manufacture: Agilent Technologies; module 1200 series, the LC systems includes vacuum degasser, quaternary pump, autosampler, column compartment and variable wavelength detector (UV–Vis/ VWD), and an external Berthold sodium iodide radiation scintillation detector.
14. HPLC column, Agilent Zorbax SB-Phenyl, 1.8 μm, 4.6 × 100 mm, part number: 828975-912
15. 25 mM phosphate buffer pH 6.73.
16. GC Agilent Technologies model 6890N with a FID detector.
17. GC headspace auto-sampler Gerstel Model MPS 2 Twister.
18. GC column, Agilent J&W GC Column DB-624, 30 m, 0.32 mm, 1.80 μm—part no. 123–1334.
19. ICH Topic Q3C (R4) Impurities: Guideline for Residual Solvents.
20. Formulation blank is 10% of ethanol in 90% of 0.9% saline for injection.
21. Jenway Model 3510 pH/mV/temperature meter.
22. Charles River Laboratories Endosafe® Portable Test System (PTS100).
23. European Pharmacopoeia, 01/2005/20601, 2.6.1. Sterility.

CHAPTER 10

SYNTHESIS OF (*R*)-[*N*-METHYL-^{11}C]PK11195

VÍTOR H. ALVES, ANTERO J. ABRUNHOSA, AND MIGUEL CASTELO-BRANCO

Institute for Nuclear Sciences Applied to Health (ICNAS), Institute for Biomedical Imaging and Life Sciences (IBILI), Faculty of Medicine, University of Coimbra, Coimbra, Portugal

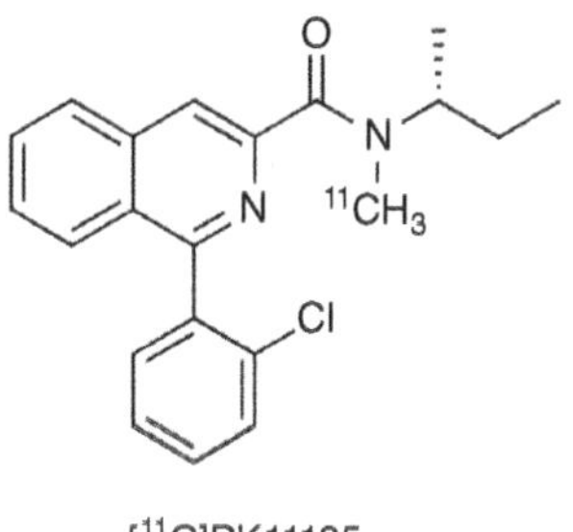

1 INTRODUCTION

Inflammation, in general, is a physiological response of the body to different types of tissue injury, triggering a cascade of coordinated cellular and biochemical events. The inflammatory process isolates the damaged tissue and promotes an immune response [1]. Likewise, neuroinflammation implicates a complex orchestrated response to various stimuli in the central nervous system (CNS) aiming toward tissue preservation and restoration. Neuroinflammatory reactions involve a cellular response by activating cells of the monocyte lineage, whether resident or circulating. Microglia are the resident cells of the monocyte lineage in the CNS, and their activation constitutes a main feature of neuroinflammation [2].

This activated stage of microglia, upregulates the expression of the 18 kDa translocator protein (TSPO, formerly known as the peripheral benzodiazepine

Radiochemical Syntheses, Volume 2: Further Radiopharmaceuticals for Positron Emission Tomography and New Strategies for Their Production, First Edition. Edited by Peter J. H. Scott.

receptor (PBR)), which can be mapped *in vivo* with positron emission tomography (PET) using selective radiolabeled TSPO ligands [3]. One of the first PET radioligands for human studies of activated microglia (TSPO overexpression) was PK11195, a lipid soluble isoquinoline carboxamide, which was enantiomerically labeled with carbon-11 ((*R*)-[*N*-methyl-^{11}C]PK11195) [4] and it remains today the most widely used radiotracer for *in vivo* imaging of TSPO [5]. With the aim of performing human studies of neuroinflammation at our institute, we worked on the optimization of the synthesis, purification, and reformulation of (*R*)-[*N*-methyl-^{11}C] PK11195 to be used for *in vivo* PET imaging studies [6–8].

2 SYNTHESIS PROCEDURES

CAUTION: All radiochemical syntheses must be carried out using appropriate equipment in a facility authorized for the use of radioactive materials. Personal protective equipment must be worn and all local radiation safety laws followed.

2.1 Production of [^{11}C]CO_2

[^{11}C]CO_2 is produced by the ^{14}N(p,α)^{11}C nuclear reaction in a 50 cm^3 aluminum target using an IBA Cyclone 18/9 cyclotron [9]. The target is loaded with high-purity scientific grade nitrogen (99.9999%) with 0.5% of oxygen of the same grade. The mixture, purchased from Praxair [10], is irradiated with a 18 MeV proton beam (35 μA for 30 min) to produce approximately 2 Ci (74 Gbq) of [^{11}C]CO_2.

2.2 Production of [^{11}C]CH_3I

Starting from in-target produced [^{11}C]CO_2, [^{11}C]CH_3I is synthesized using the chemical reaction (Fig. 1) described by Crouzel et al. in 1987 [11] on a Bioscan MeI-Plus chemistry module [12].

[^{11}C]CO_2 is trapped on a molecular sieve [13] column (Fig. 2) at room temperature (~25°C). The column is heated at 250°C to release the [^{11}C]CO_2 under a stream of nitrogen to a reaction vessel with a solution of lithium aluminum hydride in tetrahydrofuran (THF) [14] producing lithium [^{11}C]methoxide. THF is evaporated for 3 min at 115–120°C after which a white salt of lithium [^{11}C]methoxide is clearly visible in the reactor wall. In the next step, hydriodic acid [15] is added to form [^{11}C]methyl iodide that is then distilled at 115–120°C under a stream of nitrogen. After, [^{11}C]CH_3I is dried before the labeling reaction by passing it through an Ascarite [16]/phosphorus pentoxide [17] column.

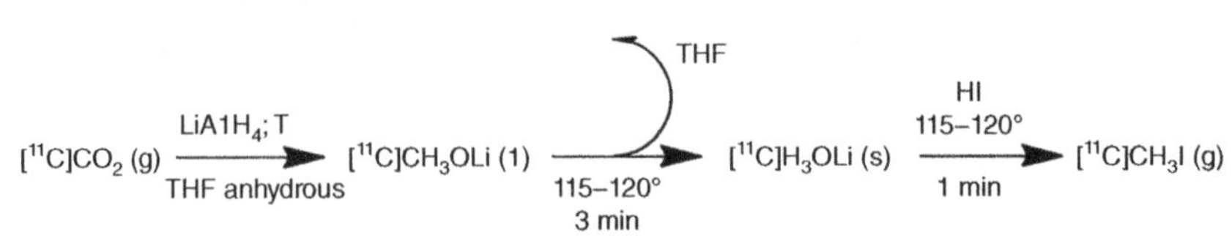

FIGURE 1 Synthesis of [^{11}C]CH_3I.

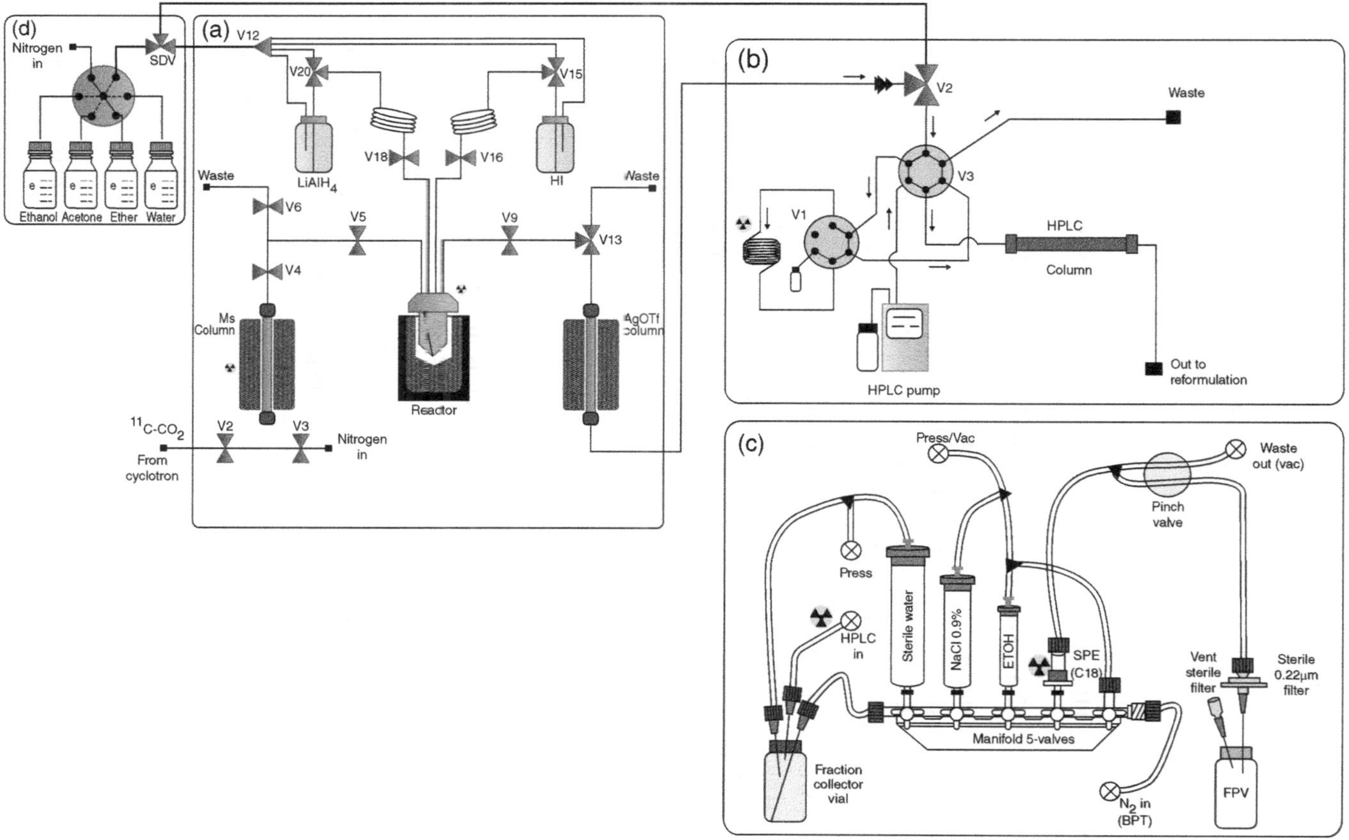

FIGURE 2 Schematic diagram of the system for preparation of (*R*)-[*N*-methyl-^{11}C]PK11195. [^{11}C]CH_3I production system (a), captive solvent and semipreparative HPLC system (b), reformulation system (c), and solvent delivery system (d).

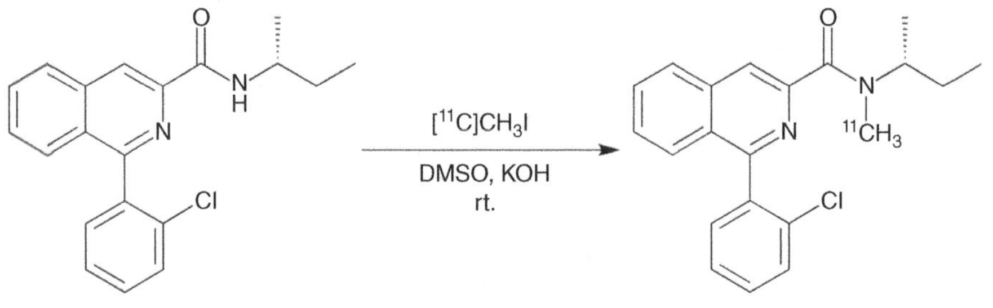

FIGURE 3 Radiosynthesis of (*R*)-[*N*-methyl-^{11}C]PK11195.

2.3 Synthesis of (*R*)-[*N*-Methyl-^{11}C]PK11195

Radiosynthesis of (*R*)-[*N*-methyl-11C]PK11195 is carried out on a specially designed stainless steel loop previously coated with the precursor solution [18, 19]. (*R*)-[*N*-Methyl-^{11}C]PK11195 is produced by [^{11}C]-methylation of (*R*)-[*N*-desmethyl] PK11195 with [^{11}C]methyl iodide using a 2 ml high-performance liquid chromatographic (HPLC) loop [18] via nucleophilic substitution (Fig. 3).

(*R*)-[*N*-Desmethyl]PK11195 (0.5–1.0 mg) salt [20] is dissolved in 150 µl of dimethyl sulfoxide (DMSO) [21], saturated with 30 mg potassium hydroxide [22], and slowly injected into the HPLC loop. Once produced, the [^{11}C]methylation agent ([^{11}C]CH_3I) is swept into the loop by the stream of nitrogen (<15 ml/min) at room temperature. After the methylation agent is quantitatively trapped in the loop, the nitrogen flow is stopped and the reaction is allowed to proceed for 5 min.

2.4 Purification and Formulation of (*R*)-[*N*-Methyl-^{11}C]PK11195

Once completed the reaction, the resulting mixture is injected onto a semipreparative HPLC column [23] and eluted using 0.1 N ammonium formate in 60% acetonitrile at a flow rate of 6 ml/min. The HPLC run is monitored using a 254 nm UV detector [24] and a radioactive detector [25]. The semipreparative chromatogram of a typical radiosynthesis of (*R*)-[*N*-methyl-^{11}C]PK11195 is shown in Fig. 4.

The product fraction of (*R*)-[*N*-methyl-^{11}C]PK11195 is collected at about 12.1 min for 1.5 min into a vial containing 10 ml of water for injection (WFI). After mixing, the solution is passed through a Sep-Pak® C18 Light cartridge [26] and washed with 30 ml of WFI. The PK11195 trapped on the cartridge is then eluted with 1.0 ml of ethanol followed by 9.0 ml of saline solution and filtered through a 0.22 µ sterile membrane filter [27].

3 QUALITY CONTROL PROCEDURES

CAUTION: All radiopharmaceuticals produced for clinical use must have local regulatory approval before use in humans. Trained personnel must carry out quality control (QC) procedures, and each dose must meet all established QC criteria before release to the clinic.

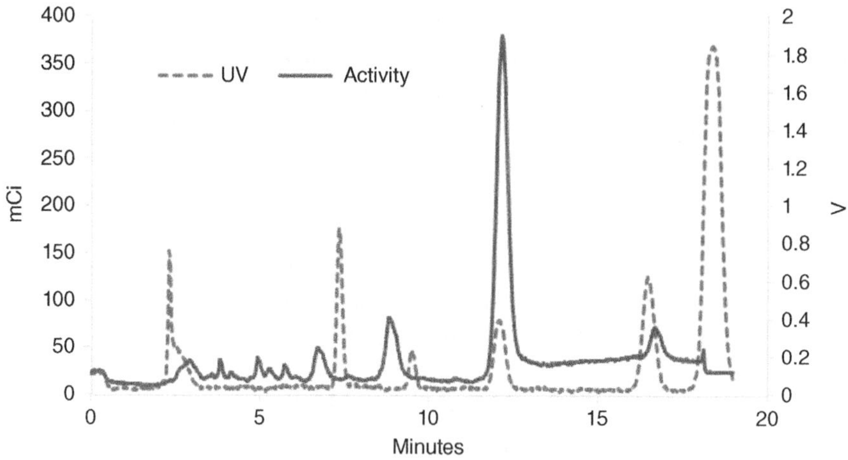

FIGURE 4 Preparative HPLC run of a typical (*R*)-[*N*-methyl-^{11}C]PK11195 purification.

TABLE 1 QC Data for Three Sequential Production Runs of (*R*)-[*N*-methyl-^{11}C]PK11195

Test	Release Criteria	RUN 1	RUN 2	RUN 3
Activity mCi (GBq)	NA	234.3 (8.7)	122 (4.5)	248.8 (9.2)
Visual inspection	Clear, colorless	Pass	Pass	Pass
Identity	RRT=0.9–1.1	1.0	1.0	1.0
Radiochemical purity (%)	≥95	99.4	99.1	100.0
Specific activity (GBq/μmol)	NA	26.36	32.24	31.49
pH	4.5–8.5	5.7	5.4	5.1
Residual solvent analysis (mg/V)	CH_3CN (<4.10)	0.01	0.02	0.02
Filter integrity—bubble point (bar)	≥3.0	Pass	Pass	Pass
Radionuclidic identity ($t_{1/2}$) (min)	18.4–22.4	20.3	20.3	20.4
Endotoxin analysis[a] (EU/V)	≤175	Pass	Pass	Pass
Sterility[a]	Sterile	Pass	Pass	Pass

NA, not applicable; V, maximum injection volume in ml.
[a] According to Ph. Eur., these tests can be performed after release for use.

Intravenous administration of a radiotracer must comply with both radiation and pharmaceutical standards not only to ensure safety but also to establish its efficiency since the *in vivo* behavior is dependent on high levels of radionuclidic, radiochemical, and chemical purity. Additionally, injectable solutions must also satisfy standards of sterility, apyrogenicity, osmolarity, and pH to ensure safe administration in humans. These parameters are detected and quantified using analytical techniques according to the guidelines for general radiopharmaceutical preparations of Ph. Eur. [28]. QC data for three sequential productions of (*R*)-[*N*-methyl-^{11}C]PK11195 using the method described herein is supplied in Table 1.

3.1 Visual Inspection

(*R*)-[*N*-methyl-^{11}C]PK11195 is stored in a clear pharmaceutical grade vial glass type I at room temperature in a lead-shielded container. The solution, inspected behind an appropriate lead glass, is clear, colorless, and without any visible particulate matter.

3.2 pH

The pH of (*R*)-[*N*-methyl-^{11}C]PK11195 product solution measured by potentiometry [29] on a small sample of the final product solution should be between 4.5 and 8.5, in accordance with the radiopharmaceutical preparations monograph of Ph. Eur. [28].

3.3 Radiochemical Identity

Radiochemical identity is determined by HPLC analysis using an Agilent 1200 Series HPLC system equipped with a multiwavelength UV detector [30] and a GABI Star NaI(Tl) radiometric detector [30]. The HPLC conditions are as follows: Zorbax Eclipse XDB-C18 column [31], 5 µm, 4.6 × 150 mm column; mobile phase consisting of 40% CH_3CN/100 mM NH_4HCO_2 at a flow rate of 2 ml/min; column oven set to 30°C; 20 µl injection volume; and UV detector operated at 254 nm. Radiochemical identity of (*R*)-[*N*-methyl-^{11}C]PK11195 is confirmed if the Rf (2.5 min) is within 10% of the reference (*R*)-[*N*-methyl-^{12}C]PK11195 standard solution [32] previously injected. The relative retention time (RRT) must be between 0.9 and 1.1:

$$\mathrm{RRT} = \frac{\text{retention time}_{[^{11}\mathrm{C}]\mathrm{PK11195}}}{\text{retention time}_{\mathrm{PK11195}}}$$

3.4 Radiochemical Purity

The radiochemical purity (RCP) of the final (*R*)-[*N*-methyl-^{11}C]PK11195 preparation is analyzed using the radioactivity detector of the HPLC system described earlier. The area of the radioactive peak corresponding to (*R*)-[*N*-methyl-^{11}C] PK11195 shall represent more than 95% of the areas of all radioactive peaks on the chromatogram, as shown in Fig. 5.

3.5 Specific Activity

The specific activity (SA) is determined by dividing the radioactivity measurement (GBq at end of synthesis) by the mass measurement (micromole) of final product solution:

$$\mathrm{SA\,(GBq/\mu mol)} = \frac{\text{activity (GBq)}}{\text{concentration PK11195}(\mu\text{g/ml}) \times \text{volume(ml)} \,/\, \text{MW PK11195}(\mu\text{g}/\mu\text{mol})}$$

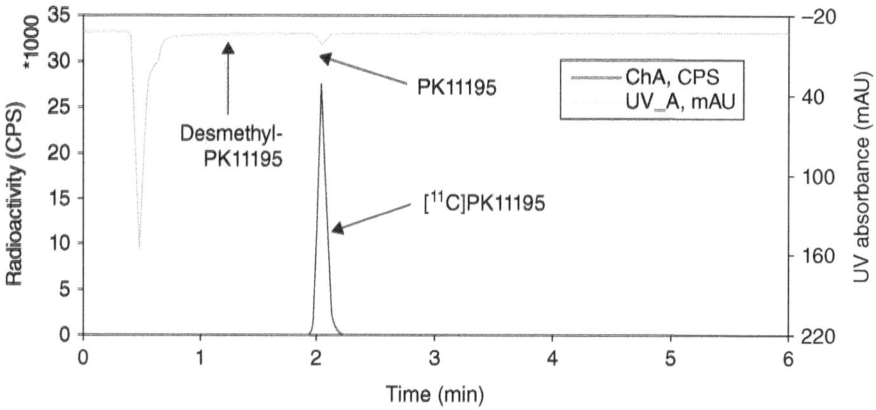

FIGURE 5 Analytical UV and radioactive HPLC trace of (*R*)-[*N*-methyl-^{11}C]PK11195.

To quantify this property, the area under (*R*)-[*N*-methyl-^{11}C]PK11195 UV mass peak is fitted to a standard calibration curve previously generated through several concentrations of *cold* PK11195 injected in the same HPLC system.

3.6 Radionuclidic Purity

The radionuclidic purity of the radioisotope is confirmed with the measure of its half-life using an aliquot of the final product and a dose calibrator [33]. The final result is compared with the characteristic half-life of ^{11}C (20.4 min) and shall be inside a margin of ±5% of this value.

3.7 Residual Solvent Analysis

Gas chromatography (GC) is used to analyze the residual solvents on the final product formulation. The levels of organic volatile impurities are analyzed through the injection of 2 µl of the final solution into a GC system [34] equipped with a flame ionization detector (FID) and previously calibrated for the residual solvents that can be present. Those are ethanol (from the formulation) and acetonitrile (from the semipreparative HPLC). According to the relevant monograph of Ph. Eur. [35], the concentration of acetonitrile on final product solution should be less than 4.1 mg/V, where V is the maximum injection volume in ml.

3.8 Sterile Filter Integrity Test

The integrity of the 0.22 µm Millex-GV sterile filter is determined after the synthesis and formulation by the *bubble point* test. After removing the final vial with the radiotracer, the needle attached to the sterile filter is submerged in water and the nitrogen pressure gradually increased to monitor for bubble formation. It is considered that the filter warrants the sterility of the final product solution if the pressure reached is higher than the filter acceptance level.

3.9 Endotoxin Analysis

The bacterial endotoxins content is analyzed using the limulus amebocyte lysate (LAL) test. This involves incubation (37°C) of the limulus lysate reagent (Charles River Laboratories Endosafe®) with a sample of our solution for 60 min to determine if there was a potential contamination above 175 EU/V (endotoxins units per maximum injectable volume). According to Ph. Eur., the apyrogenicity test can be performed after release for use.

3.10 Sterility Testing

Sterility is tested by direct inoculation of the culture medium followed by incubation for 14 days as indicated in Ph. Eur. general monograph "Sterility." According to Ph. Eur., sterility test can be performed after release for use.

WASTE DISPOSAL INFORMATION

All hazardous chemicals and toxic materials were disposed of according to Prudent Practices in the Laboratory (Washington, DC: National Academy Press, 1995).

CHEMICAL ABSTRACTS NOMENCLATURE (REGISTRY NUMBER)

(*R*)-[*N*-Desmethyl]PK11195, 1-(2-Chlorophenyl)-*N*-(1-methylpropyl)-isoquinoline-3-carboxamide (157809-85-3)
(*R*)-[*N*-Methyl]PK11195, 1-(2-Chlorophenyl)-*N*-methyl-*N*-(1-methylpropyl)-isoquinoline-3-carboxamide-(*R*) (85340-56-3)
(*R*)-[*N*-Methyl-^{11}C]PK11195, 1-(2-Chlorophenyl)-*N*-[^{11}C]methyl-*N*-(1-methylpropyl)-isoquinoline-3-carboxamide-(*R*) (157809-85-3)
Acetonitrile (75-05-8)
DMSO (67-68-5)
Ethanol (64-17-5)
Iodomethane (74-88-4)6
Potassium hydroxide (1310-58-3)

REFERENCES AND NOTES

For detailed supplier information, see Appendix 1.

1. A. Jacobs, B. Tavitian, *J. Cereb. Blood Flow Metab.* 2012, 32, 1.
2. F. Chauveau, H. Boutin, N. Van Camp, F. Dollé, B. Tavitian, *Eur. J. Nucl. Med. Mol. Imaging* 2008, 35, 2304.
3. M. Politis, P. Su, P. Piccini, *Front. Pharmacol.* 2012, 3, 96.
4. S. Venneti, B. Lopresti, C. Wiley, *Prog. Neurobiol.* 2006, 80, 308.
5. A. Ching, B. Kuhnast, A. Damont, D. Roeda, B. Tavitian, F. Dollé, *Insights Imaging* 2012, 3, 111.

6. R. Camsonne, C. Crouzel, D. Comar, M. Mazière, C. Prenant, J. Sastre, M. A. Moulin, A. Syrota, J.Labelled, Comp. Radiopharm. 1984, 21, 985.
7. P. Chakraborty, D. Maiti, T. Mangner, D. Chugani, H. Chugani, *J. Nucl. Med.* 2006, 47, 522P.
8. K. Hashimoto, O. Inoue, K. Suzuki, T. Yamasaki, M. Kojima, *Ann. Nucl. Med.* 1989, 3, 63.
9. IBA, Louvain-la-neuve, Belgium.
10. Portugal Gases, S.A, Maia, Portugal.
11. C. Crouzel, B. Langstrom, V. W. Pike, H. H. Coenen, *Int. J. Radiat. Appl. Instrum. Appl. Radiat. Isot.* 1987, 38 (8), 601–603.
12. Now Eckert & Ziegler Radiopharma, Germany.
13. Molecular sieve (13X, 100/120 mesh) was purchased from Sigma-Aldrich, pn. 2-0307.
14. Lithium aluminum hydride 0.1 M was purchased from ABX, pn. 832.0002.
15. Hydriodic acid redistilled (aqueous solution, >55%, redistilled) was purchased from ABX, pn. 830.0003.5.
16. Ascarite II® (CO_2 absorbent, ~20–30 mesh) was purchased from Sigma-Aldrich, pn. 223921.
17. Phosphorus pentoxide was purchased from Sigma-Aldrich, pn. 79610.
18. A. A. Wilson, A. Garcia, S. Houle, N. Vasdev, *J. Label Compd. Radiopharm.* 2009, 52 (11), 490–492.
19. A. A. Wilson, A. Garcia, L. Jin, S. Houle, *Nucl. Med. Biol.* 2000, 27 (6), 529–532.
20. (*R*)-[*N*-Desmethyl]PK11195 salt was purchased from ABX, pn. 1600.0001
21. Dimethyl sulfoxide (≥99.9%) was purchased from Sigma-Aldrich, pn. D8418.
22. Potassium hydroxide (puriss. p.a., Reag. Ph. Eur., ≥85%, pellets) was purchased from Sigma-Aldrich, pn. 30603.
23. Luna C18(2), 250 × 10 mm, 5 μ (Phenomenex, Torrance, United States).
24. UV K-200 detector set at 254 nm (Knauer, Berlin, Germany).
25. Reform-Plus system (Eckert & Ziegler Radiopharma, Germany).
26. Sep-Pak® C18 Light (Waters®, Milford, United States), pn. WAT023501.
27. Sterile 0.22 μm filter (Millex-GV, Millipore), pn. SLGV033RS.
28. European Pharmacopoeia 8th Edition, General Chapter on Radiopharmaceutical preparations, EDQM Council of Europe, 2013.
29. pH meter (Jenway 3510 pH; Bibby Scientific Limited, Staffordshire, United Kingdom).
30. An Agilent 1200 Series HPLC system (variable wavelength detector) with a Raytest GABI Star 1207 radiometric detector (Raytest GmbH, Straubenhardt, Germany) and a Rheodyne (IDEX H&S, Wertheim-Mondfeld, Germany) model 7125i syringe-loading sample injector valve with a 20 μl loop was used.
31. Agilent Technologies, Santa Clara, United States.
32. (*R*)-PK11195 standard salt was purchased from ABX, pn. 1610.0010.
33. ISOMED 1010 dose calibrator (Nuklear-Medizintechnik Dresden GmbH, Dresden, Germany).
34. GC Agilent 6850 (RaytestGmbH, Straubenhardt, Germany)
35. Ph. Eur., Pharmacopoeia European, the general monograph on residual solvents.

PART III

OTHER RADIOPHARMACEUTICALS

CHAPTER 11

SYNTHESIS OF OXYGEN-15 WATER ([^{15}O]H_2O)

David W. Dick and G. Leonard Watkins

Department of Radiology, University of Iowa Hospitals & Clinics, Iowa City, Iowa, USA

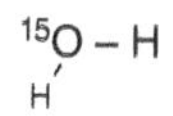

[^{15}O]Water

1 INTRODUCTION

Oxygen-15 is a cyclotron-produced, positron-emitting radionuclide with a half-life of 124 s. Depending on target conditions, the oxygen-15 produced in the cyclotron target may take on the form of [^{15}O]O_2, [^{15}O]CO, or [^{15}O]CO_2. The most common approach uses the $^{14}N(d,n)^{15}O$ nuclear reaction in the presence of up to 5% oxygen for the production of [^{15}O]O_2. Early medical studies relied on physiology to reduce [^{15}O]O_2 or [^{15}O]CO_2 to [^{15}O]H_2O so that it could be used to image tissue perfusion [1]. The ability to convert [^{15}O]O_2 to [^{15}O]H_2O using a palladium catalyst was first reported by West and Dollery in 1961 [2] and subsequently improved upon by Clark and Buckingham in 1975 [3]. The use of platinum as a catalyst for the production of [^{15}O]H_2O was first reported by Berridge in 1990, allowing a decrease in carrier oxygen in the target gas to 1% or less [4]. The use of oxygen-15 in medicine very much predates positron emission tomography (PET), having first been used in the 1950s and 1960s with gamma cameras or arrays of scintillators [5–7] with the use of [^{15}O]H_2O for PET coming into wider use in the early 1980s [8]. [^{15}O]H_2O is the gold standard for quantitative blood flow studies and still has its place in the modern clinic, even with the emergence of fMRI. The imaging studies themselves are much more technical than the production of [^{15}O]H_2O, and the reader should take care to read the relevant literature on performing [^{15}O]H_2O imaging studies [9] in order to fully understand the scope of personnel and effort required for productive imaging studies, as that information is beyond the scope of this chapter.

Radiochemical Syntheses, Volume 2: Further Radiopharmaceuticals for Positron Emission Tomography and New Strategies for Their Production, First Edition. Edited by Peter J. H. Scott.

2 SYNTHESIS PROCEDURES

CAUTION: All radiochemical syntheses must be carried out using appropriate equipment in a facility authorized for the use of radioactive materials. Personal protective equipment must be worn and all local radiation safety laws followed.

2.1 Production of [^{15}O]O_2

There are numerous production routes for oxygen-15, including ^{14}N(d,n)^{15}O, ^{16}O(p,pn)^{15}O [10], and ^{15}N(p,n)^{15}O [11]. The nitrogen-15 production route is the only available production route for low-energy proton-only cyclotrons and is hampered by the expense of enriched nitrogen-15 gas. The University of Iowa utilizes the reaction involving deuteron bombardment on nitrogen. The target body is constructed of aluminum with a volume of 35.5 ml, covered in front by a double foil assembly using 25 μm Havar foils. Irradiations are performed using 8.4 MeV deuterons. The University of Iowa Health Care GE PETtrace cyclotron has a beam current of 5–60 μA for deuterons. The target chemistry is based on the established literature methods. The system for the production of [^{15}O]O_2 using the PETtrace cyclotron is shown schematically in Figs. 1 and 2. The system uses stainless steel and Teflon tubing and electrically operated solenoid valves. The charcoal and Ascarite filters are constructed using Teflon tubes, 1/2 in. diameter × 10 in. long (Cole-Parmer), with stainless steel fittings (Swagelok). The combination trap is shown in Fig. 3. The inlet side is filled with ~9 g of activated carbon, 6–14 mesh

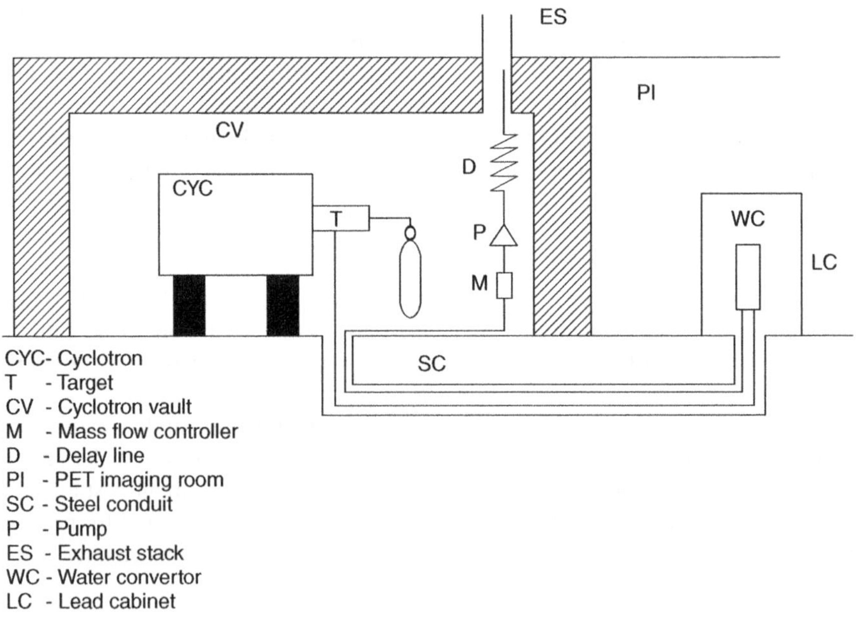

FIGURE 1 Schematic of [^{15}O] steady-state delivery system from cyclotron to [^{15}O]H_2O and returning to the waste gas system.

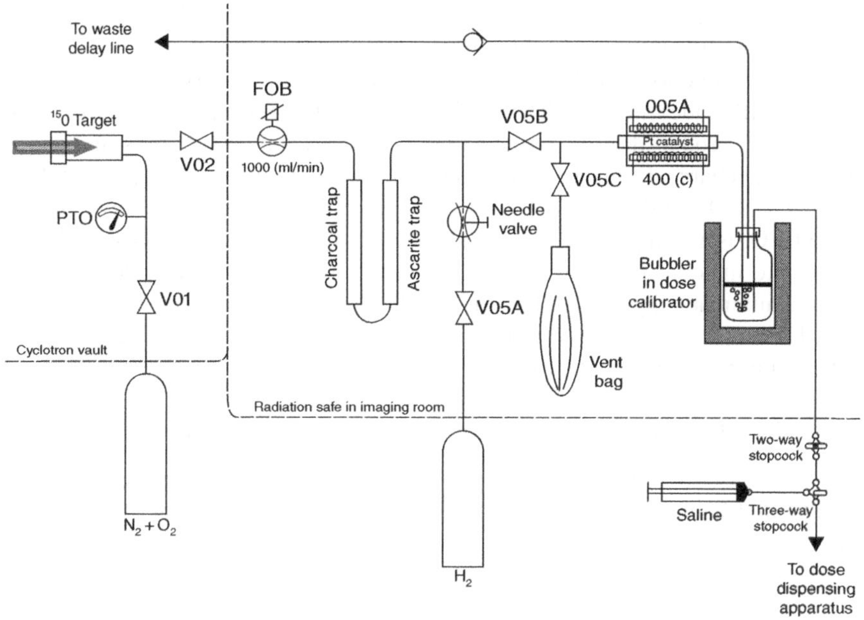

FIGURE 2 Schematic diagram of the production of [^{15}O]H_2O using the PETtrace cyclotron.

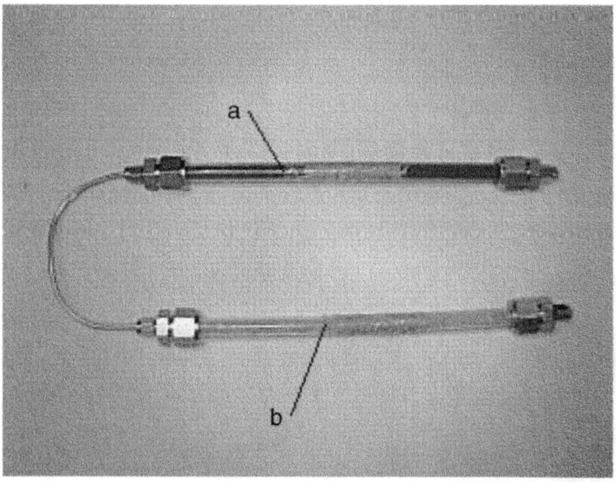

FIGURE 3 Combination trap: (a) charcoal and (b) Ascarite-II.

(Fisher Scientific). The outlet side is filled with ~19 g of Ascarite II, 8–20 mesh (Thomas Scientific). Each agent is secured in the Teflon column with ~1 cm long glass wool plugs at each end, taking care not to compromise the swaged connection. The assembly is pressure tested to ~100 psi. It is advisable to keep a freshly prepared set of traps ready for installation as soon as the pH and NH_3 levels start to increase, so as not to lose valuable time in a busy laboratory.

For the production of [^{15}O]O_2, the target is filled with 1.0% oxygen in ultrahigh-purity (UHP grade) nitrogen. The O-15 production target is pressurized to 150 psi. The [^{15}O]O_2 produced in the target along with unreacted target gas is first passed through the charcoal filter to remove several potential impurities, especially [^{13}N]NO_x, [^{13}N]N_2O, [^{15}O]NO_x, and [^{15}O]N_2O. Then, efficient removal of any [^{11}C]CO_2 or [^{15}O]CO_2 is accomplished by passing the gas stream through an Ascarite II trap. Particulate material is removed using a 60 micron filter, and then the [^{15}O]O_2 is shunted automatically to the PET imaging facility where it is converted into [^{15}O]H_2O. It is important to *tune* your tubing from the cyclotron to the imaging suite in order to minimize losses due to decay in transport [12].

2.2 Preparation, Formulation, and Administration of [^{15}O]H_2O

The convertor to synthesize [^{15}O]H_2O is composed of two main components: catalyst and bubbler (Fig. 4). [^{15}O]O_2 from the target (800 ml/min at one atmosphere), following purification and transport to the leaded-cabinet in the PET imaging suite, is mixed with research grade hydrogen (200 ml/min at one atmosphere) and passed through a glass tube containing the platinum wire catalyst. The glass tube resides in a furnace that is heated to 400°C. Under the influence of the platinum catalyst and hydrogen gas, the [^{15}O]O_2 is reduced to [^{15}O]H_2O vapor, which is subsequently bubbled into 10 ml of isotonic saline where it is trapped.

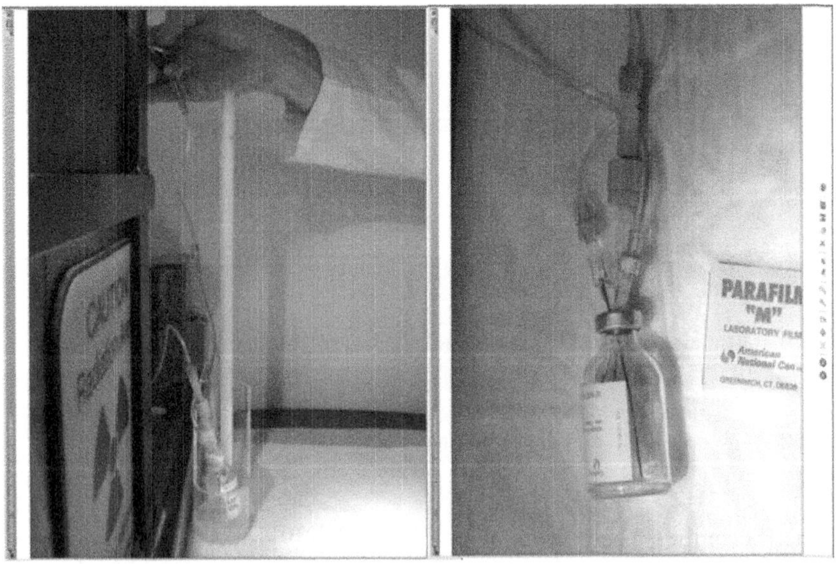

Entire apparatus Close-up: bubbler

FIGURE 4 [^{15}O] H_2O bubbler.

The distance from the furnace to the bubbler should be kept to a minimum to avoid condensation in the connecting tubing. Any untrapped radioactivity is mixed with air and returned to waste/decay line in the vault (Fig. 1). The initial sub-batch is drawn up from the bubbler and given to the radiochemistry facility for quality control (QC). The bubbler is then immediately refilled with 8 ml of saline to prepare it for subsequent administrations after all QC pre-release tests have reached a successful conclusion. The administration of the $[^{15}O]H_2O$ is done by drawing up 8 ml of isotonic saline from the bubbler. 8 ml of fresh isotonic saline is then added back to the bubbler so that subsequently produced $[^{15}O]H_2O$ will be trapped and the administration process may be repeated. Using this process, $[^{15}O]H_2O$ is continuously being prepared and sub-batches drawn as needed. Typically, in brain activation studies, an injection (sub-batch) can be performed every 12–15 min.

3 QC PROCEDURES

CAUTION: All radiopharmaceuticals produced for clinical use must have local regulatory approval before use in humans. Trained personnel must carry out QC procedures, and each dose must meet all established QC criteria before release to the clinic.

3.1 Visual Inspection

The sub-batch is visually examined. The solution must be clear, colorless, and free of particulate matter.

3.2 Radionuclidic Identity

An Atomlab 300 PET Dose Calibrator (Atomic Products Corp.) on the O-15 setting and NIST calibrated stopwatch are used to determine the half-life of the radioactive sample. Measurements are taken at 0, 5, and 10 min. Half-life is calculated for the two intermediate time points (0–5 min and 5–10 min) as well as the overall time frame. The equation used to determine the half-life is:

$$t_{1/2} = \frac{\ln(A_0/A)}{\ln(2)^* \Delta t}$$

where A_0 is the initial activity measurement, A is the later radioactivity measurement, and Δt is the elapsed time between measurements. The half-life must be in the range of 115–130 s for the product to be acceptable for administration to patients. A 124 s half-life provides convincing evidence for the radionuclidic identity of oxygen-15.

3.3 Radionuclidic Purity

A sample from the QC sub-batch is used to determine the gamma-ray energy spectrum. This should exhibit only the two expected peaks for a positron emitter at 511 keV and a possible sum peak at 1022 keV. This measurement is carried out using a NaI(Tl) well counter with output through an amplifier to a multichannel analyzer (Canberra). The instrument is calibrated with a known long-lived nuclide, [^{22}Na]sodium, standard. The sub-batch sample is twice diluted by a factor of 100 to achieve a sample with a total dilution factor of 10,000. The dilution is made in order to reduce the total activity of the sample and decrease dead time losses within the NaI(Tl) crystal. This diluted sample is placed in the NaI(Tl) well counter and a 5 min acquisition is obtained. The initial dead time should not be greater than 50%. The gamma-ray spectrum of the sample should be very similar to the [^{22}Na]sodium calibration up through 1022 keV. If no other differences are observable, then *Pass* will be recorded. The purity should *Pass* to be acceptable.

3.4 Specific Activity

The initial sub-batch vial should be weighed and the weight recorded prior to dispensing of the first sub-batch from the bubbler. Upon receipt of the QC sub-batch, the initial sub-batch vial should be immediately assayed for radioactivity and then weighed to determine the volume of liquid within the sub-batch vial. Specific activity is calculated with these two values using the following equation:

$$\text{Specific activity}\left(\frac{\text{mCi}}{\text{mol}}\right)=\frac{18.162\times\text{initial activity (mCi)}}{\text{mass of saline in vial (g)}}$$

The specific activity must be greater than 20 mCi/mol. It is usually in the 100–130 mCi/mol range.

3.5 Bacterial Endotoxin Testing

Limulus amebocyte lysate (LAL), derived from *Limulus polyphemus* amebocytes, is intended for use in the qualitative/semiquantitative detection of gram-negative bacterial endotoxins by the colorimetric method. This test is performed using the Charles River Endosafe-PTS reader and licensed PTS cartridges. Dilution of the sample is not required prior to testing. The sample value must be less than or equal to 5 EU/ml, the spike recovery value must be 50–200%, and both coefficients of variation must be less than or equal to 25% in order for the product to be considered acceptable and under the pyrogen limit.

3.6 pH Testing

pH testing is performed in order to ensure that a physiologically compatible pH product is administered to patients. pH is assessed using pH strips. The pH must be in the range of 4.5–8.0 for the product to be acceptable for administration to patients. When the pH starts increasing rapidly, it is time to change the *traps* for water production (Fig. 3).

3.7 Radiochemical and Chemical Purity

NOTE: This must be the first QC test performed in order to provide sufficient radioactivity for the radioactivity detector to obtain enough counts in a single pass to yield satisfactory results.

A sample from the QC sub-batch is removed for gas chromatographic analysis to ascertain the level of impurities in the isotonic saline (chemical purity) and the extent of radioactive impurities in relation to $[^{15}O]H_2O$ (radiochemical purity). The test is performed on a SRI 8610C gas chromatograph equipped with a thermal conductivity detector (TCD) and in tandem a radioactivity detector (CsI(Tl) scintillation crystal coupled to a Si PIN diode, low-noise preamplifier, shaping postamplifier, and analog ratemeter output (Carroll & Ramsey Associates).

NOTE: It is critical that the effluent of the first, TCD, detector be returned to the GC oven in order to maintain the water in the vapor state. The outlet tubing should exit the oven directly into a lead shield containing the radioactivity detector. The proximity of the lead shield to the oven is critical to avoid beading of water caused by condensation. The lead shield is also critical to avoid "shine through" from the sample circulating through the column. The radioactive detector is also temperature sensitive, so for best results, we have found it necessary to insert the detector immediately prior to injection of the sample.

The stationary phase is Porapak N 80/100, packed in a 6 ft × 0.25 in. o.d. × 0.210 in. i.d. stainless steel column (Alltech). Carrier gas is research grade helium and set to flow at 70 sccm. TCD temperature is set to 200°C and the oven temperature is 150°C. A 1–2 μl injection of non-radioactive water should be performed prior to QC in order to determine the retention time of water on the GC. 1–2 μl of the sub-batch is injected onto the GC and the program is run. The TCD readout is used to determine chemical purity, and the radiation detector readout is used to determine the radiochemical purity (typical example of GC chromatogram is shown in Fig. 5). To be acceptable for human use, the chemical purity should be greater than 99.5% and the radiochemical purity greater than 95.0%.

3.8 Chemical Purity: Nessler NH_3/NH_4 Test

CAUTION: Nessler's reagent is an alkaline solution of potassium tetraiodomercurate (K_2HgI_4). This reagent is highly neurotoxic and is readily absorbed through the skin. Handle with utmost care.

400 μl of the sub-batch is pipetted into a disposable glass culture tube that contains 400 μl of freshly pipetted Nessler's reagent (Aldrich), taking care to take the clear supernate. The tube is gently mixed and the resultant yellow solution is compared against a panel of known concentrations of ammonium ion made under identical conditions (400 μl ammonium ion solutions and 400 μl Nessler's reagent). To be in the acceptable range, the amount of ammonia/ammonium ion concentration must be less than 10 ppm. Normal values should be less than 3 ppm. Higher values indicate that the charcoal/Ascarite trap should be changed. Usually, the increase in ammonia/ammonium ion concentration is accompanied by a concomitant increase in pH.

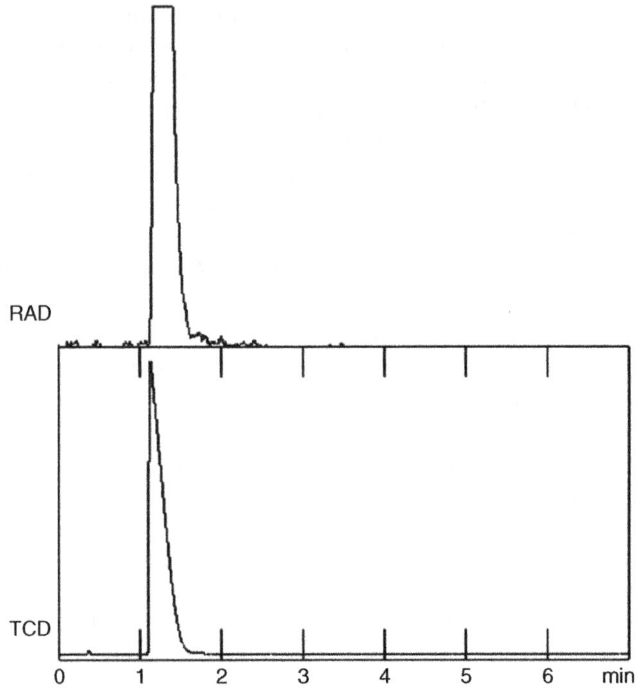

FIGURE 5 GC chromatograms for [^{15}O]H_2O.

3.9 Chemical Purity: Platinum Content

Several decayed sub-batches are combined and sent to a heavy metals lab to analyze for platinum content by inductively coupled plasma mass spectrometry (ICP-MS). The level should be less than 10 parts per billion. Note that this is not a pre-release test and only needs to be performed on an occasional basis.

3.10 Filter Integrity

The final product filter is checked for integrity using the bubble point test. The bubble point must be above the manufacturer's written specification in order for the product to be acceptable for administration to patients.

3.11 Sterility Testing

The sterility of the final product is assessed by the inoculation of the product into Trypticase soy broth and fluid thioglycolate medium. 1 ml of final product is injected into each of the two media. The media are incubated and observed daily for 14 days. The test passes if no visible growth is observed within 14 days. Note that this is a post-release test and must be initiated within 24 h of production.

WASTE DISPOSAL INFORMATION

All hazardous chemicals and toxic materials were disposed of according to Prudent Practices in the Laboratory (Washington, DC: National Academy Press, 1995).

CHEMICAL ABSTRACTS NOMENCLATURE (REGISTRY NUMBER)

Platinum (7440-06-4)
Potassium tetraiodomercurate (7783-33-7)
Water-^{15}O (24286-21-3)

REFERENCES AND NOTES

For detailed supplier information, see Appendix 1.

1. (a) C. T. Dollery, J. B. West, *Circ. Res.*, 1960, 8, 765; (b) M. M. Ter-Pogossian, J. O. Eichling, D. O. Davis, M. J.Welch, J. M. Metzger, *Radiology*, 1969, 93, 31.
2. J. B. West, C. T. Dollery, *Nature*, 1961, 189, 588.
3. J. C. Clark, P. D. Buckingham, *Short-Lived Radioactive Gases for Clinical Use*. London/Boston: Butterworths 1975.
4. M. S. Berridge, A. H. Terris, E. H. Cassidy, *Int. J. Rad. Appl. Instrum. A.*, 1990, 41, 1173.
5. M. Ter-Pogossian, W. Powers, *Radioisotopes in Scientific Research*. London: Pergamon 1958.
6. N. Dyson, P. Hugh-Jones, G. Newbery, J. Sinclair, J. West, *Br. Med. J.*, 1960, 1, 231.
7. J. West, C. Dollery, P. Hugh-Jones, *J. Clin. Investig.*, 1961, 40, 1.
8. P. Herscovitch, J. Markham, M. E. Raichle, *J. Nucl. Med.*, 1983, 24, 782.
9. R. R. Hurtig, R. D. Hichwa, D. S. O'Leary, L. L. Boles Ponto, S. Narayana, G. L. Watkins, N. C. Andreasen, *J. Cereb. Blood. Flow. Metab.*, 1994, 14, 423.
10. J. Beaver, R. Finn, H. Hupf, *Int. J. Appl. Radiat. Isot.*, 1976, 27, 195.
11. M. Sajjad, R. Lambrecht, A. Wolf, *Radiochim. Acta*, 1985, 38, 57.
12. R. D. Hichwa, R. J. Nickles, *IEEE Trans. Nucl. Sci.*, 1979, 26, 1707.

PART IV

NEW METHODS FOR THE SYNTHESIS AND QUALITY CONTROL OF RADIOPHARMACEUTICALS

CHAPTER 12

DIRECT, NUCLEOPHILIC RADIOSYNTHESIS OF [^{18}F]TRIFLUOROETHYL TOSYLATE

PATRICK J. RISS[1,2], WAQAS RAFIQUE[1], AND FRANKLIN I. AIGBIRHIO[2]

[1] *Kjemisk Institutt, Universitetet i Oslo, Oslo, Norway*
[2] *Wolfson Brain Imaging Centre, Addenbrooke's Hospital, University of Cambridge, Cambridge, UK*

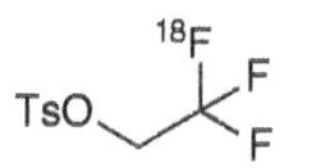

[^{18}F]trifluoroethyl tosylate

1 INTRODUCTION

The positron emitter ^{18}F is particularly well suited for PET imaging owing to almost exclusive decay via the β^+ decay branch (97%) and very low positron energy (638 keV). With a half-life of 109.7 min, multistep procedures for radiolabeling with ^{18}F are feasible. Multiple patient doses can be dispensed from a single production batch and even be shipped over moderate distances. The radionuclide is most commonly produced in high yield on standard medical cyclotrons via the $^{18}O(p,n)^{18}F$ nuclear reaction on a $H_2{}^{18}O$ liquid targets and obtained as [^{18}F]fluoride ion in high specific radioactivity, that is, a high ratio of radioactive to nonradioactive molecules (>150 GBq/μmol). In contrast to alternative production routes, this allows for PET imaging of saturable biological systems under genuine tracer conditions. Consequently, nucleophilic fluorination with $^{18}F^-$ is the chemical prelude for the majority of ^{18}F radiotracer applications. Important exemptions are cases where direct nucleophilic substitution of a leaving group is not a feasible means to introduce the ^{18}F label. In such cases, the remedial use of secondary labeling reagents for the introduction of ^{18}F-labeled prosthetic groups into suitable precursors has proved to be a valuable alternative [1–3].

Radiochemical Syntheses, Volume 2: Further Radiopharmaceuticals for Positron Emission Tomography and New Strategies for Their Production, First Edition. Edited by Peter J. H. Scott.

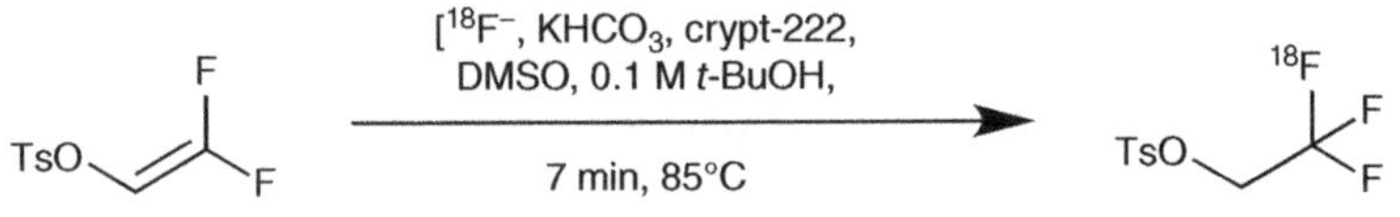

FIGURE 1 Nucleophilic radiosynthesis of 2,2,2-[^{18}F]trifluoroethyl tosylate.

2,2,2-[^{18}F]Trifluoroethyl p-toluenesulfonate is such a reagent, which facilitates introduction of an electrophilic [^{18}F]trifluoroethyl synthon ([^{18}F]F_3CCH_2), a common chemical functionality of small-molecule drugs [4, 5]. As a prosthetic group, this 2-[^{18}F]fluoro-2,2-difluoroethyl group provides great potential as a metabolically insensitive, readily available supplement to PET chemistry. Owed to the attractive characteristics of the group and its presence in many pharmaceuticals and drug candidates, two approaches for ^{18}F labeling of 2,2,2-[^{18}F]trifluoroethyl groups have been reported. The radiosynthesis of 2,2,2-[^{18}F]trifluoroethyl groups has first been completed via direct nucleophilic ^{18}F fluorination by isotopic exchange on 2,2,2-trifluoroethyl *p*-toluenesulfonate. The authors used the reagent successfully for the first time to prepare the radiotracer [^{18}F]TFMISO [5a] in low specific activity.

Shortly thereafter, we devised an efficient method to radiolabel the 2,2,2-[^{18}F] trifluoroethyl groups by direct nucleophilic addition of [^{18}F]HF to 2,2-difluorovinyl p-toluenesulfonate [6]. Herein, we provide a detailed protocol for the synthesis of 2,2,2-[^{18}F]trifluoroethyl tosylate using the latter route (Fig. 1).

2 PROCEDURES

2.1 Synthesis Procedures

CAUTION: All radiochemical syntheses must be carried out using appropriate equipment in a facility authorized for the use of radioactive materials. Personal protective equipment must be worn, and all local radiation safety laws followed.

All solvents and reagents were obtained from commercial providers such as Alfa Aesar UK Ltd. (Heysham, Morecambe, United Kingdom), Sigma-Aldrich Co. Ltd. (Poole, United Kingdom), and Fisher Scientific UK Ltd. (Loughborough, United Kingdom) and used as received. Solid-phase extraction cartridges were purchased from Waters Ltd. (Elstree, United Kingdom) and preconditioned as specified. NMR spectra were recorded on Bruker Fourier 300 (Bruker UK Ltd., Coventry, United Kingdom) spectrometer. Chemical shifts are reported downfield from a theoretical tetramethylsilane signal (0 ppm), relative to the solvent residual signal. Flash chromatography was conducted using a Gilson PLC 2020 chromatography system (Gilson Scientific Ltd., Luton, United Kingdom) and normal-phase silica gel cartridges.

2.1.1 *General Procedure for Radiolabeling*

Proton bombardment (16→3 MeV) of a silver target filled with approximately 2.5 ml $H_2{}^{18}O$ water using a beam current of 30–40 μA for 5–10 min on a GE PETtrace cyclotron afforded [^{18}F]fluoride ion. Solid-phase extraction was achieved on a Waters Accell Plus Light QMA strong anion exchanger cartridge conditioned to the $CO_2{}^2$ form using 10 ml of 1 M potassium carbonate solution and 20 ml of water. Potassium crypt-222 [^{18}F]F$^-$ cryptate complex ([K$^+$⊂K222][^{18}F]F$^-$) was obtained by elution of the trapped radioactivity using a mixture of crypt-222 (7.5 mg) in acetonitrile (600 μl) and 1 M potassium hydrogen carbonate solution (7.5 μl) followed by concentration in a stream of helium. Residual water was removed by coevaporation with anhydrous acetonitrile (3 × 1 ml). 2,2-Difluorovinyl *p*-toluenesulfonate dissolved in 0.1 M 2-methyl-2-propanol in dimethyl sulfoxide (DMSO) (1 ml) was added, and the mixture was heated to 85°C for 7 min [7, 8]. The reaction was quenched via the addition of high-performance liquid chromatography (HPLC) eluent (3 ml). The resultant sample was directly injected into radio-HPLC.

2.2 Quality Control Procedures

HPLC analyses were performed on Agilent 1100 series HPLC systems (Agilent Technologies UK Ltd., Wokingham, United Kingdom); a G1312 A gradient pump and a G1314 variable wavelength UV detector were used. Thallium-doped sodium iodide detectors NaI(Tl) of 1" diameter (Bioscan, Inc., Washington, DC, United States) with Flow-Count B-FC-4000 analog/digital interfaces were used for radioactivity detection. LabLogic Laura 3 and Laura 4 software (LabLogic Systems Ltd., Sheffield, United Kingdom) were used for data acquisition and evaluation. Quality control was performed using radio-HPLC and radio-TLC. Analytical radio-HPLC was conducted using a Chromolith RP18e (5 μm) 0.4 mm × 100 mm column (Merck KGaA, Darmstadt, Germany) at a flow rate of 3 ml/min (water/acetonitrile; 1:1).

3 DISCUSSION

We present a straightforward protocol for the radiosynthesis of no-carrier-added 2,2,2-[^{18}F]trifluoroethyl p-toluenesulfonate as a labeling reagent for ^{18}F labeling. The reagent was obtained in a single radioactive step in moderate to good yields. The conditions were optimized for the use of potassium 4,7,13,16,21,24-hexaoxa-1, 10-diazabicyclo[8.8.8]hexacosan (crypt-222) cryptate [^{18}F]fluoride ion complex, the most commonly used method for radiolabeling with [^{18}F]fluoride ion. This reactive [^{18}F]fluoride ion source can be readily obtained from the target using solid-phase extraction and azeotropic removal of remaining water. This process has been automated on a variety of commercial systems, which facilitate straightforward transfer of the methodology into general application.

The labeling precursor 2,2-difluorovinyl p-toluenesulfonate is reacted with said [^{18}F]fluoride ion complex obtained from $KHCO_3$ and crypt-222 in DMSO at 85°C in the presence of 2-methyl-2-propanol to suppress formation of a side product (Fig. 2). At 85°C, [^{18}F]fluoride is rapidly incorporated into the labeling precursor, and the desired product can be obtained in high yield within 7 min. Prolonged reaction times result in lower RCY of 2,2,2-[^{18}F]trifluoroethyl *p*-toluenesulfonate.

Rapid HPLC purification was achieved using a low precursor concentration of unlabeled precursor, which is desirable to achieve demanding conditions for separation. Highest radiochemical yields were obtained using 2 ± 0.4 mg of precursor in 800 ml of DMSO, equivalent to a molar concentration of 1 mM. An appropriate

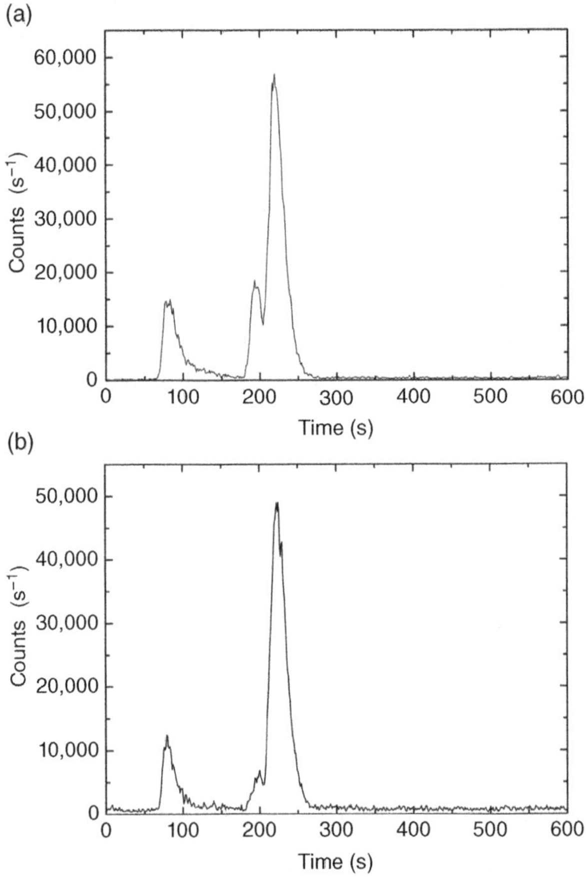

FIGURE 2 Example of a typical chromatogram of [^{18}F]**1** from reaction mixture in the absence (a) and presence (b) of *t*-BuOH. HPLC conditions are described in the text. Radio-TLC was conducted with 1 μl samples diluted to 1 MBq/ml in $CHCl_3$. A BAS-*IP* MS storage phosphor screen 35 cm × 43 cm was used for radio-TLC (Fisher Scientific UK Ltd., Loughborough, United Kingdom). Screens were evaluated using a Duerr CR 35 NDT (Raytest Isotopenmessgeraete GmbH, Straubenhardt, Germany) and Raytest AIDA QWBA software.

protic additive is required to achieve maximum RCY. We resorted to 2-methyl-2-propanol, which has a pK_a value similar to water in DMSO (pK_a 31.4) [9]. But contrary to water, this alcohol is tolerated in much higher, hence easily controllable concentration, without a detrimental effect on ^{18}F labeling.

A routine production batch afforded HPLC-purified 2,2,2-[^{18}F]trifluoroethyl *p*-toluenesulfonate in a yield of about 50% using a batch of approximately 12 GBq [^{18}F]fluoride ion for the radiosynthesis. Typical specific activity of the final after trifluoroalkylation with [^{18}F]**1** product was >150 MBq/nmol.

WASTE DISPOSAL INFORMATION

All hazardous chemicals and toxic materials were disposed of according to Prudent Practices in the Laboratory (Washington, DC: National Academy Press, 1995).

CHEMICAL ABSTRACTS NOMENCLATURE (REGISTRY NUMBER)

2,2,2-[^{18}F]trifluoroethyl p-toluenesulfonate [1300731-45-6]
2,2-Difluorovinyl p-toluenesulfonate [185739-14-4]
4,7,13,16,21,24-Hexaoxa-1,10-diazabicyclo[8.8.8]hexacosan (crypt-222) [23978-09-8]
DMSO [67-68-5]
Potassium hydrogen carbonate [298-14-6]
2-Methyl-2-propanol [75-65-0]
Acetonitrile [75-05-8]

REFERENCES AND NOTES

For detailed supplier information, see Appendix 1.

1. (a) P. W. Miller, N. J. Long, R. Vilar, A. D. Gee, *Angew. Chem. Int. Ed.*, 2008, 47, 8998; (b) L. Cai, S. Lu, V. W. Pike, *Eur. J. Org. Chem.*, 2008, 2008, 2853.
2. (a) H. J. Wester in *Handbook of Nuclear Chemistry*, Vol 4 (Eds: A. Vértes, S. Nagy, Z. Klencsár), Kluwer Academic Publishers, Amsterdam, 2003, 167; (b) P. J. Riss, F. Roesch, *Org. Biomol. Chem.*, 2008, 6, 4567.
3. G. J. Meyer, S. L. Waters, H. H. Coenen, A. Luxen, B. Maziere, B. Langström, *Eur. J. Nucl. Med.*, 1995, 22, 1420.
4. (a) H. L. Yale, *J. Med. Pharm. Chem.*, 1959, 1, 121; (b) A. G. M. Barret, M. R. Crimmin, M. S. Hill, P. B. Hitchcock, P. A. Procopiou, *Angew. Chem. Int. Ed.*, 2007, 46, 6339; (c) D. O'Hagan, C. Isanbor, *J. Fluor. Chem.*, 2006, 127, 303; (d) T. Furuya, A. S. Kamlet, T. Ritter, *Nature*, 2011, 473, 470. (e) S. Purser, P. R. Moore, S. Swallow, V. Gouverneur, *Chem. Soc. Rev.*, 2008, 37, 320.
5. (a) M. Suehiro, G. Yang, G. Torchon, E. Ackerstaff, J. Humm, J. Koutcher, O. Ouerfelli, *Bioorg. Med. Chem.*, 2011, 19, 2287; (b) M. V. Fawaz, A. F. Brooks, M. E. Rodick, G. M. Carpenter, X. Shao, T. J. Desmond, P. Sherman, C. A. Quesada, B. G. Hockley, M. R. Kilbourn, R. L. Albin, K. A. Frey, P. J. H. Scott, *ACS Chem. Neurosci.*, 2014, 5, DOI: 10.1021/cn500103u; (c) H. Watanabe, N. Chopra, D. Laver, H. S. Hwang, S. S. Davies, D. E. Roach, H. J. Duff, D. M. Roden, A. A. M. Wilde,

B. C. Knollmann, *Nat. Med.*, 2009, 15, 380; (d) A. van Oeveren, M. Motamedi, N. S. Mani, K. B. Marschke, F. J. López, W. T. Schrader, A. Negro-Vilar, L. Zhi, *J. Med. Chem.*, 2006, 49, 6143; (e) F. Mauler, J. Mittendorf, E. Horváth, J. De Vry, *J. Pharmacol. Exp. Ther.*, 2002, 302, 359; (f) L. G. Miller, W. R. Galpern, J. J. Byrnes, D. J. Greenblatt, *Pharmacol. Biochem. Behav.*, 1992, 43, 413.

6. (a) P. J. Riss, F. I. Aigbirhio, *Chem. Commun.*, 2011, 47, 11873; (b) P. J. Riss, V. Ferrari, L. Brichard, P. Burke, R. Smith, F. I. Aigbirhio, *Org. Biomol. Chem.*, 2012, 10, 6980.
7. T. M. Gøgsig, L. S. Søbjerg, A. T. Lindhardt (neé Hansen), K. L. Jensen, T. Skrydstrup, *J. Org. Chem.*, 2008, 73, 3404.
8. Sigma-Aldrich 41647 DMSO puriss., absolute, over molecular sieve (H_2O ≤0.005%), ≥99.5% (GC) was used.
9. (a) W. M. Olmstead, Z. Margolin, F. G. Bordwell, *J. Org. Chem.*, 1980, 45, 3295. (b) D. W. Kim, H.-J. Jeong, S.T. Lim, M.-H. Sohn, J.A. Katzenellenbogen, D.Y. Chi, *J. Org. Chem.*, 2008, 73, 957; (c) C. M. Sharts, W. A. Sheppard, *Org. React.*, 1974, 21, 125.

CHAPTER 13

SYNTHESIS OF [^{18}F]*N*-METHYL LANSOPRAZOLE VIA GENERATION OF A [^{18}F]TRIFLUOROMETHYL GROUP

Allen F. Brooks, Melissa E. Rodnick, Garret M. Carpenter, and Peter J. H. Scott

Department of Radiology, University of Michigan School of Medicine, Ann Arbor, Michigan, USA

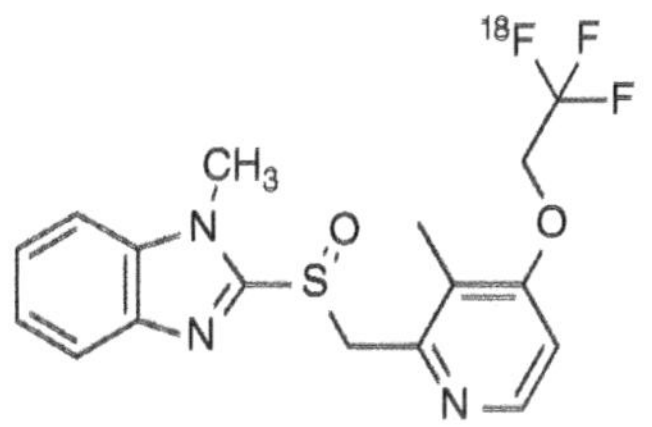

[^{18}F]*N*-methyl lansoprazole ([^{18}F]NML)

1 INTRODUCTION

As part of our PET imaging R & D program, we have an interest in developing radiopharmaceuticals for imaging tau in neurodegenerative tauopathies such as Alzheimer's disease (AD), progressive supranuclear palsy (PSP), frontotemporal dementia (FTD), and corticobasal degeneration (CBD). Our efforts in this area have focused upon repurposing lansoprazole, an FDA-approved proton pump inhibitor, for tau imaging following a report by Rojo and coworkers that, it has low nanomolar affinity for tau and selectivity for tau over amyloid [1]. We have recently reported analogs of lansoprazole labeled with carbon-11 or fluorine-18 and are in the process of translating our lead compound, [^{18}F]*N*-methyl lansoprazole, into clinical imaging studies [2, 3]. Development of a synthesis of [^{18}F]*N*-methyl lansoprazole in our laboratory concentrated on adaptation of recent chemistry reported

Radiochemical Syntheses, Volume 2: Further Radiopharmaceuticals for Positron Emission Tomography and New Strategies for Their Production, First Edition. Edited by Peter J. H. Scott.

FIGURE 1 Radiosynthesis of [18F]*N*-methyl lansoprazole.

by Riss and coworkers (see Chapter 12 of this book as well as references [4, 5]). Riss was able to generate [^{18}F]trifluoroalkyl tosylates, labeled at the trifluoromethyl group, by reacting the corresponding gem-difluoroenol ether with [^{18}F]fluoride, potassium carbonate, and Kryptofix-2.2.2 (K222), and then use them as fluorine-18-labeled prosthetic groups. We were able to adapt this methodology to the synthesis of [^{18}F]*N*-methyl lansoprazole (Fig. 1) and demonstrated that the radiotracer has subnanomolar affinity for aggregated tau as well as imaging characteristics suitable for future applications in clinical imaging of tau [3].

2 SYNTHESIS PROCEDURES

CAUTION: All radiochemical syntheses must be carried out using appropriate equipment in a facility authorized for the use of radioactive materials. Personal protective equipment must be worn and all local radiation safety laws followed.

2.1 Production of [^{18}F]Fluoride

[^{18}O]H_2O (1.5 ml) [6] was loaded into the [^{18}F]fluoride target [7] of a General Electric (GE) PETtrace cyclotron [8]. The target was bombarded (60 µA beam for 15 min) to generate ~900 mCi (33.3 GBq) of [^{18}F]fluoride by the ^{18}O(p,n)^{18}F nuclear reaction.

2.2 Azeotropic Drying of [^{18}F]Fluoride

The [^{18}F]fluoride was delivered to a GE TRACERlab FX_{FN} synthesis module [8] as a solution in [^{18}O]H_2O (1.5 ml). This solution was passed through a QMA-Light Sep-Pak® cartridge [9] to trap the [^{18}F]fluoride and recycle the [^{18}O]H_2O. The [^{18}F]fluoride was then eluted into the TRACERlab FX_{FN} reaction vessel using a solution of aqueous potassium carbonate (3.5 mg in 0.5 ml H_2O) [10]. A solution of K222 (15 mg in 1 ml MeCN) [11] was added and the reaction mixture was azeotropically dried, initially at 80°C under vacuum for 4 min and subsequently at 60°C with both vacuum and argon flow for an additional 4 min.

2.3 Synthesis of [^{18}F]*N*-Methyl Lansoprazole

Following drying of the [^{18}F]fluoride, the TRACERlab reactor was cooled to 90°C and a solution of enol ether precursor **1** (3.5 mg) in DMSO (0.95 ml) and aq. satd. ammonium chloride (5 µl) was added to the reaction vessel [12]. The reaction was

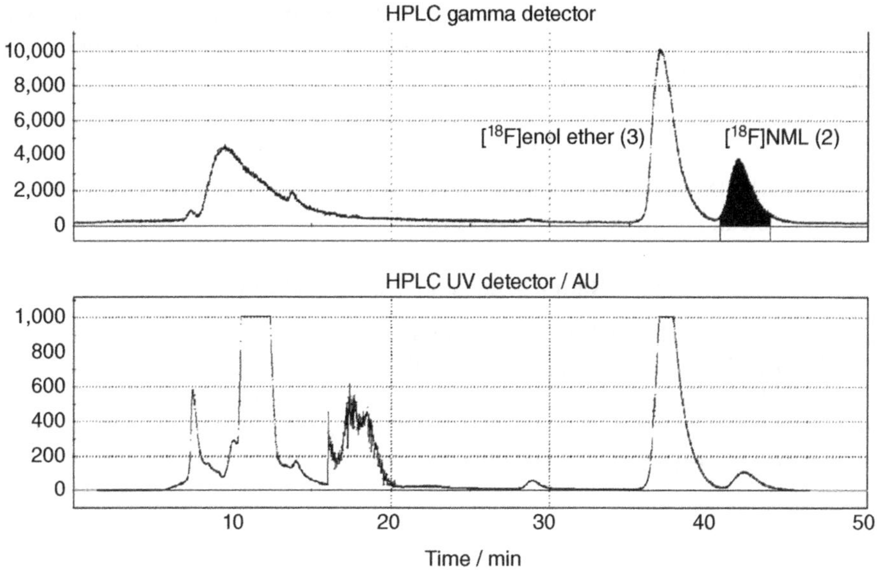

FIGURE 2 Representative semipreparative HPLC trace from the purification of [18F]*N*-methyl lansoprazole.

stirred for 3 min at 90°C, cooled to 50°C, and quenched by the addition of HPLC solvent (3.5 ml, 30% acetonitrile). The diluted reaction mixture was purified by semi-preparative HPLC (column: Phenomenex Luna PFP(2) 250 × 10 mm [13]; mobile phase: 65% water in 35% acetonitrile (v/v); flow rate, 2.0 ml/min). The fraction corresponding to [^{18}F]*N*-methyl lansoprazole (typically eluting out at 44 min—see Fig. 2 for a representative semi-preparative HPLC trace) was collected into a dilution flask containing sterile water (50 ml). The resulting solution was transferred through a C18 Sep-Pak [14] to collect the desired product. The Sep-Pak was subsequently rinsed with 10 ml of sterile water and eluted with dehydrated ethanol for injection (0.5 ml) into the collection vial containing saline (4.5 ml of 0.9% for injection, USP). Finally, the Sep-Pak was rinsed with an additional 5 ml of saline and the resulting isotonic formulated dose (10 ml) was transferred through a Millex-GV sterile filter [15] into a sterile 10 ml dose vial [16] to yield [^{18}F]*N*-methyl lansoprazole (typically 27–60 mCi, 3–7% non-decay-corrected radiochemical yield at end of synthesis, based upon 900 mCi of [^{18}F]fluoride, $n = 4$).

3 QUALITY CONTROL PROCEDURES

CAUTION: All radiochemicals produced for clinical use must have local regulatory approval (e.g., FDA, EMEA, MHRA, PFSB, etc.) prior to human use. Quality control procedures must be carried out by trained personnel and each dose must meet all established quality control criteria before release to the clinic.

Quality control (QC) procedures for [^{18}F]*N*-methyl lansoprazole, based upon the current requirements for research use radiopharmaceuticals outlined in the US Pharmacopeia [17], are summarized in the following:

3.1 Visual Inspection

Doses of [^{18}F]*N*-methyl lansoprazole were examined behind a PET L-block and had to be clear, colorless, and free of particulate matter.

3.2 Radiochemical and Chemical Purity

Radiochemical purity and dose concentration were determined using Shimadzu LC-2010A HT system equipped with a Bioscan B-FC-1000 radiation detector (column: Phenomenex Luna PFP(2) (150×4.6 mm); mobile phase: 30% acetonitrile/70% H_2O; UV wavelength: 254 nm; flow rate: 2.0 ml/min; retention time, ~12.5 min (see Fig. 3 for a representative analytical HPLC trace)). Coinjection of

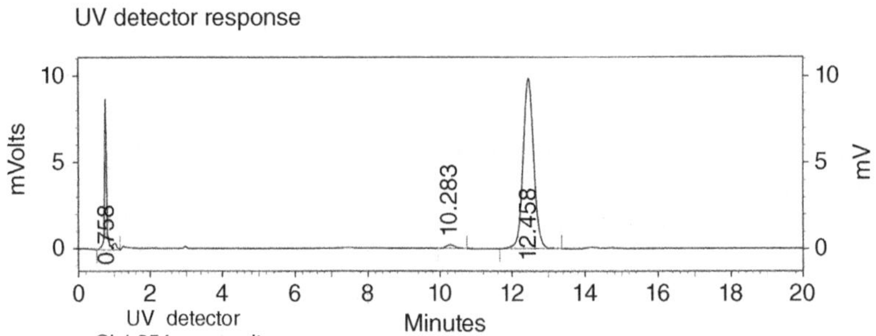

UV detector
Ch1-254 nm results

Retention time	Area	Area %
0.76	37,352	15.62
10.28	3,548	1.48
12.46	19,8291	82.90

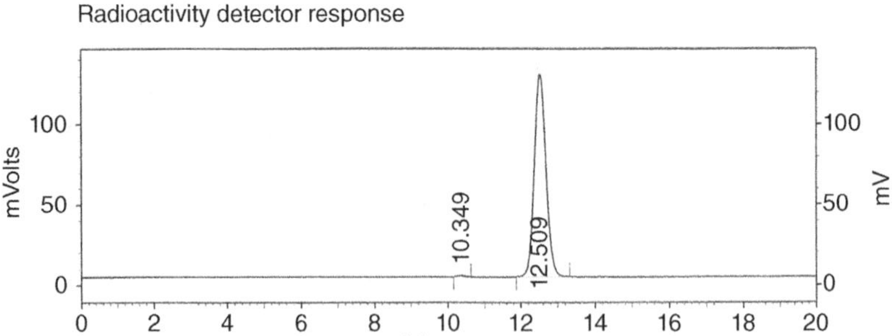

RAD results

Retention time	Area	Area %
10.35	10,893	0.39
12.51	2,756,186	99.61

FIGURE 3 HPLC analysis of [18F]*N*-methyl lansoprazole.

the dose with unlabeled reference standard was performed to confirm compound identity, and radiochemical purity was greater than or equal to 95%.

3.3 Dose pH

The pH of [^{18}F]*N*-methyl lansoprazole doses was analyzed by applying a small amount of the dose to colorpHast® pH 2.0–9.0 nonbleeding pH-indicator strips [18] and determined by visual comparison to the scale provided. Dose pH must be 4.5–7.5, and all formulated doses of [^{18}F]*N*-methyl lansoprazole had a pH of 5.0.

4 DISCUSSION

Radiofluorination of precursor **1** [12] in anhydrous DMSO (950 μl) spiked with 5 μl of saturated ammonium chloride proceeded in good radiochemical conversion to generate [^{18}F]*N*-methyl lansoprazole **2** as well as radiolabeled enol ether **3** as a by-product (ratio of **2**:**3** = 1:3). Attention was then turned to identifying a suitable HPLC method for purification of **2** from the crude reaction mixture. This initially proved challenging given the structural similarities of **2** and **3** and, not surprisingly, traditional reverse-HPLC stationary phases (e.g., C18) proved unsuitable. As a key difference between the two products is fluorine content, a perfluorophenyl-capped matrix was investigated as the stationary phase (Luna PFP(2) [13]). This was found to be the method of choice, and baseline separation between **2** and **3** was obtained in both the semi-preparative (Fig. 2) and analytical cases (Fig. 3). Following purification by semi-preparative HPLC (see Section 2.3), the fraction corresponding to [^{18}F]*N*-methyl lansoprazole **2** was collected and transferred into sterile water. The product was then reformulated into ethanolic saline using a C18 Sep-Pak [14], and transferred through a 0.22 μm sterile filter [15] into a sterile 10 ml dose vial [16] to yield [^{18}F]*N*-methyl lansoprazole.

WASTE DISPOSAL INFORMATION

All hazardous chemicals and toxic materials were disposed of according to Prudent Practices in the Laboratory (Washington, DC: National Academy Press, 1995).

CHEMICAL ABSTRACTS NOMENCLATURE (REGISTRY NUMBER)

Acetonitrile (75-05-8)
Ammonium chloride (12125-02-9)
H-Benzimidazole, 2-[[[4-[(2,2-difluoroethenyl)oxy]-3-methyl-2-pyridinyl]methyl] sulfinyl]-1-methyl (1613520-79-8)
Carbonic acid and potassium salt (1:2) (584-08-7)
Dimethyl sulfoxide (DMSO) (67-68-5)
4,7,13,16,21,24-Hexaoxa-1,10-diazabicyclo[8.8.8]hexacosane (23978-09-8)
Lansoprazole (103577-45-3)
[^{18}F]*N*-Methyl lansoprazole (1613520-80-1)

REFERENCES AND NOTES

For detailed supplier information, see Appendix 1.

1. L. E. Rojo, J. Alzate-Morales, I. N. Saavedra, P. Davies, R. B. Maccioni, *J. Alzheimer's Dis.* 2010, 19, 573.
2. X. Shao, G. M. Carpenter, T. J. Desmond, P. Sherman, C. A. Quesada, M. V. Fawaz, A. F. Brooks, M. R. Kilbourn, R. L. Albin, K. A. Frey, P. J. H. Scott, *ACS Med. Chem. Lett.* 2012, 3, 936.
3. M. V. Fawaz, A. F. Brooks, M. E. Rodnick, X. Shao, G. M. Carpenter, T. J. Desmond, P. Sherman, C. A. Quesada, B. G. Hockley, M. R. Kilbourn, R. L. Albin, K. A. Frey, P. J. H. Scott, *ACS Chem. Neurosci.* 2014, 5, 718.
4. P. J. Riss, F. I. Aigbirhio, *Chem. Commun.* 2011, 47, 11873.
5. P. J. Riss, V. Ferrari, L. Brichard, P. Burke, R. Smith and F. I. Aigbirhio. *Org. Biomol. Chem.*, 2012, 10, 6980.
6. Virgin [^{18}O]H_2O purchased from ABX, Amic, Rotem, or Medical Isotopes and used as received.
7. GE Silver high-yield [^{18}F] target.
8. GE Healthcare, United States.
9. QMA-Light Sep-Pak® cartridges were purchased from Waters (part no. WAT023525) and conditioned with 10 ml ethanol, 10 ml water, 10 ml 0.5 M sodium bicarbonate and a further 10 ml water prior to use.
10. Potassium carbonate purchased from Aldrich (part no. 209619) and used as received. Sterile water purchased from Hospira (part no. 0409-4887-50) and used as received.
11. Kryptofix 222 was purchased from Acros (part no. 29195-0010) and used as received. Anhydrous acetonitrile was purchased from Acros (part no. 61096-1000) and used as received.
12. ^{18}F]*N*-Methyl lansoprazole precursor and reference standard were prepared as previously described (see Ref. [3].).
13. Phenomenex. Semipreparative HPLC column maintenance. Following completion of a radiosynthesis, careful maintenance of the Luna PFP(2) semi-preparative HPLC column is recommended by Phenomenex to maximize the lifetime of the column. The column is conditioned as follows: flow rate, 3.0 ml/min; oven temperature, 40°C. Run each of the following for 60 min each: 100% water, 100% THF, 100% MeCN, 65:35 MeCN/H_2O. The column is then stored under 65:35 MeCN/H_2O until the next use.
14. Waters C18 1 cm^3 Sep-Pak (part no. WAT054955).
15. Millipore, Billerica MA, United States.
16. HollisterStier Laboratories, LLC.
17. US Pharmacopeia <823>. *Radiopharmaceuticals for Positron Emission*. USP 32–NF 27. 2009.
18. EMD Chemicals, Inc., United States (part no. 9578-3).

CHAPTER 14

[^{18}F]FLUORINATION OF (MESITYL)(ARYL)IODONIUM SALTS

Naoko Ichiishi[1], Allen F. Brooks[2], Joseph J. Topczewski[1], Melissa E. Rodnick[2], Melanie S. Sanford[1], and Peter J. H. Scott[2]

[1] *Department of Chemistry, University of Michigan, Ann Arbor, Michigan, USA*
[2] *Department of Radiology, University of Michigan School of Medicine, Ann Arbor, Michigan, USA*

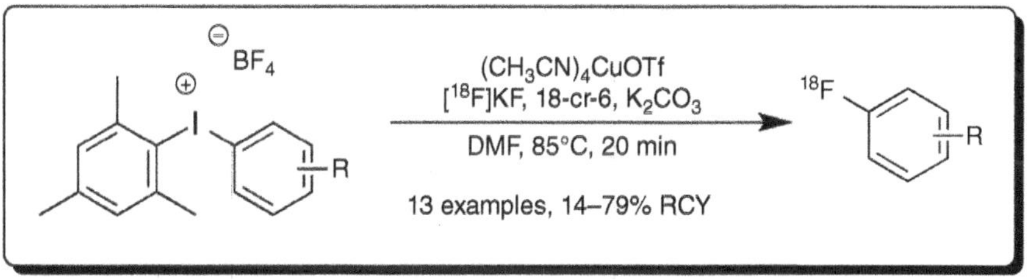

1 INTRODUCTION

The ability to prepare [^{18}F]fluoroarenes using high specific activity nucleophilic [^{18}F]fluoride has long been of interest to the fluorine-18 radiochemistry community, but has not been without its challenges. While a range of procedures have been developed for the direct nucleophilic [^{18}F]fluorination of activated aromatic rings containing electron-withdrawing groups and a suitable leaving group ($-NMe_3^+$, $-NO_2$, F, Cl, $-SAr_2^+$, etc.), robust methods for the corresponding nucleophilic fluorination of more electron rich systems are lacking. Two avenues of research have been explored to address this need. Firstly, diaryliodonium salts have been used as [^{18}F]fluoroarene precursors [1] for the radiosynthesis of electron-rich and electron-neutral arenes. Despite promising proof-of-concept studies, these precursors have not seen widespread clinical application because of difficulties associated with their synthesis and stability [2]. Moreover, the fluorination of an electron-neutral or electron-rich substrate often demands high temperatures and proceeds in poor to average radiochemical yields. These conditions result in limited regioselectivity and functional group tolerance. A second strategy exploits transition-metal catalysts (and/or reagents)

Radiochemical Syntheses, Volume 2: Further Radiopharmaceuticals for Positron Emission Tomography and New Strategies for Their Production, First Edition. Edited by Peter J. H. Scott.

to accelerate rates of reaction, improve selectivity of reactions, and enhance reactivity. Several groups have developed complementary transition-metal mediated processes enabling radiofluorination of a variety of (hetero)arenes [3, 4].

We sought a general, mild, and high-yielding procedure for the fluorination of diverse aromatic substrates by merging these two approaches. Recently, we demonstrated that copper catalyzes the fluorination of stable and synthetically accessible (mesityl)(aryl)iodonium salts [5]. The scope of this work was then expanded to include the use of fluorine-18, using [^{18}F]KF and commercially available $(MeCN)_4CuOTf$ [6].The larger mesityl group directs oxidative addition and fluorination to the smaller aryl group on iodine, independent of its electronic properties, offering a general, practical, and rapid radiofluorination of electronically diverse arenes. The process consistently affords good selectivity for a single radiolabeled product. The synthesis of ten diverse [^{18}F]fluoroarenes has been demonstrated (Fig. 1), as well as a small series of clinically relevant tracers including a protected version of 6-[^{18}F]fluoroDOPA (Fig. 2).

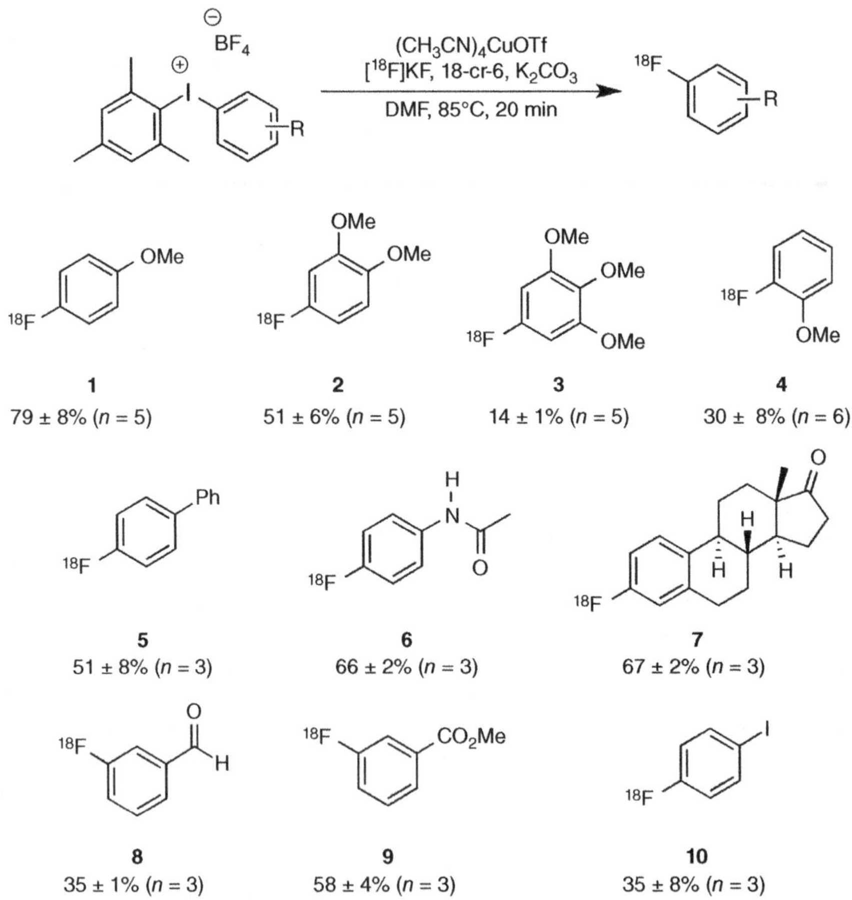

FIGURE 1 Radiofluorination of electronically diverse arenes.

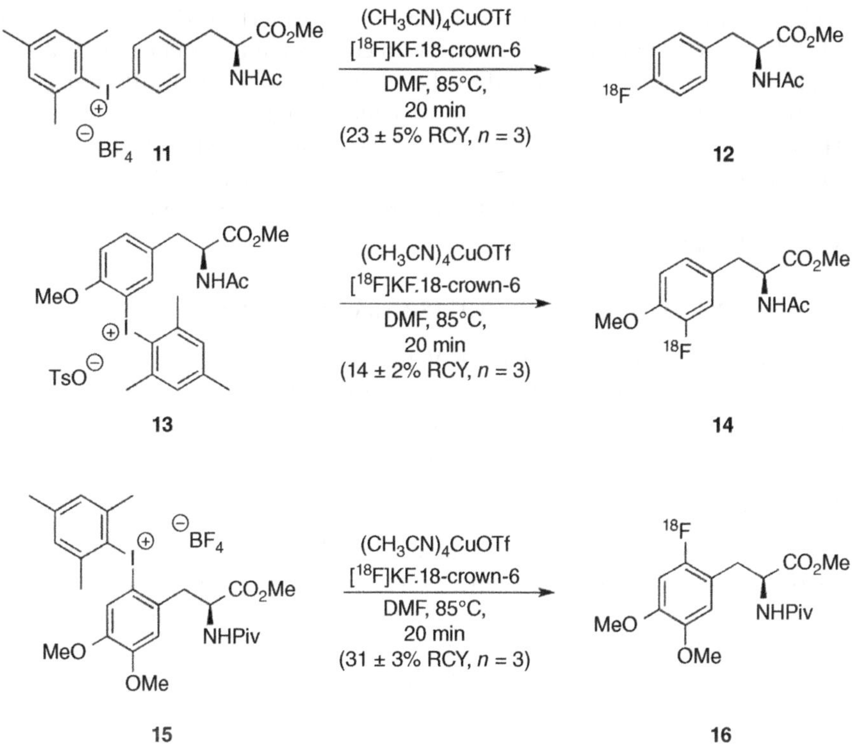

FIGURE 2 Synthesis of clinically relevant radiotracers.

2 SYNTHESIS PROCEDURES

CAUTION: All radiochemical syntheses must be carried out using appropriate equipment in a facility authorized for the use of radioactive materials. Personal protective equipment must be worn and all local radiation safety laws followed.

2.1 Production of [^{18}F]Fluoride

[^{18}O]H_2O (2 ml) [7] was loaded into the [^{18}F]fluoride target [8] of a General Electric PETtrace cyclotron [9]. The target was bombarded (40 μA beam for 2 min to generate ~ 150 mCi (5.55 GBq) of [^{18}F]fluoride for manual runs or 40 μA beam for 30 min to generate ~1.5 Ci (55.5 GBq) of [^{18}F]fluoride for full-scale automated runs) by the ^{18}O(p,n)^{18}F nuclear reaction.

2.2 Synthesis of (Mesityl)(Aryl)Iodonium Salts

[Ar–I–Mes]BF_4 substrates were prepared using literature procedures from either the corresponding aryl boronic acid or aryl stannane (commercially available or synthesized in-house) [5]. For example, in a typical procedure, the aryl boronic acid (1 eq.) was dissolved in CH_2Cl_2 (0.075 mM). The resulting solution was cooled to 0°C, BF_3·Et_2O (1.1 eq.) was added, and the mixture was stirred for 10 min.

A solution of 2-(diacetoxyiodo)mesitylene (1.1 eq.) in CH_2Cl_2 (0.33 M) was then added dropwise. The reaction was then warmed to room temperature and stirred for 2 h. The reaction was quenched by the addition of saturated $NaBF_4$ (aqueous, ~30–50 ml). After 30 min of vigorous stirring, the aqueous layer was extracted with CH_2Cl_2 (2×30–50 ml). The combined organic layers were dried ($MgSO_4$), filtered, and concentrated under vacuum. The residue was triturated with Et_2O at rt, collected via filtration, washed with Et_2O, and dried under vacuum overnight, to afford the [Ar–I–Mes]BF_4 precursors.

2.3 Synthesis of [^{18}F]KF·18-Crown-6·K_2CO_3 Complex

[^{18}F]KF was prepared using a GE TRACERlab FX_{FN} automated radiochemistry synthesis module [9, 10]. The [^{18}F]fluoride was delivered to the synthesis module in a 1.5 ml bolus of [^{18}O]water and trapped on a QMA-Light Sep-Pak® [11] to remove [^{18}O]water. [^{18}F]Fluoride was then eluted into the reaction vessel using aqueous potassium carbonate (3.5 mg in 0.5 ml of water). A solution of 18-crown-6 (15 mg in 1 ml of acetonitrile) was added to the reaction vessel, and the resulting solution was dried by azeotropic distillation to give dry [^{18}F]KF·18-crown-6·K_2CO_3. Evaporation was achieved by heating the reaction vessel to 100°C and drawing vacuum for 4 min. After this time, the reaction vessel was subjected to an argon stream and simultaneous vacuum draw for an additional 4 min. Finally, *N,N*-dimethylformamide (8 ml) was added to the dried reagent, and the resulting solution was transferred to a sterile vial for use in further reactions (~30 mCi (1.11 GBq) of prepared ^{18}F reagent was transferred).

2.4 Preparation of Stock Solution of Tetrakis(Acetonitrile)Copper(I) Triflate

A stock solution of tetrakis(acetonitrile)copper(I) triflate (CuOTf, [12]) was prepared by dissolving 14.3 mg in anhydrous DMF (1 ml). Aliquots of this solution were used for several reactions.

2.5 General Procedures for Manual Synthesis of ^{18}F-Labeled Compounds

Solid [Mes-I-Ar]BF_4 (6 µmol) was weighed into a 4 ml amber glass vial containing a stir bar and dissolved in DMF (0.35 ml). A 150 µl aliquot of the CuOTf stock solution (6 µmol, see Section 2.4) was added to the vial containing [Mes-I-Ar]BF_4. The reaction vial was sealed under an atmosphere of ambient air with a PTFE/silicone septum cap, and then the solution was thoroughly mixed using a vortex shaker [13]. A 250 µl aliquot of [^{18}F]KF·18-crown-6·K_2CO_3 complex (typically 300–900 µCi (11.1–33.3 MBq), prepared as described in Section 2.3) was added to the reaction vial via syringe [14], and the vial was heated in an aluminum block with stirring at 85°C for 20 min. After this time, the reaction was cooled to room temperature. A 100 µl aliquot was withdrawn from the vial and added to 400 or 900 µl of CH_2Cl_2 in a 4 ml vial (choice of volume was dependent on activity to minimize exposure).

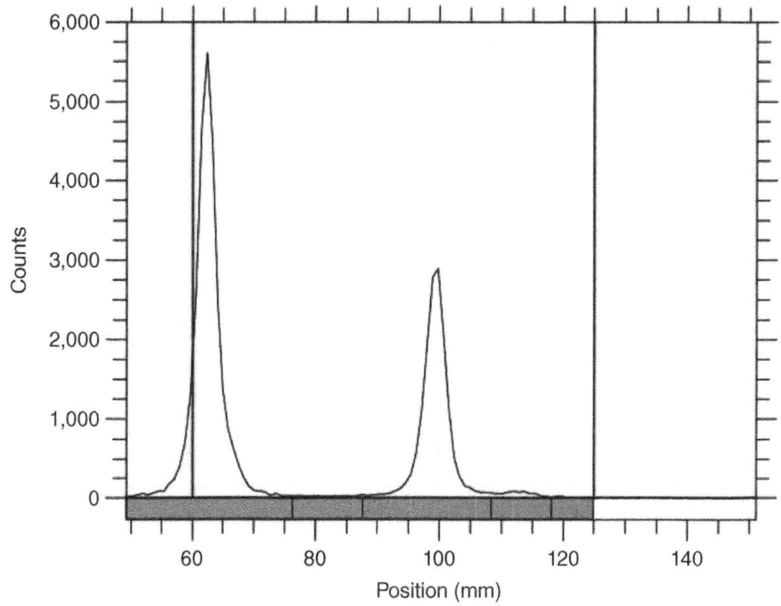

Replicate	TLC yield
1	33.9% (1.5% F-Mes)
2	30.4% (1.4% F-Mes)
3	27.6% (1.6% F-Mes)
Average	31%
Standard deviation	3%

F-Mes = [^{18}F]fluoromesitylene

FIGURE 3 Representative radio-TLC trace of compound **16.**

The CH_2Cl_2 mixture was shaken by hand and then used for radio-TLC analysis (Section 3.2) to obtain radiochemical conversions (RCC) [15]. In addition, a 100 µl aliquot of the reaction solution was used for radio-HPLC analysis (Section 3.1) by diluting the sample into 50/50 MeCN/H_2O (300 µl total volume). The RCC was determined by dividing the integrated area under the fluoroarene spot by the total integrated area of the TLC plate (see Fig. 3 for a typical TLC trace) [16].

2.6 General Procedures for Automated Synthesis of ^{18}F-Labeled Compounds

The production-scale synthesis of radiolabeled arenes was conducted using a GE TRACERlab FX_{FN} automated radiochemistry synthesis module [9]. The synthesis module was preloaded with a solution of the [Mes-I-Ar]BF_4 precursor (18 µmol) and tetrakisacetonitrile copper(I) triflate (8.0 mg, 20 μmol) in DMF (0.75 ml) to be added from a single automated port prior to ^{18}F delivery. [^{18}F]Fluoride (1.5 Ci (55.5 GBq)) was produced as described in Section 2.1, delivered to the synthesis module (in a 1.5 ml bolus of [^{18}O]water) and trapped on a QMA-Light Sep-Pak [11] to remove [^{18}O]water. [^{18}F]Fluoride was eluted into the reaction vessel using aqueous K_2CO_3 (3.0 mg in 0.5 ml of water). A solution of 18-crown-6 (5 mg in 1 ml of acetonitrile) was added to the reaction vessel, and the resulting solution was dried by heating the reaction vessel to 100°C and drawing vacuum for 4 min. After this time, the reaction vessel was subjected to an argon stream and simultaneous vacuum draw for an additional 4 min to give dry [^{18}F]KF·18-crown-6·K_2CO_3. The reaction vessel was cooled to 50°C, DMF (0.75 ml) was added, and the resulting mixture was stirred for 1 min. The preloaded solution of [Mes-I-Ar]BF_4 precursor and tetrakisacetonitrile copper(I) triflate was added to the synthesis module reactor via automated injection, and the vessel was sealed and heated at 85°C for 20 min. The reaction vessel was then cooled to 50°C and DMF (8.5 ml) was added. The resulting solution (10 ml) was transferred to a sterile vial for analysis (radio-TLC and radio-HPLC—see Section 3). The RCC was determined by dividing the integrated area under the fluoroarene spot by the total integrated area of the TLC plate (see Fig. 3 for a typical TLC trace) [16].

2.7 ^{19}F Reference Standards

Authentic ^{19}F reference standards of the test compounds (**1–10, 12, 14,** and **16**) were purchased directly [17, 18], or prepared according to literature procedures [5]. In the latter case, all spectroscopic data matched literature values.

3 QUALITY CONTROL PROCEDURES

CAUTION: All radiochemicals produced for clinical use must have local regulatory approval (e.g., FDA, EMEA, MHRA, PFSB, etc.) prior to human use. Quality control procedures must be carried out by trained personnel and each dose must meet all established quality control criteria before release to the clinic.

3.1 Radio-HPLC Analysis

Two general sets of HPLC conditions were used. A gradient method (Section 3.1.1) was used for all manual scale reactions. For the reactions conducted with higher levels of radioactivity (>1 Ci (37 GBq) of initial activity), an isocratic method (Section 3.1.2) was used to determine the specific activity of the product. A known volume of the crude reaction mixture was transferred to a vial, and the activity of the vial was counted using a dose calibrator [19]. The activity in the vial was then multiplied by the RCC to obtain the activity of the product in the vial (Ci/ml). An aliquot of the sample was then injected onto the HPLC using an isocratic method (Section 3.1.2). The UV peak corresponding to the radiofluorinated product was determined by overlaying the UV and radioactivity traces. The UV area was then used to calculate the concentration of the product in mmol/ml [20]. Dividing the activity concentration (Ci/ml) by the HPLC-derived concentration of product (mmol/ml) provided the specific activity in Ci/mmol. This reflects an end-of-synthesis (EOS) specific-activity.

3.1.1 HPLC Method 1

Condition: 5–95% gradient of (MeCN + 0.05% TFA) in (H_2O + 0.05% TFA)
Flow rate: 1 ml/min
Column: Luna C18 column 150 × 4.6 mm, 5 μm
Method:

0–3 min	5% MeCN	Isocratic
3–20 min	5%–95% MeCN	Linear increase
20–30 min	5% MeCN	Isocratic

3.1.2 HPLC Method 2

Condition: 40% (MeCN + 0.05% TFA) in (H_2O + 0.05% TFA)
Flow rate: 1.5 ml/min
Column: Luna C18 column 150 × 4.6 mm 5 μm

3.2 Radio-TLC Analysis

Radio-TLC analysis was performed using a Bioscan AR 2000 Radio-TLC scanner [21] with EMD Millipore TLC silica gel 60 plates (3.0 cm wide × 6.5 cm long) [22]. The RCC was determined by dividing the integrated area under the fluoroarene spot by the total integrated area of the TLC plate, as shown in Fig. 3 for Compound **16**.

4 DISCUSSION

A copper-catalyzed radiofluorination of (mesityl)(aryl)iodonium precursors [23] using [^{18}F]KF and commercially available $(MeCN)_4$CuOTf has been developed [24]. The radiofluorination conditions were applied to a diverse series of (mesityl)(aryl)iodonium tetrafluoroborate salts (Fig. 1). The transformation is performed under ambient conditions without the requirement for a drybox or extensive drying of reagents and glassware. In each case, modest to excellent RCC was observed, and products were obtained in high specific activity [25]. The RCY was very reproducible [26], and all of the reactions were highly selective for a single ^{18}F-containing product [27]. Reaction mixtures were homogeneous, and the only radiochemical by-product appeared to be unreacted ^{18}F sequestered as inorganic fluoride salts.

This new protocol can be used for the radiofluorination of electron-rich arenes such as those that contain one or more electron-donating methoxy substituents (**1**–**4**). Remarkably, even the highly electron-rich product 1-[^{18}F]fluoro-3,4,5-trimethoxybenzene (**4**) can be prepared using this method. Steric factors did affect the reaction yields. For example, 4-[^{18}F]fluoroanisole was obtained in 79% RCC, but the corresponding 2-[^{18}F]fluoroanisole was only formed in 30% RCC. Radiofluorination of electron-neutral and electron-deficient aryl rings (**5–10**) also proceeded in high yield and excellent selectivity and enabled the radiosynthesis of a number of commonly utilized PET prosthetic groups (e.g., [^{18}F]fluorobenzaldehyde (**8**), 4-methyl [^{18}F]fluorobenzoate (**9**), and [^{18}F]fluoroiodobenzene (**10**)). The methodology was also tolerant of a range of functional groups, and compatibility was demonstrated with amide NH bonds (**6**), ketones (**7**), aldehydes (**8**), esters (**9**), and aryl iodides (**10**).

Finally, the radiofluorination of several molecules of clinical relevance was investigated (Fig. 2). (Mesityl)(aryl)iodonium salts derived from aromatic amino acids (**11, 13** and **15**) were prepared and subjected to the radiofluorination protocol to yield protected versions of [^{18}F]fluoro-$_{L}$-phenylalanine (**12**), the more electron-rich 3-[^{18}F]fluorotyrosine derivative **14** (in this case, the tosylate salt of the precursor was employed), and a protected version of 6-[^{18}F]fluoroDOPA (**16**) (Fig. 2). Without any substrate-specific optimization, the radiolabeled products were obtained in acceptable yields, and further development of fully automated syntheses for clinical use is ongoing in our laboratories at this time.

WASTE DISPOSAL INFORMATION

All hazardous chemicals and toxic materials were disposed of according to Prudent Practices in the Laboratory (Washington, DC: National Academy Press, 1995).

CHEMICAL ABSTRACTS NOMENCLATURE (REGISTRY NUMBER)

Acetonitrile (75-05-8)
Carbonic acid and potassium salt (1:2) (584-08-7)
Dichloromethane (75-09-2)
N,N-Dimethylformamide (68-12-2)
1,4,7,10,13,16-Hexaoxacyclooctadecane (17455-3-9)
Tetrakisacetonitrile copper(I) triflate (CAS number not assigned)
2,2,2-Trifluoroethanoic acid (76-05-1)

REFERENCES AND NOTES

For detailed supplier information, see Appendix 1.

1. For a review of radiochemistry with diaryliodonium salt precursors, see M. S. Yusubov, D. Y. Svitich, M. S. Larkina and V. V. Zhdankin, ARKIVOC, 2013, 364.
2. In our hands, (2-thienyl)(aryl)iodonium derivatives exhibit only modest stability (e.g., [(4-OMePh)(2-thienyl)I]Br was found to decompose within 2 weeks under ambient conditions).
3. For transition metal-catalyzed radiofluorination of arenes, see (a) E. Lee, A. S. Kamlet, D. C. Powers, C. N. Neumann, G. B. Boursalian, T. Furuya, D. C. Choi, J. M. Hooker, T. Ritter, *Science*, 2011, 334, 639; (b) A. Kamlet, C. N. Neumann, E. Lee, S. M. Carlin, C. K. Moseley, N. Stephenson, J. M. Hooker, T. Ritter, *PLoS One*, 2013, 8, e59187; (c) E. Lee, J. M. Hooker, T. Ritter, *J. Am. Chem.*

Soc., 2012, 134, 17456; and (d) M. Tredwell, S. M. Preshlock, N. J. Taylor, S. Gruber, M. Huiban, J. Passchier, J. Mercier, C. Génicot, V. Gouverneur, *Angew. Chem.*, 2014, 53, 7751.

4. For metal-catalyzed C-sp^3 radiofluorination, see (a) J. J. Topczewski, T. J. Tewson, H. M. Nguyen, *J. Am. Chem. Soc.*, 2011, 133, 19318; (b) E. Benedetto, M. Tredwell, C. Hollingworth, T. Khotavivattana, J. M. Brown, V. Gouverneur, *Chem. Sci.*, 2013, 4, 89; (c) T. J. A. Graham, R. F. Lambert, K. Ploessl, H. F. Kung, A. G. Doyle, *J. Am. Chem. Soc.*, 2014, 136, 5291; and (d) X. Huang, W. Liu, H. Ren, R. Neelamegam, J. M. Hooker, J. T. Groves, *J. Am. Chem. Soc.*, 2014, 136, 6842.
5. (a) N. Ichiishi, A. J. Canty, B. F. Yates, M. S. Sanford, *Org. Lett.*, 2013, 15, 5134; (b) J-H. Chun, V. W. Pike, *J. Org. Chem.*, 2012, 77, 1931; (c) F. Basuli, H. Wu, G. L. Griffiths, *J. Label. Compd. Radiopharm.*, 2011, 54, 224; (d) M. E. Kieffer, K. V. Chuang, S. E. Reisman, *Chem. Sci.*, 2012, 3, 3170; (e) Phipps, R. J.; Grimster, N. P.; Gaunt, M. J. *J. Am. Chem. Soc.*, 2008, 130, 8172; and (f) A. McKillop, D. Kemp, *Tetrahedron*, 1989, 45, 3299.
6. N. Ichiishi, A. F. Brooks, J. J. Topczewski, M. E. Rodnick, M. S. Sanford, P. J. H. Scott, *Org. Lett.*, 2014, 16, 3224.
7. Virgin [^{18}O]H_2O purchased from ABX, Amic, Rotem, or Medical Isotopes and used as received.
8. GE Silver high-yield [^{18}F]fluoride target.
9. GE Healthcare, United States.
10. All loading operations were conducted under ambient atmosphere, and argon was used as a pressurizing gas during automated sample transfers.
11. QMA-Light Sep-Pak® cartridges were purchased from Waters (part no. WAT023525) and conditioned with 10 ml ethanol, 10 ml water, 10 ml 0.5 M sodium bicarbonate, and a further 10 ml water prior to use.
12. Sigma-Aldrich, Part No. 685038; the copper catalyst is stable in solution over at least 4 h as 77% RCC of **1** was obtained using a solution of $(CH_3CN)_4CuOTf$ in DMF that was allowed to stand for 4 h under ambient conditions prior to use.
13. Barnstead® Thermolyne Type 16700.
14. On a typical day, several reactions (4–20) were set up together. Due to this, the time of mixing and time of incubation at room temperature prior to heating varied slightly from day to day. However, the results of the radiofluorination appear to be insensitive to this variation.
15. The reaction mixture was diluted to obtain more reproducible TLC results. Undiluted samples of the reaction showed the same RCC; however, broadening was observed as a result of the DMF, and this made accurate integration more difficult. Radio-TLCs where counted immediately after being developed. This was particularly critical when the fluoroarene was volatile (e.g., 4-fluoroanisole), because the apparent RCC was found to decrease as a function of time due to the product evaporating off of the TLC plate.
16. The RCC reported here do not reflect losses during the preparation of [^{18}F]KF·18-crown-6·K_2CO_3.
17. Acros Organics/Thermo Fisher Scientific, United States.
18. Oakwood Products, Inc., Unites States.
19. Capintec-CRC-15R dose calibrator, Capintec, Inc., United States.
20. Determined by linear regression analysis of appropriate fluoroarene standards. A standard curve was generated from standard solutions, each run in duplicate (140 μM–1.1 μM).
21. Bioscan, Inc., United States (now Eckert and Ziegler).
22. EMD Millipore, Unites States.
23. (Mesityl)(aryl)iodonium salts are colorless, free-flowing solids that are shelf stable for months under ambient conditions.
24. For all of the substrates examined, ≤2% of the corresponding fluoroarene product was observed in the absence of Cu catalyst, confirming that Cu is vital for accelerating the reaction rate as well as controlling selectivity.
25. Obtaining high specific activity products using BF_4 salts suggests that isotopic exchange is not an issue under the mild conditions and short reaction times employed.
26. For the model 4-methoxy system (synthesis of **1**) the RCY typically did not vary more than 5% even over >35 trials on different days using different solvent sources.
27. F-DOPA was the outlier where ≤2% fluoromesitylene by-product was detected.

CHAPTER 15

Pd(IV)-MEDIATED FLUORINATION OF ARENES WITH [^{18}F]F$^-$ FOR PET IMAGING

Nathan J. Schauer, Stephen M. Carlin, Hong Ren, and Jacob M. Hooker

Athinoula A. Martinos Center for Biomedical Imaging, Department of Radiology, Massachusetts General Hospital, Harvard Medical School, Charlestown, Massachusetts, USA

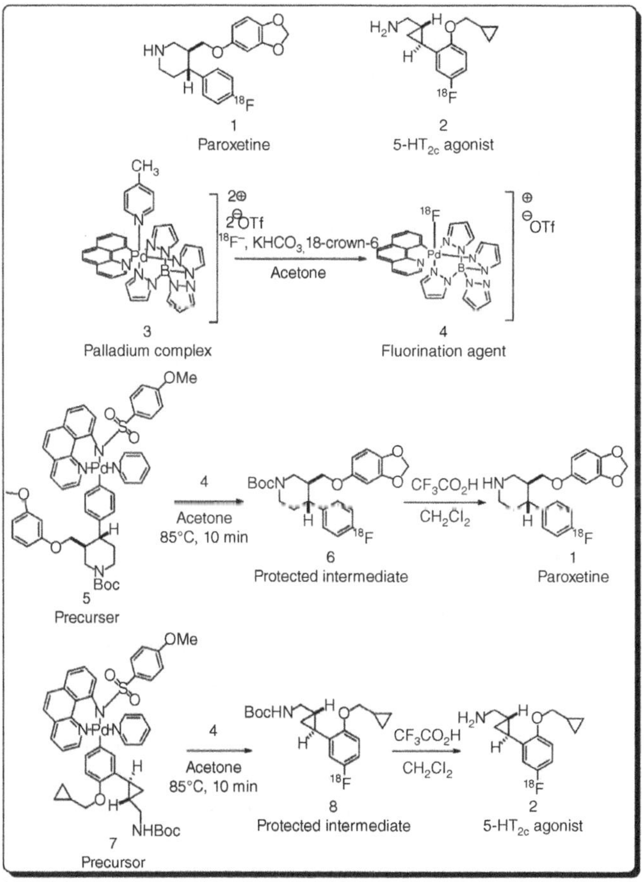

Radiochemical Syntheses, Volume 2: Further Radiopharmaceuticals for Positron Emission Tomography and New Strategies for Their Production, First Edition. Edited by Peter J. H. Scott.

1 INTRODUCTION

Positron emission tomography (PET) imaging provides a noninvasive method for preclinical and clinical investigation of biological systems and chemical interactions. Fluorine-18 (^{18}F) is a clinically relevant radioisotope for PET imaging, as evidenced by the impact of [^{18}F]fluorodeoxyglucose ([^{18}F]FDG) in clinical diagnosis in oncology. Despite the prevalence of PET imaging utilizing ^{18}F in research and clinical settings, access to ^{18}F PET tracers remains a significant challenge. ^{18}F-labeling is typically achieved with nucleophilic [^{18}F]fluoride ([$^{18}F]F^-$) or less typically with electrophilic [^{18}F]fluorine gas ([$^{18}F]F_2$). Fluorine gas is challenging to handle in a process-controlled environment due to its powerful oxidizing behavior, necessitating the use of specialized equipment. In addition, liberation of [$^{18}F]F_2$ requires the addition of nonradioactive [$^{19}F]F_2$ as a carrier. This not only reduces the maximum possible radiochemical yield to 50% but also substantially reduces the specific activity (the ratio of ^{18}F to ^{19}F) of [$^{18}F]F_2$. High radiotracer-specific radioactivity is desired in order to maximize occupancy of the biological target with ^{18}F; saturation of the target with PET-inactive ^{19}F could mask low and possibly even high density binding sites that would be revealed at high specific activity. Electrophilic fluorination with [$^{18}F]F_2$ in higher specific activity than is usual has been achieved, but the scope of such methods are narrow. For these reasons, electrophilic fluorination with [$^{18}F]F_2$ is impractical, limiting access to certain PET tracers and molecules that may have potential as radiotracers.

[$^{18}F]F^-$, readily obtained from proton bombardment of [^{18}O]-enriched water, provides a means of radiolabeling with ^{18}F in high specific activity but also comes with its own challenges. Aqueous fluoride must first be dried in the presence of a phase-transfer catalyst before it is introduced into the organic phase to undergo substitution. Suitable drying conditions can readily be achieved when using high concentrations of [$^{19}F]F^-$ but are more difficult to optimize for radiochemistry with [$^{18}F]F^-$, which utilizes nanomole quantities of fluoride and is more sensitive to the presence of a small amount of water. In addition, radiosynthetic routes for introducing ^{18}F into arenes utilizing [^{18}F] fluoride have traditionally been limited to nucleophilic aromatic substitution reactions, which require the presence of strong electron-withdrawing substituents on the aromatic compounds. Thus, the labeling of electron neutral or rich arenes presents a challenging chemical transformation. Finally, radiolabeling of complex PET tracers must occur late stage in order for the radiotracer to be sufficiently radioactive due to the short half-life of ^{18}F ($t_{1/2} = 110$ min), necessitating rapid regioselective and functional-group-tolerant methods. Recent work by Ritter and Hooker has addressed some of these obstacles by developing late-stage palladium-based electrophilic fluorination reagents that afford aryl [^{18}F]fluorides in sufficient yields for PET imaging, are stable to established drying conditions, and are tolerant of various functional groups. This chapter provides a brief history of the development of palladium-mediated electrophilic fluorination and an overview of the Ritter Group's reagents and their translation to radiochemistry for PET imaging in collaboration with the Hooker lab.

2 HISTORY

Research into the use of palladium catalysts to form carbon–fluorine bonds has been ongoing for over a decade. In 2002, Grushin reported the synthesis and characterization of the first soluble palladium fluoride complexes [1]. Further work by Grushin and Yandulov demonstrated the feasibility of aryl fluoride reductive elimination from palladium(II) [2, 3], and in 2009, Buchwald demonstrated the first synthesis of aryl fluorides using reductive elimination on a Pd(II) catalyst [4]. Sanford pioneered the use of oxidative methods in Pd(II)-catalyzed aryl fluorinations, and in 2006, her lab became the first to successfully form a palladium-mediated aryl fluoride [5]. However, the application of these early methods remained limited due to harsh reaction conditions and a narrow substrate scope, particularly for the late-stage fluorination desired for radiochemistry.

In 2008, Ritter and coworkers reported a regioselective and high-yielding two-step fluorination reaction of palladium aryl complexes with the electrophilic fluorination reagent F-TEDA (Selectfluor). Substrate screening of the reaction demonstrated its functional group tolerance and its broad scope, extending to electron-rich, electron-poor, heterocyclic, and *ortho*-substituted arenes [6]. Further studies by Ritter et al. showed that F-TEDA can oxidize palladium(II) aryl complexes, accessing a high-valence palladium(IV) aryl fluoride intermediate that affords aryl fluorides via reductive elimination [7], and that discrete palladium(IV) complexes could be isolated and applied to form carbon–fluorine bonds [8]. These methods formed the basis of the design of a fluorination reagent that replaces the function of F-TEDA with fluoride, enabling the synthesis of ^{18}F radiotracers in high specific activity.

In 2011, Ritter and Hooker reported a new fluoride-derived electrophilic fluorination reagent and its application for PET imaging in an automated synthesis. The fluorination reagent (**4**) is a high-valence palladium complex. [^{18}F]Paroxetine (**1**) and a 5-HT_{2C} agonist (**2**) were chosen as preliminary PET tracers to be produced with electrophilic fluorination from reagent **4**. In 2013, Hooker and Ritter reported an automated synthesis for these radiotracers using Eckert and Ziegler (E&Z) automated synthesis modules and Modular-Lab in a hot cell [9]. The procedure for the automated production of reagent **4** and radiotracers **1** and **2** is reproduced below.

3 PROCEDURES

3.1 Synthesis Procedures

CAUTION: All radiochemical syntheses must be carried out using appropriate equipment in a facility authorized for the use of radioactive materials. Personal protective equipment must be worn, and all local radiation safety laws followed.

3.1.1 Production of [^{18}F]Fluoride

[^{18}O]H_2O [10] was loaded into the [^{18}F]fluoride target of a Siemens Eclipse HP 11 MeV cyclotron. The target was bombarded at 52 μA for 40 min to generate ~1.5 Ci of [^{18}F]fluoride by the $^{18}O(p,n)^{18}F$ nuclear reaction.

3.1.2 Preparation of E&Z Automated Synthesis Unit

An E&Z automated synthesis unit in a lead-shielded hot cell was loaded with the following components prior to delivery of [^{18}F]fluoride: a vial containing 5 mg $KHCO_3$ in 800 µl acetonitrile and 200 µl deionized water [11], a vial containing 13.1 mg 18-crown-6 in 500 µl acetonitrile [12], a vial containing 1 ml dry acetonitrile [13], a vial containing 1.5 ml acetone, a vial containing 10 mg Pd(IV) complex **3** in 500 µl acetone [14], a vial containing 1 ml 2-butanone [15], and a vial containing 10 mg Pd(II)-complexed precursor **5** or **7** [16]. In addition to these solvents, two reaction vessels were loaded with magnetic stir bars and connected to the synthesis unit. A Macherey-Nagel Chromafix 30-PS-HCO_3 cartridge that had been primed with 2 ml 5 mg/ml $KHCO_3$ in deionized water and washed with 20 ml deionized water [17] was affixed to the unit's [^{18}F]fluoride inlet line. A glass syringe was filled with 10 mg cotton and 25 mg Janda-Jel polypyridine [18] and then loaded between the synthesis unit's reaction vessels. Finally, a Sep-Pak Plus Waters Accell QMA cartridge was conditioned with 1 ml acetone, washed with 20 ml deionized water, and affixed to the synthesis unit's outlet line [19]. The line from the QMA cartridge was connected to a vial in a lead-shielded container so that the reaction mixture could be transported to a lead-shielded hot cell for manual manipulation.

3.1.3 Operation of E&Z Automated Synthesis Unit Using Modular-Lab Software

Once the automated synthesis unit was fully prepared, the [^{18}F]fluoride was delivered from the cyclotron and trapped on the Chromafix cartridge [20]. Using the Modular-Lab software, the 1 ml $KHCO_3$ solution in 8:2 acetonitrile/water was pushed through the Chromafix cartridge into the first reaction vessel to elute the trapped [^{18}F]fluoride. This initial solution was heated to 108°C and blown dry with nitrogen to evaporate the vessel's contents. 500 µl acetonitrile was added to the vessel, and it was again heated to 108°C under nitrogen until evaporation. The last 500 µl acetonitrile was added to the vessel, and after heating at 108°C under nitrogen, a white precipitate remained on the bottom of the vessel. At this point, 500 µl of the acetone was added to the vessel, and it was again heated to 108°C under nitrogen to leave a glassy film on the vial's surface. The vessel was cooled to room temperature and purged with nitrogen. Then, the 500 µl solution containing **3** was added to the first reaction vessel and allowed to react at room temperature for 7.5 min [21]. After the reaction, the mixture was passed through the Janda-Jel and into the second reaction vessel. Subsequently, the 1 ml 2-butanone was added to the first reaction vessel and pushed through the Janda-Jel to wash off any excess product. For the second reaction step, precursor **5** or **7** was added to the reaction vessel, which was heated to 85°C for 10 min. After the reaction, the mixture was pushed through the QMA cartridge and into the intermediate vial. The second reaction vessel was then washed with the final 1 ml acetone, which was sent through the QMA cartridge and into the intermediate vial.

3.1.4 Manual Deprotection and Preparative HPLC Purification

Prior to completion of the previous syntheses, the lead-shielded hot cell had been prepared with the following components: a 1 in. plug of silica gel in a glass syringe [22], a solution of 500 µl ethyl acetate with 500 µl hexanes [23], a solution of 500 µl trifluoroacetic acid with 500 µl dichloromethane [24], and a solution of 75 µl deionized water with 25 µl acetonitrile. A preparative HPLC column [25] had been equilibrated with isocratic mobile phase [26] and was prepared for injection of crude product.

Upon completion of the second reaction step, the intermediate vial was transferred to the lead-shielded hot cell, and the 2:1 mixture of 2-butanone/acetone was dried down under a steady nitrogen stream [27]. After drying, the 1:1 solution of ethyl acetate/hexanes was added to the vial, and its contents were pushed through the silica gel into a third reaction vessel [28]. Again, a steady nitrogen stream was used to dissolve the solvents, and after this drydown, the 1:1 solution of TFA/CH_2Cl_2 was added to the reaction vessel. Nitrogen flow was immediately applied for a final drydown. After the final product was dried, the 100 µl acetonitrile/water solution was added to the reaction vessel. A glass syringe with a blunt-tipped needle was used to draw up the crude product and manually inject it onto the preparative HPLC [29]. Using a three-way stopcock attached to the HPLC outlet [30], a fraction of the HPLC eluent containing the desired product was collected into a 28-mL glass test tube [31].

3.1.5 Final Product Reformulation

Prior to HPLC fraction collection, a basic buffer with 0.1 M K_2CO_3 and 0.1 M $KHCO_3$ in deionized water was prepared. A Grace Extract-Clean SPE 50 mg/1.5 ml column was fitted with a syringe adapter [32], conditioned with 1 ml of the basic buffer, and then washed with 20 ml deionized water. A solution of 0.1 M NaOAc buffered to pH 5 was also prepared.

After collection from the HPLC, deionized water was added to the test tube to dilute its contents to ~25 ml. The test tube's contents were then pulled into a syringe and pushed through the SPE column into a waste beaker. 10 ml of deionized water was then pushed through the SPE column using a 10 ml Luer adapter syringe, followed by a low pressure nitrogen push to wash off any residual solvent. The activity on the Sep-Pak was measured using a Capintec CRC-25 PET ion chamber [33]. A 10 ml Luer adapter syringe preloaded with 0.5 ml ethanol was used to elute the product from the SPE column into a 20 ml scintillation vial. After elution, the vial was capped and measured in the ion chamber [34]. 1 ml of the pH 5 NaOAc solution was added to the vial, followed by 3.5–8.5 ml sterile injectable saline [35, 36]. The final formulation was then loaded into a 10 ml syringe and passed through a sterile Millex LG 0.2 µm 25 mm syringe-driven filtering unit into a 10 ml sterile vial [37], followed by a low pressure nitrogen push [38]. The final activity was measured and recorded, and 200 µl of the sample was retained for quality control.

3.2 Quality Control Procedures

CAUTION: All radiopharmaceuticals produced for clinical use must have local regulatory approval before use in humans. Trained personnel must carry out QC procedures, and each dose must meet all established QC criteria before release to the clinic.

3.2.1 Analytical HPLC

Twenty microliters of formulated product was loaded into a 300 μl HPLC vial and added to the autosampler of an Agilent 1100 Series HPLC connected to a Carol and Ramsey Associates Model 105-S radioactivity detector. A second vial had been charged with 20 μl of a 1 mg/ml solution of authentic reference standard in saline and run to verify chemical identity [39].

A calibration curve had been previously determined by injecting known masses of authentic reference standard and integrating the UV signals corresponding to those masses [40]. By fitting the UV signal of the sample (not coinjected with reference standard) to this calibration curve, estimates for injected mass and specific activity could be obtained [41].

3.2.2 Residual Palladium

Serial dilutions of Palladium were prepared and analyzed using an Agilent 7500a ICP-MS [42]. A lutetium control was also analyzed to ensure constancy of measurements at low concentration. Two samples of **2** were analyzed by ICP-MS to determine the concentration of Pd in the final product [43].

4 DISCUSSION

The successful fluorination of these *p*-substituted arenes by electrophilic aromatic substitution is a major accomplishment from a radiochemistry perspective, as the fluorination of substituted aryl halides is an area of active interest. However, this reaction requires a well-planned approach in order to ensure that the synthesis is achieved with minimal exposure to the operator.

One of the major challenges for large-scale radiosynthesis is the need for automated synthesis units in order to limit the operator's exposure. Because this is essentially a three-step synthesis (fluorination of **3**, fluorination of **5** or **7**, deprotection of **6** or **8**) followed by two purification steps (preparative HPLC and Sep-Pak reformulation), few automated synthesis units on the market are capable of performing all of the necessary steps for a large-scale synthesis using **3**. At the investigator's facility, the first two steps were addressed by preloading the required components for labeling of **3** and **5** or **7**, and the final reaction step was performed in a lead-shielded hot cell using manual manipulation. In order to minimize the dose during this third reaction step, excess [^{18}F]fluoride was sequestered on a QMA cartridge prior to handling of the eluted dose, and care was taken to minimize exposure time by preparing the necessary reagents outside of the hot cell. Similar precautions were taken during the HPLC purification and reformulation of the final product.

Bringing this synthesis method to human use requires a fully automated synthesis because of the high levels of radiation involved in large-scale production and translation to larger facilities making multiple doses with a single run. An additional E&Z reaction vessel could in principle be used to perform the final reaction step remotely. In addition, the investigators have previously used an Explora LC unit for remote HPLC loading and peak collection. For the use of the LC unit in the investigator's lab, a larger volume than 100 µl would be required for loading the 5 ml loop on that system, but this process could again be achieved in principle by drawing the crude product onto the HPLC loop using a 5 ml syringe [44]. As with other radiotracers that this lab has prepared, the HPLC collection fraction could be loaded directly into a vial attached to an Explora FM unit preloaded with the requisite solvents for automated formulation. Thus, while scale-up and automation provide a set of logistical challenges, they can in principle be overcome with careful planning and the requisite equipment.

A larger hurdle for large-scale reactions using these Pd(II) precursors is their tendency to decompose in solution. Precursors **5** and **7** must be added to the second reaction vessel in dry, powdered form, and this makes them more difficult to work with remotely.

WASTE DISPOSAL INFORMATION

All hazardous chemicals and toxic materials were disposed of according to Prudent Practices in the Laboratory (Washington, DC: National Academy Press, 1995).

CHEMICAL ABSTRACTS NOMENCLATURE (REGISTRY NUMBER)

Acetone (67-64-1)
18-Crown-6 (17455-13-9)
Dichloromethane (75-09-2)
Potassium bicarbonate (298-14-6)
Trifluoroacetic acid (76-05-1)

REFERENCES AND NOTES

For detailed supplier information, see Appendix 1.

1. V. V. Grushin, *Chem. Eur. J.* 2002, 8, 1006.
2. D. V. Yandulov, N.T. Tran, *J. Am. Chem. Soc.* 2007, 129, 1342.
3. V. V. Grushin, W. J. Marshall, *Organometallics*. 2007, 26, 4997.
4. D. A. Watson, M. Su, G. Teverovskiy, Y. Zhang, J. Garcia-Fortanet, T. Kinzel, S. L. Buchwald, *Science*. 2009, 325, 1661.
5. K. L. Hull, W. Q. Anani, M. S. Sanford, *J. Am. Chem. Soc.* 2006, 128, 7134.
6. T. Furuya, H. M. Kaiser, T. Ritter. *Angew. Chem. Int. Ed.* 2008, 47, 5993.
7. T. Furuya, D. Benitez, E. Tkatchouk, A. E. Storm, P. Tang, W. A. Goddard, T. Ritter. *J. Am. Chem. Soc.* 2010, 132, 3793.

8. T. Furuya, T. Ritter. *J. Am. Chem. Soc.* 2008, 130, 10060.
9. A. S. Kamlet, C. N. Neumann, E. Lee, S. M. Carlin, C. K. Moseley, N. Stephenson, J. M. Hooker, T. Ritter, *PLoS One*. 2013, 8(3).
10. $[^{18}O]H_2O$ (>97% O-18 enriched) was obtained from ISOFLEX USA. 2.4 ml was loaded into the cyclotron's silver-bodied F-18 target.
11. $KHCO_3$ (≥99%) and acetonitrile were obtained from Sigma-Aldrich, and deionized water was prepared on site using a Millipore Milli-Q Integral Water Purification System. $KHCO_3$ was dried at 23°C for 24 h under dynamic vacuum (10^{-4} Torr) prior to use.
12. 18-Crown-6 (1,4,7,10,13,16-hexaoxacyclooctadecane) was obtained from Sigma-Aldrich.
13. Dry acetonitrile was obtained from Sigma-Aldrich. It was distilled over P_2O_5 prior to use.
14. HPLC-grade acetone was obtained from Sigma-Aldrich and distilled over B_2O_3 prior to use. The procedure for synthesizing **3** is beyond the scope of this report, but it can be found in Ref. [9].
15. 2-Butanone was obtained from Sigma-Aldrich and distilled over B_2O_3 prior to use.
16. The synthesis procedures for **5** and **7** can be found in Ref. [9].
17. The Chromafix cartridge was obtained from ISOFLEX USA. A 5 mg/ml $KHCO_3$ stock solution had previously been prepared. While this technique was not employed for the original syntheses of **1** and **2**, the investigators have found that the elution efficiency of [^{18}F]fluoride from the ion-exchange cartridge can be enhanced by removing the septum's cartridge with a 23 gauge needle, emptying 20 mg PS-HCO_3 into a waste vial, and replacing the septum.
18. Cotton was washed with acetone and water and dried at 150°C prior to use. Janda-Jel polypyridine (100–200 mesh, extent of labeling ~8 mmol/g loading, 1% cross-linked) was obtained from Sigma-Aldrich and dried at 23°C for 24 h under dynamic vacuum (10^{-4} Torr) prior to use.
19. The QMA cartridge was obtained from Waters.
20. The Eclipse HP cyclotron uses He as a push gas for delivery of [^{18}F]fluoride. Excess [^{18}O]water was collected in a vial and retained for recycling.
21. The reaction temperature was set at 23°C. The initial solution containing **3** was a translucent orange brown. Over the course of the reaction, the solution mixture turned opaque.
22. Silica gel was obtained from Fisher.
23. Ethyl acetate was obtained from Sigma-Aldrich. Hexanes were obtained from Fisher.
24. Trifluoroacetic acid and dichloromethane were obtained from Sigma-Aldrich.
25. Preparative HPLC was performed on an Agilent 1100 series HPLC connected to a Carol and Ramsey Associates Model 105-S radioactivity detector. A Macherey-Nagel VP 250/10 Nucleosil 100-5 C18 Nautilus column was used for purification.
26. For **1**, the isocratic mobile phase contained 0.1% trifluoroacetic acid a 29:71 solution of acetonitrile: water. For **2**, the isocratic mobile phase contained 0.1% trifluoroacetic acid in 32:68 acetonitrile: water. The flow rate for both preparative HPLC methods was 5 ml/min.
27. A nitrogen outlet within the hot cell was connected to a 1/8 in. PTFE tubing line fitted with a 4 in. × 22 gauge needle. The needle was bent so that it could rest on the rim of the open vial, which was placed in a vial holder on a hot plate at room temperature.
28. The contents of the intermediate vial were drawn up in the same syringe used to add the ethyl acetate/hexanes solution, then quickly added to the third reaction vial.
29. 100 µl glass syringes were obtained from Hamilton. The HPLC injector was obtained from Rheodyne and equipped with a 2 ml loop.
30. The three-way stopcock was obtained from Smiths Medical.
31. Test tubes were obtained from VWR. Retention time of **1** was 13.5–14.5 min and retention time of **2** was 10.7–11.7 min, although these values may shift depending on the specific HPLC system, columns, and mobile phases that are used. In both cases, a large gamma signal will indicate that collection should begin. A large gamma signal at the solvent front (2–3 min) indicates the presence of unbound [^{18}F]fluoride.
32. To ensure that the syringe adapter stays in place, it should be wrapped tightly several times with Parafilm.
33. The lead-shielded hot cell is outfitted with this ion chamber. The SPE column should not be removed from the hot cell to measure activity.

34. Investigators should ensure that sufficient activity is present in the vial at this stage. While the reporters aimed to obtain at least 10 mCi for injection into a nonhuman primate, depending on the weight and species of the animal to be studied, a range of minimum activities may be required. If additional activity is required (and residual activity remains trapped on the SPE cartridge, more ethanol can be used to elute from the cartridge, but the final concentration of ethanol should be ≤10%, so care should be taken to ensure that the final formulated dose is not too dilute Ref. [35].
35. Varying levels of saline were added depending on the activity in the final dose. For lower levels of activity (in the reporters' case, ~10 mCi), the final dose was diluted to 5 ml in order to keep the formulation's concentration sufficiently high. For higher activity products, more saline can be used for formulation. The PET scanning team should be consulted as to the total volume that can be injected into the animal to be studied. For rats, a hard limit is typically set at 1 ml, while for nonhuman primate studies, as much as 10 ml may be injectable.
36. Injectable saline was obtained from Hospira.
37. Sterile filters were obtained from Millex. Sterile vials were obtained from VWR.
38. The nitrogen pressure must be kept below the manufacturer-specified bubble point for the sterile filter, which is 13 psi for the LG filters used in the investigator's work.
39. For analytical HPLC, an Agilent XDB-C18 5 μm, 4.6 × 150 mm column was used. The mobile phase ran on a 12 min gradient from 5% acetonitrile to 95% acetonitrile in 0.1% TFA, with 3 min of posttime prior to the next injection. The injection volume was 10 μl for each sample, and gamma peak of the sample injection was compared to the UV peak at 280 nm of the standard injection to verify consistent retention time.
40. For **1**, mol amounts of 7, 26, 65, 260, 651, and 2603 pmol were used to determine the calibration curve. For **2**, the standards were 29, 72, 286, 716, and 2863 pmol. These standards were prepared by serial dilution in saline.
41. In the experiments conducted by the investigators, 89 μCi of **1** did not display a significant UV signal, while 122 μCi of **2** had a UV absorbance of 5.1 at 280 nm. These values were used to back-calculate mole amounts of <7 pmol for **1** and 120 pmol for **2** and specific activities of >13 Ci/μmol for **1** and 1.0 Ci/μmol for **2**. Both of these values are within the typical injection limits for most known radiotracers.
42. Pd standards were prepared by serial dilution. The concentrations used for calibration were 0.020, 0.10, 0.50, 2.0, 10, and 50 ppb.
43. The samples of **2** had an MS signal lower than the smallest serial dilution, corresponding to a concentration <0.020 ppb, which is within the USP injection limit for palladium.
44. The standard method of loop-loading in the investigators' lab is to affix a 4 in. 22 gauge needle to a line coming from the HPLC loop outlet and a 5 ml Luer adapter syringe to a line coming from the same loop's inlet. With the loop prefilled with HPLC mobile phase and set to the load position, the needle is dropped into a conical vial containing the crude product (at least 3 ml solution), and the syringe is drawn back to pull the sample onto the loop. HPLC mobile phase will fill the syringe to give an estimate of the total volume of solution that has been drawn onto the loop. Once the syringe has been filled to the appropriate volume, the loop's contents are injected onto the HPLC using the Explora LC software.

CHAPTER 16

SILICON FLUORIDE ACCEPTORS (SIFAs) FOR PEPTIDE AND PROTEIN LABELING WITH ^{18}F

Ralf Schirrmacher[1], Alexey Kostikov[1], Carmen Wängler[2], Klaus Jurkschat[3], Vadim Bernard-Gauthier[1], Esther Schirrmacher[1], and Björn Wängler[4]

[1] McConnell Brain Imaging Centre, Montreal Neurological Institute, McGill University, Montreal, Canada
[2] Biomedical Chemistry, Department of Clinical Radiology and Nuclear Medicine, Medical Faculty Mannheim, Heidelberg University, Mannheim, Germany
[3] Lehrstuhl für Anorganische Chemie II, TU University of Dortmund, Dortmund, Germany
[4] Molecular Imaging and Radiochemistry, Department of Clinical Radiology and Nuclear Medicine, Medical Faculty Mannheim, Heidelberg University, Mannheim, Germany

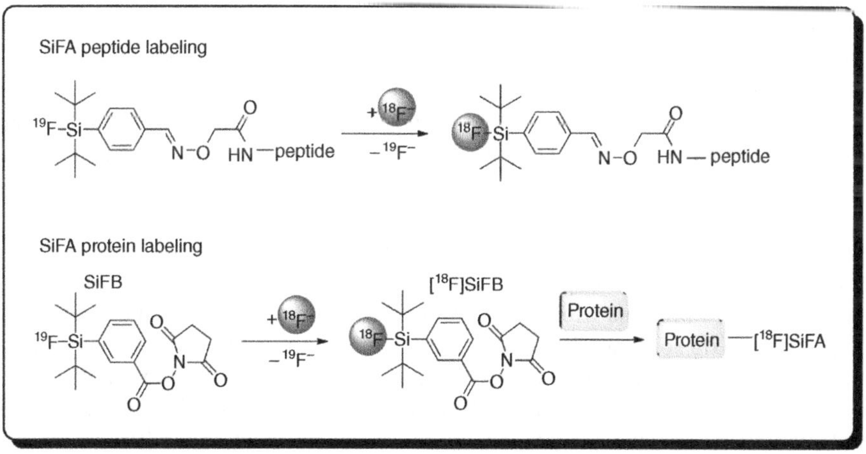

Radiochemical Syntheses, Volume 2: Further Radiopharmaceuticals for Positron Emission Tomography and New Strategies for Their Production, First Edition. Edited by Peter J. H. Scott.

1 INTRODUCTION

The radionuclide fluorine-18 (^{18}F) is one of the most popular radionuclides for the development of radiopharmaceuticals for positron emission tomography (PET) imaging of cancer [1], neurologic diseases [2], and various other biological targets [3]. ^{18}F can be comfortably produced via a medium energy cyclotron in high specific activities via the $^{18}O(p,n)^{18}F$ pathway [4]. The importance of ^{18}F in comparison to other cyclotron-produced positron-emitting radionuclides stems from its almost ideal physical decay properties such as a low positron energy and a comfortable half-life of 110 min allowing for even complex syntheses [5]. Despite these positive qualities, the number of commercially available ^{18}F-radiotracers is quite limited, and 2-[^{18}F]fluorodeoxyglucose ([^{18}F]FDG) has been dominating the market [6]. The reason is the limited chemistry available to introduce nucleophilic $^{18}F^-$ into biologically active compounds [7]. Small molecules can be easily labeled with $^{18}F^-$ in one or two steps usually, but larger molecules like peptides and proteins most often require complex multistep syntheses not suitable for a viable commercial distribution [6b, 8]. In the light of the fact that peptide and protein-based imaging agents have become very popular in molecular imaging, it is obvious that radiochemists should seek for novel routes to prepare those labeled compounds with simpler protocols. Over the last decade, radiochemists have started to report on novel, rather *unorthodox*, labeling methods primarily for peptides and proteins based on the formation of boron–^{18}F, silicon–^{18}F, and aluminium–^{18}F bonds [9]. These efforts have been driven by the requisition for radiochemical protocols providing labeled peptides and proteins in a minimum number of steps, preferably a one-step reaction with $^{18}F^-$. The inherent problem with direct peptide and protein ^{18}F-labeling without prosthetic group chemistry is due to the multitude of functional groups present in those compounds. Only a few conventional methods (characterized by the formation of a carbon–^{18}F bond) to label multifunctional compounds have been reported so far and their general applicability remains to be proven [10]. Due to the general nonselectivity of conventional radiolabeling methods and the presence of interfering acidic protons in many functional groups, a protecting group strategy is necessary to achieve the radiolabeling step. From a practical point of view, it appears tedious to protect each reactive side chain in a peptide, and for proteins with numerous functional groups, it would be chemically impossible to do so. In addition, almost all nucleophilic radiofluorinations require the so-called *naked* fluoride ions in nonaqueous solvents not suitable for nucleophilic F-18 chemistry, but boron–^{18}F as well as aluminium–^{18}F and silicon–^{18}F radiochemistry have the advantage to work in aqueous solution too [9e–g]. In case of Si–^{18}F formation, the labeling efficiency in aqueous media depends highly on the nature of the substance. Some Si–^{18}F-labeled peptides and even small molecules can be labeled in high yields at elevated temperatures, conditions that commonly result in the formation of radioactive and nonradioactive by-products necessitating a complex purification. Other compounds like proteins are not amenable to such harsh labeling conditions at all. The bond energy of the SiF bond is one of the highest in nature surpassing the C–F bond by 90 kJ. It seems therefore reasonable to exploit this chemical stability for the formation of radioactive Si–^{18}F compounds. However, the highly polar character of the SiF bond unfortunately entails a very low tolerance for water, and especially, bases lead to a quick hydrolysis and disintegration. Thus, the development of a suitable Si–^{18}F radiochemistry had to address this issue. The pioneering work finally resulting in several viable labeling approaches was performed by three groups led by (i) D. Perrin [9e], (ii) by R. Schirrmacher and K. Jurkschat [9f], and (iii) by S. Ametamey and U. Klar [11]. Perrin's approach from 2005 entailed the reaction of biotinylated (aminopropyl) triethoxysilane with aqueous $^{18}F^-$ forming [^{18}F]tetrafluorosiliconates (coined *ate* salts). Those seminal findings launched an extensive research program aimed at providing Si-based PET tracers for *in vivo* imaging and extended simultaneously into boron–$^{18}F^-$ radiochemistry. Ametamey and Klar in 2008 observed that simple silanols and alkoxysilanes can be efficiently converted into the corresponding ^{18}F-fluorosilanes under aqueous as well as nonaqueous conditions [11, 12]. The research team observed a strong structural influence on the Si–^{18}F bond rigidity finally leading to just a few suitable building blocks of sufficient hydrolytic stability [13]. However, the first suitable hydrolysis-resistant SiF building block displaying an adequate *in vivo* stability was reported in 2006 by Schirrmacher and Jurkschat as well as by P. Blower [9a, f]. It became obvious from those studies that the stability of the SiF bond toward hydrolysis is imparted by bulky groups such as *tert*-butyl moieties attached to the Si atom. Despite becoming hydrolytically stable in the sense that the Si atom is efficiently shielded from a nucleophilic attack of water, smaller anions such as hydroxy- and radioactive [^{18}F]fluoride are still able to pass the hydrophobic barrier provided by the *tert* butyl groups and replace nonradioactive fluoride. This unusual isotopic exchange was named after its building block silicon fluoride acceptor (SiFA), and it proceeds with an unexpected efficiency. A very large, experimentally determined preexponential factor in the Arrhenius equation for the isotopic exchange in comparison to the formation of a carbon–^{18}F bond (e.g., formation of 2-[^{18}F]fluoroethyltosylate) results in a very fast rate constant for the isotopic exchange even at very low temperatures (–20°C). Since the isotopic exchange takes place in a dipolar aprotic solvent such as acetonitrile, the calculated Gibbs free energy was far lower than in the gas phase resulting in a less favored pentacoordinate transition state. The pentacoordinate siliconate resulting from the attack of the $^{18}F^-$ anion is hence not favored. So far, the SiFA technology has been utilized primarily to radioactively label compounds of higher molecular weight such as peptides and proteins although a first attempt by Wängler et al. reported the successful integration of the SiFA unit into smaller organic brain imaging agents [12, 14]. Furthermore, novel PET imaging agents for hypoxia have been published recently combining the SiFA group with nitroimidazole [15]. The biggest structure-related drawback of the SiFA labeling technology is the inherently high lipophilicity of the SiFA building block. This characteristic is very detrimental for *in vivo* applications since highly lipophilic compounds are usually accumulated in the liver, result in a high plasma protein binding and thus background accumulation, and do not reach the intended biological target *in vivo*. The key to a successful PET imaging agent based on SiFA technology lies therefore in reducing the SiFA lipophilicity by either making the SiFA moiety more hydrophilic or by adding hydrophilic compounds to the peptide to compensate for the high SiFA lipophilicity. This problem has no impact on protein labeling as a result of the protein's high molecular weight and thus diminished SiFA influence on the overall lipophilicity [14f, g, 16]. Most recently, significant improvements have been made regarding peptide labeling, reducing the lipophilicity of peptide SiFA conjugates by using hydrophilic auxiliaries such as PEG spacers and hydrophilic amino acids or sugar-derivatized amino acids [14d, e, 17]. The following procedures describe the labeling of peptides and proteins using different SiFA strategies that have been validated many times in independent laboratories.

2 PROCEDURES

2.1 SiFA–Aoa–Aux–Peptides

The most convenient SiFA building block for peptide modification and subsequent single step ^{18}F-labeling is p-(di-tert-butylfluorosilyl)benzaldehyde (SiFA-A), an aldehyde moiety containing a SiFA compound that can be chemoselectively conjugated to aminooxy-derivatized peptides via click chemistry [9f, 14b, 17, 18]. The reaction conditions for the formation of the resulting oxime are compatible with all common peptide side chains, and the coupling is performed at pH4 in aqueous solution, preventing any solubility issues. The peptides are usually synthesized via Fmoc solid-phase peptide synthesis on a resin with *bis* Boc-protected aminooxy acetic acid (Aoa) being the last building block in the sequence. Between the actual peptide sequence and the *bis*Boc-aminooxy acetic acid, hydrophilic auxiliaries (Aux) such as Fmoc-NH-PEG$_x$-COOH, Fmoc-Asp(OtBu)-OH, and Fmoc-Asn-(Ac$_3$AcNH-β-Glc)-OH can be added to the sequence in an unspecified combination depending on the final targeted lipophilicity of the SiFA–peptide conjugate. The fully protected peptides are then cleaved from the resin, and side chains are deprotected using a standard protocol. The peptide–Aux–Aoa conjugate is purified using gradient reverse-phase HPLC, freeze-dried, and coupled solution phase to SiFA-A in aqueous solution at pH4. A second HPLC purification is necessary to obtain the final labeling precursor peptide–Aux–Aoa–SiFA for ^{18}F-labeling.

The ^{18}F-labeling of the conjugate is easily achieved using oxalic acid-assisted isotopic exchange (Fig. 1) [17]. This method requires the use of $^{18}F^-$ dried according to the Munich protocol because the amount of base ($[K^+\subset 2.2.2]OH^-$ [19], prepared by freeze-drying an equimolar aqueous solution of Kryptofix 2.2.2® and KOH) has to be known in order to add the appropriate amount of oxalic acid. The Munich ^{18}F drying protocol requires no azeotropic drying, is based entirely on the use of strong anion exchange (SAX) cartridges, and does not need any sophisticated equipment. The entire drying setup has minimal space requirements and can be easily integrated into commercial synthesis boxes alongside with the actual peptide labeling reactor.

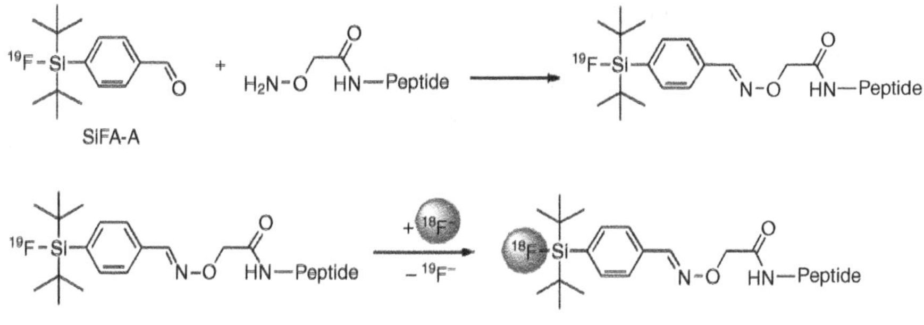

FIGURE 1 Conjugation of SiFA-A to a peptide and subsequent ^{18}F-radiolabeling via isotopic exchange.

2.2 Protein Labeling Using SiFA

The most convenient way of labeling proteins is the use of radiolabeled active esters. Those secondary labeling precursors nonselectively target nucleophilic moieties in the unmodified protein and transfer the radiolabeled tag primarily to amino groups found at the *N*-terminus and in lysine (Lys) side chains. In stark contrast to the actual protein labeling with active esters, which is an easy one-pot reaction followed by a single size exclusion chromatography, the synthesis of the ^{18}F-labeled active esters themselves is usually a major impediment due to the relatively complicated radiochemistry required to obtain these labeling synthons. The most prominent example is undoubtedly *N*-succinimidyl-4-[^{18}F]fluorobenzoate ([^{18}F]SFB) introduced by Zalutsky and coworkers [20]. The synthesis of this NHS ester has been subject of many investigations and despite improvements having been made, the synthesis still requires an HPLC purification and highly skilled personnel. To make protein labeling available to a larger number of potential users, the preparation of the active ester has to become a kit-like procedure, both convenient and efficient. As a comfortable substitute for [^{18}F]SFB, the SiFA-based active ester *N*-succinimidyl 3-(di-tert-butyl[^{18}F]fluorosilyl)benzoate ([^{18}F]SiFB) was developed featuring both the SiFA moiety for easy ^{18}F-labeling via isotopic exchange as well as an NHS ester for easy protein labeling [16a, b]. Usually an active ester disintegrates during the radiolabeling process and has thus to be formed after the ^{18}F introduction, adding an additional complexity to the overall preparation. In case of [^{18}F]SiFB, however, the active ester integrity is not affected by the mild isotopic exchange reaction conditions of the SiFA unit as long as the base from the Munich labeling cocktail is neutralized with oxalic acid. The necessary amount of oxalic acid to be added to the basic $^{18}F^-$ solution (aliquots or full batch) for labeling SiFB was determined experimentally (cf. Synthesis Procedures). No HPLC is required for purification. The crude labeling reaction is simply diluted with 0.01 M H_3PO_4 and loaded onto a C18 cartridge followed by washing the cartridge with water to remove any unreacted ^{18}F- and Kryptofix 2.2.2. The [^{18}F]SiFB can then be eluted from the C18 cartridge with diethyl ether and the ether removed in a stream of inert gas. The respective protein is then added in borate buffer (pH 9) for final radiolabeling. The ^{18}F-labeled protein is purified by size exclusion HPLC with sodium phosphate buffer (0.1 M, pH 7.2) as an eluent. The collected labeled protein fraction is suitable for injection after sterile filtration.

2.3 Synthesis Procedures

CAUTION: All radiochemical syntheses must be carried out using appropriate equipment in a facility authorized for the use of radioactive materials. Personal protective equipment must be worn, and all local radiation safety laws followed.

2.3.1 Peptide Labeling Using SiFA-A

Production of [^{18}F]Fluoride Aqueous [^{18}F]fluoride (10–20 GBq) is prepared by bombarding 2.2 ml of a highly enriched (97%) [^{18}O]water target with protons (35 µA) following the ^{18}O(p,n)^{18}F nuclear reaction using an IBA cyclotron (Cyclone 18/9) for 15 min. After finishing the irradiation, the $^{18}F^-$/[^{18}O] H_2O solution is transferred to an automated synthesis module (HotBox III, Scintomics, Germany) using a stream of He, passing through a SAX cartridge (Sep-Pak Light (46 mg) Accell Plus QMA). The $^{18}F^-$ is trapped on the cartridge.

Preparation of [$K^+ \subset$2.2.2]OH Elution Cocktail for the Munich Protocol 41 mg Kryptofix 2.2.2 (110 µmol) are dissolved in 1 M KOH (100 µl). To this solution, water (1 ml) is added and the mixture is lyophilized to dryness overnight. The salt is redissolved in anhydrous acetonitrile (0.5 ml) yielding a stock solution immediately before use.

Drying of $^{18}F^-$ Via the Munich Method The $^{18}F^-$ loaded SAX cartridge is washed with 5 ml of anhydrous acetonitrile and dried with He for 3 min. The trapped $^{18}F^-$ is eluted with [$K^+ \subset$2.2.2]OH elution cocktail (0.5 ml) and is ready for ^{18}F-radiolabeling of the peptides and SiFB.

Synthesis of Peptides for SiFA-A Conjugation Peptides described herein are not further specified but are generally synthesized by Fmoc solid-phase peptide synthesis (SPPS) following a standard protocol. The *N*-terminal auxiliary modifications to the original peptides are outlined in the following and have been proven to be applicable for Tyr^3-octreotate derivatives (TATE derivatives) and other peptides such as cRGD and bombesin. The final coupling and Boc deprotection of *bis*Boc-Aoa-OH on the modified peptides were done using standard conditions. Only the deprotection of the Fmoc-Asn-(Ac_3AcNH-β-Glc)-OH side chain protecting groups (if applicable) and the last conjugation of SiFA-A to the *N*-terminal aminooxy group of the peptide are performed solution phase.

Solid-Phase Coupling of Hydrophilic Auxiliaries (Aux) to Peptides Via Fmoc Strategy The following protected auxiliaries introducing a varying degree of hydrophilicity to the original peptide can be applied in variable combinations depending on the desired grade of overall hydrophilicity: Fmoc-NH-PEG_x-COOH, Fmoc-Asp(OtBu)-OH, and Fmoc-Asn(Ac_3AcNH-β-Glc)-OH.

The coupling uses the desired peptide (synthesized by Fmoc SPPS) bound on resin providing a free terminal amino group as well as a fourfold excess of the auxiliaries, HBTU (3.9 eq.) and *N,N*-diisopropylethylamine (4 eq.) in DMF at room temperature. The standard coupling protocol can be used.

Cleavage of the Aoa–Aux–Peptide from the Resin and Purification For most peptides prepared by the standard Fmoc protocol, the cleavage from the solid support is achieved by treatment of the resin-bound peptide with TFA/triisopropylsilane/H_2O 95:2.5:2.5 for approximately 1 h at room temperature. The crude peptide is precipitated by the addition of diethyl ether, filtered and dried before the final HPLC purification.

HPLC Purification of the Aoa–Aux–Peptide The crude peptide (2–3 mg; the amount depends on the column size and quality of separation on the column) from Section "Cleavage of the Aoa-Aux-Peptide from the Resin and Purification" is dissolved in a minimum amount of water:acetonitrile 1:1 and injected onto a preparative HPLC system (e.g., Waters SymmetryPrep C18 HPLC column (300 × 19 mm, 7 µm)). The mobile phase is a gradient of 0–60% (vol/vol) water containing 0.1% (vol/vol) TFA to acetonitrile also containing 0.1% (vol/vol) TFA. The flow rate of the HPLC system is about 8 ml/min (this can vary in the individual laboratory setup). The peak corresponding to the product is collected and the final product lyophilized.

Note: If Fmoc-Asn(Ac_3AcNH-β-Glc)-OH is used as an auxiliary during the SPPS, the OH groups of the carbohydrate moiety have to be deprotected by treating the peptide with sodium methanolate solution in methanol (0.5 M) at pH 12–13 at room temperature for 30–60 min. The reaction mixture is acidified using neat TFA (ca. 1 ml), if required again purified via HPLC and lyophilized.

Coupling of SiFA-A to the Aoa–Aux–Peptide The Aoa–Aux–peptide is dissolved in phosphate buffer (0.5 M, pH = 4), and SiFA-A (5 eq. dissolved in a minimum amount of acetonitrile) is added. The reaction is allowed to stand at room temperature for 15 min. The final product SiFA–Aoa–Aux–peptide is obtained after preparative HPLC purification and lyophilized for ^{18}F-radiolabeling. The peptide can be aliquotized and stored in a freezer for up to 1 year (tested for several peptides such as RGD and TATE derivatives).

Radiolabeling of SiFA–Aoa–Aux–Peptide Cf. Sections "Preparation of [K^+⊂2.2.2]OH Elution Cocktail for the Munich Protocol" and "Drying of $^{18}F^-$ Via the Munich Method" for the $^{18}F^-$ Munich drying procedure to obtain [K^+⊂2.2.2][^{18}F]F^- without azeotropic drying. Oxalic acid (25 µl, 1 M in anhydrous acetonitrile) is added to the [K^+⊂2.2.2][^{18}F]F/[K^+⊂2.2.2]OH eluate solution from Section "Preparation of [K^+⊂2.2.2]OH Elution Cocktail for the Munich Protocol". The full batch of $^{18}F^-$ is used for the labeling of the SiFA–Aoa–Aux–peptide.

Note: The amount of oxalic acid varies according to the peptide to be labeled. The specific conditions herein refer to the labeling of a Tyr^3-octreotate derivative

(SiFA-Aoa-Asn(AcNH-β-Glc)-(Asp)$_2$-PEG-TATE) derivative. For other peptides, the amount might slightly vary.

To 0.5 ml of this solution, the SiFA–Aoa–Aux–peptide in anhydrous DMSO (10–25 µl of a 1 mmol/l stock solution) is added (full $^{18}F^-$ batch). The reaction is allowed to proceed at room temperature for 5 min.

Note: Reaction times might require optimization for individual peptides.

Purification of [^{18}F]SiFA–Aoa–Aux–Peptide A Sep-Pak C18 Light cartridge is conditioned with ethanol (10 ml) followed by water (10 ml).

Note: The conditioning has to be slow to ensure proper swelling of the material, and no air should be passed through the cartridge after finishing the conditioning.

HEPES buffer (9 ml, 0.1 M, pH 2) is added to the reaction mixture before being slowly passed through the preconditioned Sep-Pak Light cartridge. The loaded cartridge is washed with water for injection (ca. 10 ml), the final labeled [^{18}F]SiFA–Aoa–peptide is eluted from the solid phase using ethanol (200–500 µl), and the isotonic saline solution is added to obtain the final injectable solution after sterile filtration.

Note: The volume of ethanol depends on the nature of the peptide conjugate and has to be optimized for each individual case. The concentration of ethanol in the final solution must not exceed 10%.

2.3.2 Synthesis of [^{18}F]SiFB for Protein Labeling

Oxalic acid in anhydrous acetonitrile (150 µl, 0.5 M, 75 µmol) is added to the SAX eluate containing $^{18}F^-$ and [$K^+ \subset$ 2.2.2]OH in acetonitrile (cf. Section "Drying of $^{18}F^-$ Via the Munich Method"). To this solution, SiFB (200 nmol) in acetonitrile (40 µl) is added at room temperature and allowed to stand for 30 min without stirring.

Note: A reaction time of 30 min is only needed if a full batch of $^{18}F^-$ in 0.5 ml is used. Lower volumes of $^{18}F^-$ in acetonitrile require shorter reaction times as well as adjusted amounts of oxalic acid. The optimal ratio for [$K^+ \subset$ 2.2.2]OH/oxalic acid is 4:3.

The reaction mixture is diluted with H_3PO_4 (5 ml, 0.01 M) passed through a C18 cartridge, and unreacted $^{18}F^-$ is removed by washing with water (5 ml). Radiolabeled [^{18}F] SiFB is eluted from the cartridge with diethyl ether (3 ml) into the reaction vessel (e.g., Eppendorf vial). The ether is removed under a gentle stream of inert gas (nitrogen or argon) within 2 min.

Protein Labeling with [^{18}F]SiFB The protein (1–3 mg, depending on molecular weight) is dissolved in borate buffer (ca. 1 ml, 0.5 M, pH 9), added to the dried [^{18}F]SiFB and allowed to react at room temperature for 30 min (Fig. 2).

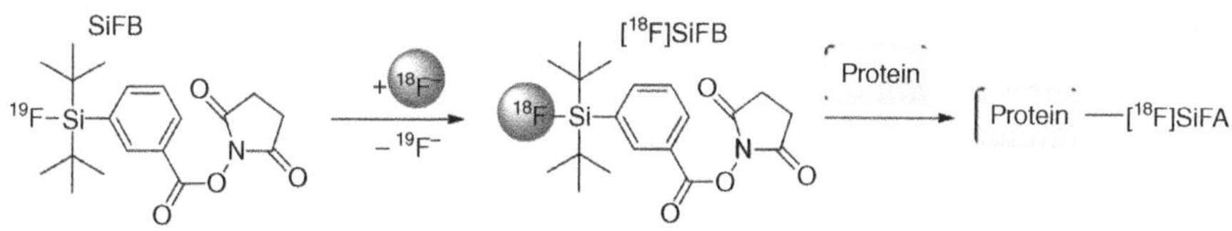

FIGURE 2 Synthesis of [^{18}F]SiFB and subsequent protein labeling (e.g., RSA).

Purification of [^{18}F]SiFB-Labeled Proteins (e.g., RSA) The crude reaction mixture is injected onto a size exclusion HPLC system. The radioactivity peak corresponding to the labeled protein fraction (R_f = 13.5–15 min in our individual laboratory setup) is collected and can be directly used for animal PET experiments after sterile filtration. The following HPLC system has been used for preparation as well as quality control: Phenomenex Biosep-SEC-S4000 HPLC column, 0.1 M sodium phosphate buffer (pH 7.2) as mobile phase, and flow rate of 0.7 ml/min.

2.4 Quality Control Procedures

CAUTION: All radiopharmaceuticals produced for clinical use must have local regulatory approval before use in humans. Trained personnel must carry out QC procedures, and each dose must meet all established QC criteria before release to the clinic.

2.4.1 Visual Inspection

[^{18}F]SiFA–Aoa–peptides The [^{18}F]SiFA–Aoa–peptide dose is examined behind a lead-glass window. The solution must be clear, colorless, and free of any particulate matter.

[^{18}F]SiFB-Labeled Proteins The [^{18}F]SiFB-labeled proteins in phosphate buffer are examined behind a lead-glass shielding. The solution must be clear, colorless, and free of any particulate matter.

2.4.2 Radiochemical Identity and Specific Activity

[^{18}F]SiFA–Aoa–peptides To prove radiochemical identity of the final injectable peptide solution, an aliquot is injected into an analytical radio-HPLC system, and the UV and radioactivity channels are analyzed (Fig. 3). The specific activity (SA) can either be determined experimentally using a UV calibration curve or mathematically by measuring the radioactivity and dividing it by the amount of peptide precursor used in the synthesis.

[^{18}F]SiFB-Labeled Proteins The average amount of protein used for labeling is 1–3 mg (e.g., RSA; 1–3 mg (15–45 nmol)). The average specific activity was calculated to be greater than 18.5 GBq/μmol (>500 Ci/mmol). The average number of SiFA moieties attached to one protein is 2 units.

2.4.3 *Analytical HPLC*

[^{18}F]SiFA–Aoa–Peptides
Column: Chromolith performance RP18e HPLC column (100×4.6 mm) (Fig. 3)
Solvent: 0–100% water including 0.1% (vol/vol) TFA to acetonitrile 0.1% (vol/vol) TFA over 5 min
Flow: 4 ml/min

[^{18}F]SiFB-Labeled Proteins
Column: Phenomenex Biosep-SEC-S4000 (Fig. 4)
Solvent: 0.1 M sodium phosphate buffer (pH 7.2)
Flow: 0.7 ml/min

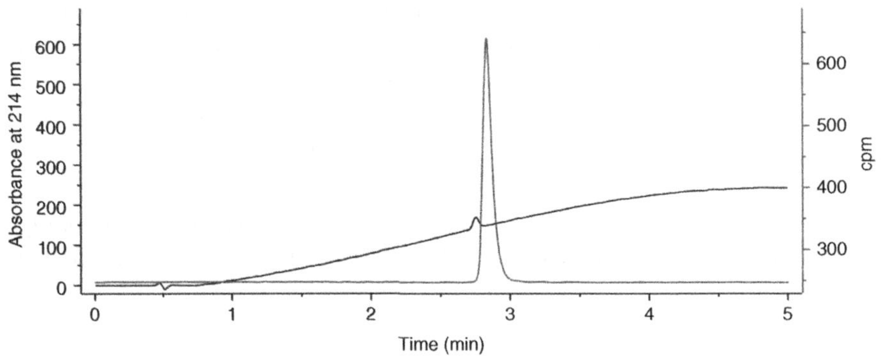

FIGURE 3 Analytical HPLC run (UV and radioactivity channel) of the cartridge-purified [^{18}F] SiFA–Aoa–peptide (e.g., TATE).

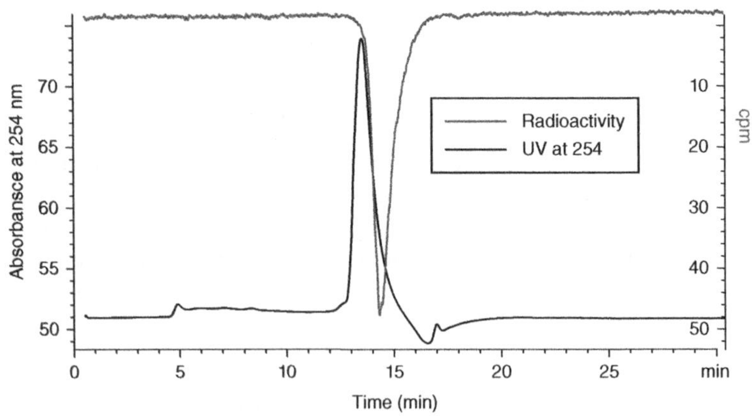

FIGURE 4 Analytical HPLC run (UV and radioactivity channel) of HPLC-purified [^{18}F]SiFB-labeled protein (e.g., RSA).

3 DISCUSSION

The SiFA radiochemistry has evolved over the last few years into a new powerful tool for the development of novel PET tracers based on peptides and proteins. The initial limitations from the high lipophilicity of the original SiFA building block were overcome by introducing a positive charge into the SiFA moiety as well as adding hydrophilic auxiliaries [14d, e, 16a, b, 17]. Those refinements led to positive *in vivo* imaging results in the field of oncologic nuclear imaging of tumors using SiFA-derivatized peptides [14e]. The labeling of peptides is straightforward because the SiFA building block SiFA-A is simple to synthesize and all common peptides prepared by the Fmoc strategy are easily derivatized with the complementary aminooxy group for solution-phase click chemistry. The coupling of hydrophilic auxiliaries to the peptides counterbalances the unfavorably high lipophilicity of the SiFA moiety and can be used to fine-tune the characteristics of the final peptide conjugate toward an optimized *in vivo* behavior. The final labeling of the SiFA–Aoa–Aux–peptides is an easy one-pot reaction leading only to the desired labeled peptide and unreacted $^{18}F^-$, which can be removed easily by solid-phase extraction chemistry. The radiochemical yields of the ^{18}F-labeling of well-known peptides such as octreotate, RGD, and bombesin with the hydrophilic auxiliaries and equipped with the SiFA building block are usually in the range of 50% [17].

Besides peptides, the ^{18}F-labeling of proteins such as antibodies, Fab or scFv antibody fragments, or cellular proteins such as annexin for apoptosis imaging has steadily gained interest over the last decade. Numerous secondary labeling synthons for protein labeling directly at their nucleophilic side chains have been reported in the literature [9b, 21]. Most of these prosthetic group transferring agents entail a cumbersome, lengthy synthesis and demand special equipment as well as superior experience and skills from the operating personnel. This has limited the use of radiolabeled proteins as imaging agents for PET significantly. To alleviate this situation, the labeling synthon [^{18}F]SiFB, mimicking the current gold standard for ^{18}F protein labeling, [^{18}F]SFB was developed [16a]. It provides access to a simple one-pot labeling of a highly reactive ^{18}F bearing active ester for protein labeling [16b]. Normally, an active ester does not survive the usual conditions for nucleophilic $^{18}F^-$ labeling. The SiFA labeling protocol however is based on a very mild isotopic exchange reaction, which works even at low temperatures. This keeps the crucial *N*-succinimidyl ester moiety intact for subsequent protein labeling. The synthesis is simple and the purification relies on straightforward HPLC or solid-phase extraction techniques (using a simple Nap10 size exclusion column). The labeling efficiency is about 50% (nondecay corrected), and the labeling yields for proteins vary between 19 and 36%, which is comparable to [^{18}F]SFB [20c]. In addition to the technical simplicity and easy potential implementation into contemporary automated synthesis modules, the overall synthesis time for [^{18}F]SiFB is significantly reduced in comparison to [^{18}F]SFB. Since labeling yields of proteins are comparable and the higher lipophilicity of the transferred SiFA moiety to the protein does not significantly affect the properties of the protein itself, the use of [^{18}F]SiFB is a promising alternative to [^{18}F]SFB.

WASTE DISPOSAL INFORMATION

All hazardous chemicals and toxic materials were disposed of according to Prudent Practices in the Laboratory (Washington, D.C.: National Academy Press, 1995).

CHEMICAL ABSTRACTS NOMENCLATURE

[^{18}O]H_2O (Huayi Isotopes)
Kryptofix 2.2.2 (Aldrich, cat. No. 291110)
$^{18}F^-$ from $^{18}O(p,n)^{18}F$ nuclear reaction using a cyclotron
Potassium hydroxide pellets (Aldrich, cat. no. 221473)
Acetonitrile anhydrous (Aldrich, cat. No. 271004)
Deionized water (Baxter, cat. No. JF7623)
Ethanol anhydrous (Trace Grade. VWR cat. no CABDH6058-4)
Oxalic acid (Aldrich, cat. No. 241172)
Diethyl ether (Fisher, cat. No. E-138-4)
Sodium phosphate dibasic heptahydrate ($Na_2HPO_4 \cdot 7H_2O$, Aldrich, cat. No. S9390)
Sodium phosphate monobasic monohydrate ($NaH_2PO_4 \cdot H_2O$, Aldrich, cat. No. S9638)
Sodium tetraborate decahydrate ($Na_2B_4O_7 \cdot 10H_2O$, Aldrich, cat. No. S9640)
Hydrochloric acid (HCl, Aldrich, cat. no. 258148)
Rat serum albumin (Aldrich, cat. no. A6272), as protein example
Phosphoric acid (Fisher, cat. no. 351290)

REFERENCES AND NOTES

For detailed supplier information, see Appendix 1.

1. (a) M. P. S. Dunphy, J. S. Lewis, *J. Nucl. Med.*, 2009, 50, 106S; (b) M. E. Juweid, B. D. Cheson, *N. Engl. J. Med.*, 2006, 354, 496.
2. (a) C. Catana, A. Drzezga, W. D. Heiss, B. R. Rosen, *J. Nucl. Med.*, 2012, 53, 1916; (b) T. Jones, E. A. Rabiner, *J. Cereb. Blood. Flow. Metab.*, 2012, 32, 1426; (c) L. Zimmer, A. Luxen, *Neuroimage*, 2012, 61, 363.
3. (a) E. V. Garcia, *J. Nucl. Cardiol.*, 2012, 19, S19; (b) M. D. Majmudar, M. Nahrendorf, *J. Nucl. Med.*, 2012, 53, 673; (c) K. Kubota, K. Ito, M. Morooka, R. Minamimoto, Y. Miyata, H. Yamashita, Y. Takahashi, A. Mimori, *Ann. N Y Acad. Sci.*, 2011, 1228, 29; (d) S. Hermann, A. Starsichova, B. Waschkau, M. Kuhlmann, C. Wenning, O. Schober, M. Schafers, *J. Nucl. Cardiol.*, 2012, 19, 609.
4. (a) J. C. Clark, D. J. Silvester, *Int. J. Appl. Radiat. Isot.*, 1966, 17, 151; (b) M. Guillaume, A. Luxen, B. Nebeling, M. Argentini, J. C. Clark, V. W. Pike, *Appl. Radiat. Isot.*, 1991, 42, 749.
5. H. H. Coenen, *Ernst Schering Research Foundation Workshop* 2007, Springer: Berlin 15.
6. (a) K. Hamacher, H. H. Coenen, G. Stocklin, *J. Nucl. Med.*, 1986, 27, 235; (b) C. Y. Shiue, M. J. Welch, *Radiol. Clin. North Am.*, 2004, 42, 1033.
7. L. S. Cai, S. Y. Lu, V. W. Pike, Eur. J. Org. Chem., 2008, 2853.
8. (a) S. M. Ametamey, M. Honer, P. A. Schubiger, *Chem. Rev.*, 2008, 108, 1501; (b) P. W. Miller, N. J. Long, R. Vilar, A. D. Gee, *Angew. Chem. Int. Ed.*, 2008, 47, 8998.
9. (a) G. E. Smith, H. L. Sladen, S. C. G. Biagini, P. J. Blower, *Dalton Trans.*, 2011, 40, 6196; (b) R. Schirrmacher, C. Wangler, E. Schirrmacher *Mini-Rev. Org. Chem.*, 2007, 4, 317; (c) A. R. Studenov,

M. J. Adam, J. S. Wilson, T. J. Ruth, *J. Label. Compd. Radiopharm.*, 2005, 48, 497; (d) M. S. Rosenthal, A. L. Bosch, R. J.Nickles, S. J. Gatley, *Int. J. Appl. Radiat. Isot.*, 1985, 36, 318; (e) R. Ting, M. J. Adam, T. J. Ruth, D. M. Perrin, *J. Am. Chem. Soc.*, 2005, 127, 13094; (f) R. Schirrmacher, G. Bradtmoller, E. Schirrmacher, O. Thews, J. Tillmanns, T. Siessmeier, H. G. Buchholz, P. Bartenstein, B. Waengler, C. M. Niemeyer, K. Jurkschat, *Angew. Chem. Int. Ed.*, 2006, 45, 6047; (g) W. J. McBride, R. M. Sharkey, H. Karacay, C. A. D'Souza, E. A. Rossi, P. Laverman, C. H. Chang, O. C. Boerman, D. M. Goldenberg, *J. Nucl. Med.*, 2009, 50, 991.

10. (a) D. E. Olberg, J. M. Arukwe, D. Grace, O. K. Hjelstuen, M. Solbakken, G. M. Kindberg, A. Cuthbertson, *J. Med. Chem.*, 2010, 53, 1732; (b) T. Ramenda, J. Steinbach, F. Wuest, *Amino Acids*, 2013, 44, 1167; (c) J. Becaud, L. J. Mu, M. Karramkam, P. A. Schubiger, S. M. Ametamey, K. Graham, T. Stellfeld, L. Lehmann, S. Borkowski, D. Berndorff, L. Dinkelborg, A. Srinivasan, R. Smits, B. Koksch, *Bioconjug. Chem.*, 2009, 20, 2254.
11. L. J. Mu, A. Hohne, R. A. Schubiger, S. M. Ametamey, K. Graham, J. E. Cyr, L. Dinkelborg, T. Stellfeld, A. Srinivasan, U. Voigtmann, U. Klar, *Angew. Chem. Int. Ed.*, 2008, 47, 4922.
12. A. Hoehne, L. Mu, M. Honer, P. A. Schubiger, S. M. Ametamey, K. Graham, T. Stellfeld, S. Borkowski, D. Berndorff, U. Klar, U. Voigtmann, J. E. Cyr, M. Friebe, L. Dinkelborg, A. Srinivasan, *Bioconjug. Chem.*, 2008, 19, 1871.
13. A. Hohne, L. Yu, L. J. Mu, M. Reiher, U. Voigtmann, U. Klar, K. Graham, P. A. Schubiger, S. M. Ametamey, *Chem.-Eur. J.*, 2009, 15, 3736.
14. (a) L. Iovkova-Berends, C. Wängler, T. Zoller, G. Hofner, K. T. Wanner, C. Rensch, P. Bartenstein, A. Kostikov, R. Schirrmacher, K. Jurkschat, B. Wangler, *Molecules*, 2011, 16, 7458; (b) E. Schirrmacher, B. Wangler, M. Cypryk, G. Bradtmoller, M. Schafer, M. Eisenhut, K. Jurkschat, R. Schirrmacher, *Bioconjug. Chem.*, 2007, 18, 2085; (c) L. Iovkova, D. Konning, B. Wangler, R. Schirrmacher, S. Schoof, H. D. Arndt, K. Jurkschat, Eur. J. Inorg. Chem., 2011, 2238–2246; (d) J. Chin, A. P. Kostikov, B. Wangler, R. B. Lennox, R. Schirrmacher, *J. Label. Compd. Radiopharm.*, 2011, 54, S485; (e) C Wangler, B Waser, A Alke, L Iovkova, HG Buchholz, S Niedermoser, K Jurkschat, C Fottner, P Bartenstein, R Schirrmacher, JC Reubi, HJ Weste, B. Wangler *Bioconjug. Chem.*, 2010, 21, 2289; (f) B. Wangler, G. Quandt, L. Iovkova, E. Schirrmacher, C. Wangler, G. Boening, M. Hacker, M. Schmoeckel, K. Jurkschat, P. Bartenstein, R. Schirrmacher, *Bioconjug. Chem.*, 2009, 20, 317; (g) P. Rosa-Neto, B. Waengler, L. Iovkova, G. Boening, A. Reader, K. Jurkschat, E. Schirrmacher, *Chembiochem.*, 2009, 10, 1321.
15. P. Bohn, A. Deyine, R. Azzouz, L. Bailly, C. Fiol-Petit, L. Bischoff, C. Fruit, F. Marsais, P. Vera, *Nucl. Med. Biol.*, 2009, 36, 895.
16. (a) A. P. Kostikov, J. Chin, K. Orchowski, S. Niedermoser, M. M. Kovacevic, A. Aliaga, K. Jurkschat, B. Wangler, C. Wangler, H. J. Wester, R. Schirrmacher, *Bioconjug. Chem.*, 2012, 23, 106; (b) A. P. Kostikov, J. S. Chin, K. Orchowski, E. Schirrmacher, S. Niedermoser, K. Jurkschat, L. Iovkova-Berends, C. Wangler, B. Wangler, R. Schirrmacher, *Nat. Protoc.*, 2012, 7, 1956; (c) B. Wangler, A. P. Kostikov, S. Niedermoser, J. S. Chin, K. Orchowski, E. Schirrmacher, L. Iovkova-Berends, K. Jurkschat, C. Wangler, R. Schirrmacher, *Nat. Protoc.*, 2012, 7, 1964.
17. C. Wangler, S. Niedermoser, J. S. Chin, K. Orchowski, E. Schirrmacher, K. Jurkschat, L. Iovkova-Berends, A. P. Kostikov, R. Schirrmacher, B. Wangler, *Nat. Protoc.*, 2012, 7, 1946.
18. C. Wangler, R. Schirrmacher, P. Bartenstein, B. Wangler, *Curr. Med. Chem.*, 2010, 17, 1092.
19. S. H. Wessmann, G. Henriksen, H. J. Wester, *Nuklearmed-Nucl. Med.*, 2012, 51, 1.
20. (a) G. Vaidyanathan, M. R. Zalutsky, *Nucl. Med. Biol.*, 1992, 19, 275–281; (b) G. Vaidyanathan, M. R. Zalutsky, *Bioconjug. Chem.* 1994, 5, 352; (c) G. Vaidyanathan, M. R. Zalutsky, *Nat. Protoc.*, 2006, 1, 1655.
21. B. Wangler, R. Schirrmacher, P. Bartenstein, C. Wangler, *Mini-Rev. Med. Chem.*, 2011, 11, 968.

CHAPTER 17

SOLID PHASE SYNTHESIS OF [^{18}F]-2-FLUORO-2-DEOXY-D-GLUCOSE: A RESIN-LINKER-VECTOR (RLV) APPROACH

LYNDA J. BROWN[1], IMTIAZ KHAN[2], HARRY J. WADSWORTH[2], ALEXANDER JACKSON[2], NIANCHUN MA[1], NICOLAS MILLOT[2], SUE M. CHAMPION[2], DENIS R. BOUVET[2], ALEX M. GIBSON[2], AND RICHARD C. D. BROWN[1]

[1] *Department of Chemistry, University of Southampton, Southampton, UK*
[2] *GE Healthcare, The Grove Centre, Amersham, Buckinghamshire, UK*

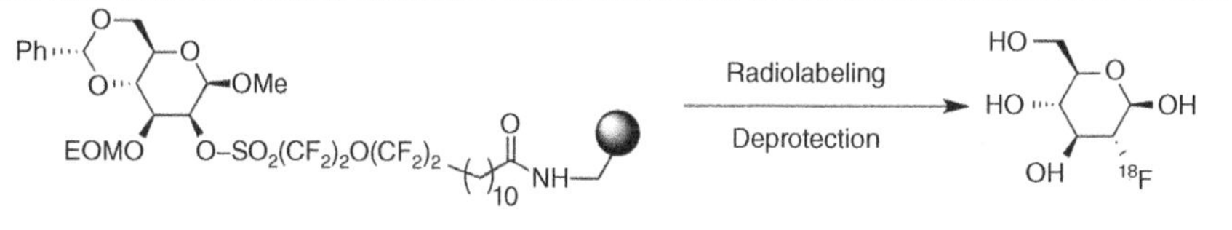

1 INTRODUCTION

Positron emission tomography (PET) is a powerful imaging tool used to locate and assess physiological abnormalities [1]. [^{18}F]2-Fluoro-2-deoxy-D-glucose ([^{18}F]FDG, **9**) is the most widely used radiochemical tracer in clinical PET applications, using the positron-emitting radionuclide ^{18}F (half-life ~ 110 min) [2]. Traditional methods of preparation are based on nucleophilic fluoridation of excess tetra-*O*-acetyl-2-*O*-trifluoromethanesulfonyl D-mannose with limiting [^{18}F]fluoride ion affording [^{18}F]FDG as a crude mixture with radiochemical yields typically around 60% [3]. In this chapter, we present a resin-linker-vector (RLV) approach to the synthesis of protected [^{18}F]FDG using [^{18}F]fluoride ion. The radiotracer [^{18}F]FDG is obtained in high radiochemical yield and radiochemical purity (>95%), following acidic cleavage of the protecting groups [4].

Radiochemical Syntheses, Volume 2: Further Radiopharmaceuticals for Positron Emission Tomography and New Strategies for Their Production, First Edition. Edited by Peter J. H. Scott.

2 PROCEDURES

2.1 Synthetic Procedures

2.1.1 *Methyl-4,6-O-benzylidene-3-O-ethoxymethyl-β-D-glucopyranoside (2)*

To methyl-4,6-*O*-benzylidene-β-D-glucopyranoside [5] (**1**, 847 mg, 3.00 mmol) [6] in anhydrous THF (40 ml) [7] was added NaH (60% in mineral oil, 192 mg, 4.80 mmol) [8] and the effervescing mixture stirred, under argon, for 30 min at rt (CAUTION! Liberation of explosive gas H_2). Chloromethylethyl ether (425 mg, 0.417 ml, 4.27 mmol) [9] was added dropwise, and the reaction stirred at rt for 20 h. The reaction was quenched by careful addition of methanol (0.5 ml), then concentrated *in vacuo*. The crude product was dissolved in CH_2Cl_2 (150 ml), washed with water (150 ml), and the aqueous phase extracted with CH_2Cl_2 (100 ml) [10]. The combined organic solution was washed with water (50 ml), brine, dried (anhydrous Na_2SO_4), and concentrated *in vacuo*. Purification by silica gel column chromatography (hexane–EtOAc, 2 : 1) afforded the title compound **2** (white solid, 500 mg, 49%) [11].

2.1.2 *Methyl-4,6-O-benzylidene-3-O-ethoxymethyl-β-D-mannopyranoside (3)*

A solution methyl-4,6-*O*-benzylidene-3-*O*-ethoxymethyl-β-D-glucopyranoside (**2**, 14.0 g, 41.1 mmol) in DMSO (168 ml) and Ac_2O (84.5 ml, 0.943 mol) was stirred at rt for 24 h [12]. The reaction was diluted with EtOAc (1 l) and washed with saturated aqueous K_2CO_3 solution (600 ml). The organic layer was separated, dried (anhydrous Na_2SO_4), and concentrated to dryness *in vacuo*, giving the crude ketone as a white solid (20 g). The crude ketone (20 g) was redissolved in MeOH (200 ml) and cooled to −20°C before $NaBH_4$ (1.69 g, 44.4 mmol) was added portionwise over 10 min, with stirring (CAUTION! Liberation of explosive gas H_2) [13]. The reaction mixture was allowed to warm to 25°C and then stirred at this temperature for a further 48 h. The reaction was concentrated *in vacuo* to give a gum, which was partitioned between EtOAc (250 ml) and saturated aqueous K_2CO_3 solution (150 ml). The organic layer was separated, re-extracting the aqueous with EtOAc (2 × 100 ml). The combined organic solution was dried (anhydrous Na_2SO_4) and concentrated *in vacuo*. Purification of the resulting off-white solid by silica gel column chromatography (hexane–EtOAc, 1 : 1) afforded the title compound **3** (white solid, 10.62 g, 76%) [14].

2.1.3 *Methyl-4,6-O-benzylidene-3-ethoxymethyl-2-(5-iodooctafluoro-3-oxapentanesulfonate)-β-D-mannopyranoside (4)*

To a solution of methyl-4,6-*O*-benzylidene-3-*O*-ethoxymethyl-β-D-mannopyranoside (**3**, 315 mg, 0.925 mmol) in anhydrous THF (10 ml) at rt was added NaHMDS (1.0 M solution in THF, 1.1 ml, 1.10 mmol) and the reaction stirred, under argon, at rt for 20 min [15]. 5-Iodooctafluoro-3-oxapentanesulfonyl fluoride (437 mg,

1.11 mmol) was added dropwise, and the reaction stirred for a further 45 min at rt [16]. The reaction was concentrated *in vacuo*, redissolved in Et_2O (100 ml), washed with saturated aqueous K_2CO_3 solution (100 ml) and water (100 ml) and the aqueous phase re-extracted with Et_2O (2 × 50 ml). The combined organic phase was washed with brine, dried (anhydrous $MgSO_4$), and concentrated *in vacuo*. Purification by silica gel column chromatography (hexane–EtOAc, 3 : 1) afforded the title compound **4** (white solid, 540 mg, 78%).

2.1.4 Methyl-4,6-O-benzyledine-3-ethoxymethyl-2-(3-oxa-12,12,13,13,15,15,16,16-octafluoro-10-iodo-hexadecanoic acid-16-sulfonate)-β-D-mannopyranoside (5)

To the iodide **4** (300 mg, 0.402 mmol) and undec-10-enoic acid (89 mg, 0.483 mmol) in MeCN (2 ml) at rt was added H_2O (1 ml), $NaHCO_3$ (41 mg, 0.488 mmol) and $Na_2S_2O_4$ (85%, 97 mg, 0.48 mmol) and the reaction stirred at rt for 30 min [17]. The reaction was concentrated *in vacuo*, redissolved in Et_2O (150 ml), washed with water (100 ml), and the aqueous phase re-extracted with Et_2O (50 ml). The combined organic solution was washed with brine (100 ml), dried ($MgSO_4$), and concentrated *in vacuo*. Purification by silica gel column chromatography, eluting with (CH_2Cl_2–CH_3OH, 98 : 2), afforded the title compound **5** (colourless oil, 200 mg, 53%).

2.1.5 Methyl-4,6-O-benzylidene-3-ethoxymethyl-2-(3-oxa-12,12,13,13,15,15,16,16-octafluoro-hexadecanoic acid-16-sulfonate)-β-D-mannopyranoside (6)

To acid **5** (190 mg, 0.203 mmol) in Et_2O (2 ml) was added zinc powder (41 mg, 0.627 mmol) [18] and AcOH (1 ml) and the reaction heated at reflux, under argon, for 3 h (bath temp = 80°C) [19]. The reaction was allowed to cool to rt and filtered through celite, washing with Et_2O (3 × 30 ml). The filtrate and combined washings were concentrated *in vacuo*. Purification by silica gel column chromatography, eluting with (EtOAc–hexane, 1 : 3), afforded the title compound **6** (colourless oil, 97 mg, 59%).

2.1.6 Methyl-4,6-O-benzylidene-3-ethoxymethyl-2-(3-oxa-12,12,13,13,15,15,16,16-octafluoro-hexadecanoic acid-16-sulfonate)-β-D-mannopyranoside polystyryl amide (7)

To a suspension of amino-methylated polystyrene (50 mg, 0.075 mmol) [20] and acid **6** (80 mg, 0.098 mmol, 1.3 equiv) in anhydrous CH_2Cl_2 (1 ml) at rt was added *N,N*-diisopropylethylamine (34 µl, 0.195 mmol, 2.6 equiv), followed by diphenylphosphoryl chloride (26 mg, 0.098 mmol, 1.3 equiv) [21]. The reaction was stirred gently, under argon, at rt for 18 h [22]. The resin was removed by filtration, washed with CH_2Cl_2 (3 × 10 ml), CH_3OH (3 × 10 ml), Et_2O (3 × 10 ml), and dried *in vacuo*, at 40°C for 48 h. This gave the title resin **7** (pale yellow solid, 103 mg, 90%) [23].

2.2 Radiochemical Procedures

CAUTION: All radiopharmaceuticals produced for clinical use must have local regulatory approval before use in humans. Trained personnel must carry out QC procedures, and each dose must meet all established QC criteria before release to the clinic.

2.2.1 *The Manual Radiosynthesis of [^{18}F]-2-Fluoro-2-deoxy-D-glucose (9) from Resin-Bund-β-D-mannose Derivative 7*

Labeling (Compound 8) A carbon glass reaction vessel was placed in a brass heater, the pot lid tightened down and the whole system was leak tested [24] (Figs. 1 and 2). Kryptofix [2.2.2] (22 mg, 0.058 mmol) in CH_3CN (300 µl) and K_2CO_3 (8 mg, 0.058 mmol) in water (300 µl) were transferred using a plastic syringe (1 ml) into the carbon glass reaction vessel [25]. The [^{18}F]fluoride ion (185–370 MBq) in water (0.5–2 ml) was added and the vessel heated to 125°C. After 15 min, three aliquots of CH_3CN (0.5 ml) were added at 1 min intervals. [^{18}F] Fluoride ion was dried up to 40 min in total. The heater was cooled to rt, the pot lid removed and CH_3CN (200 µl) was added. The pot lid was replaced and the lines were capped off with stoppers. The heater was set at 100°C for 10 min and the [^{18}F]

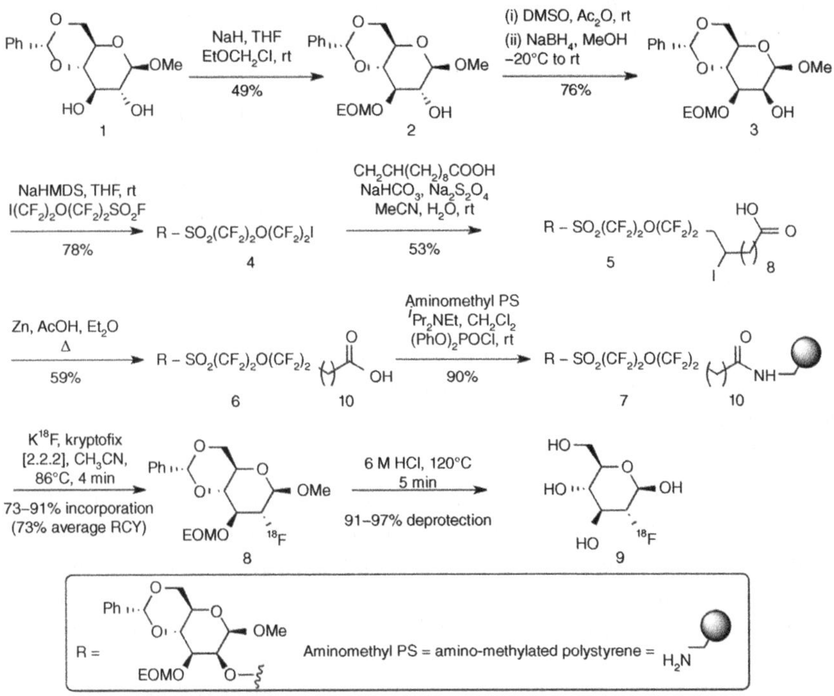

FIGURE 1 Solid phase synthesis of [^{18}F]-2-fluoro-2-deoxy-D-glucose: A resin-linker-vector (RLV) approach.

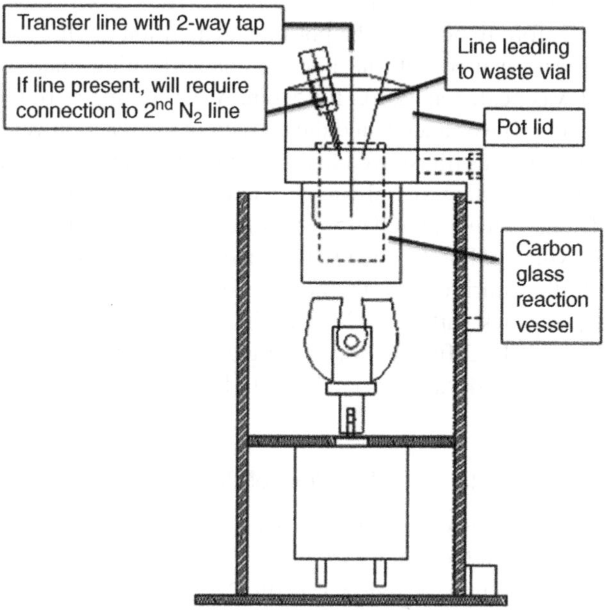

FIGURE 2 Glassy carbon reactor.

fluoride ion redissolved. After cooling to rt, the [^{18}F]fluoride ion solution (200 µl) was transferred by plastic syringe (1 ml) to a second carbon glass reaction vessel, equipped with a stirrer bar [26], containing the resin **7** (20–25 mg). This carbon glass vessel was transferred to an ion chamber and the labeling activity measured. The carbon glass vessel was replaced in the brass heater and the capped pot lid tightened. The gently stirred reaction was heated to 86°C for 4 min before cooling with compressed air. The pot lid was removed, CH_3CN (1.0 ml) was added and the activity in the reaction vessel was measured. The resin was mixed and drawn up into a plastic syringe and then passed through a sintered syringe and into a collection vial. The reaction vessel was washed with a further volume of CH_3CN (0.5 ml) and passed through the sintered syringe. The activity in the collection vial, of the resin, and of the sintered syringe was measured. The reaction mixture was diluted with water (1 ml), passed through a C_{18} sep-pak that had been preconditioned with EtOH (5 ml) and water (10 ml) [27]. This solution was then passed through a silica sep-pak [28] that had been preconditioned with Et_2O (5 ml). The activities of all solutions were measured at each stage and incorporation yields were consistently in the range 73–91% (reverse phase HPLC, average 80–85%) [29].

*Deprotection ([^{18}F]2-Fluoro-2-deoxy-D-glucose ([^{18}F]FDG, **9**))* The solution of [^{18}F]-**8**, after treatment through the sep-pak cartridges, was transferred into a third carbon glass reaction vessel and the activity measured. The reaction vessel was placed in the brass heater and the pot lid was tightened. Lines from the pot were connected to the nitrogen supply and waste vial and then the solution heated at 100°C for 10 min to evaporate the solvent. The vessel was cooled with compressed air before removal of the pot lid and addition of HCl (6 M, 0.5 ml). A second pot lid was secured with no lines to enable heating under pressure. The reaction vessel was heated at 120°C for 5 min. After cooling to rt, the pot lid was removed, water (1 ml) was added, and the activity in the reaction vessel measured. The reaction mixture was transferred to a collection vial and once again activity in the vial and empty reaction vessel was measured. Reaction mixture (100 µl) was added to the Dionex eluent [NaOH (50–52%)aq. (2 ml) : 600 ml H_2O] (4–5 ml) to give a solution with a pH of 7 ± 2 ready for ion exchange and reverse phase HPLC analysis. For further purification, the neutralized solution was flushed through a conditioned C_{18} sep-pak cartridge [27] and the eluate analysed. The deprotection proceeded giving [^{18}F]-2-fluoro-2-deoxy-D-glucose (**9**) with an average radiochemical yield of 73% (decay corrected) with 91–97% of the activity retained (ion exchange HPLC) [30].

2.3 Quality Control Procedures

Low-activity labeling experiments were carried out on the resin **7** manually using K^{18}F (185–350 MBq) and Kryptofix [2.2.2] in CH_3CN (200 µl), and radiochemical yields were established by reversed-phase HPLC analysis of the protected product **8** using UV and γ-detection. [^{18}F] Incorporation between 73 and 91% (53–86% RCY) was achieved within 3–4 min, and no further improvement was observed from longer reaction times (10–15 min). Incorporation gradually improved with increasing temperature up to 86°C; beyond this, there was no significant effect. Reaction volume proved to be important, and increasing the volume to 1 ml resulted in significantly reduced incorporation (42%). High activity labeling studies (5.18–6.16 Ci) were also conducted, and incorporation yields in the range of 68–77% were realized.

Figure 3 shows C_{18} RP-HPLC chromatograms of the of crude tetracetoxy protected [^{18}F]FDG from conventional solution-phase synthesis (left) and from the

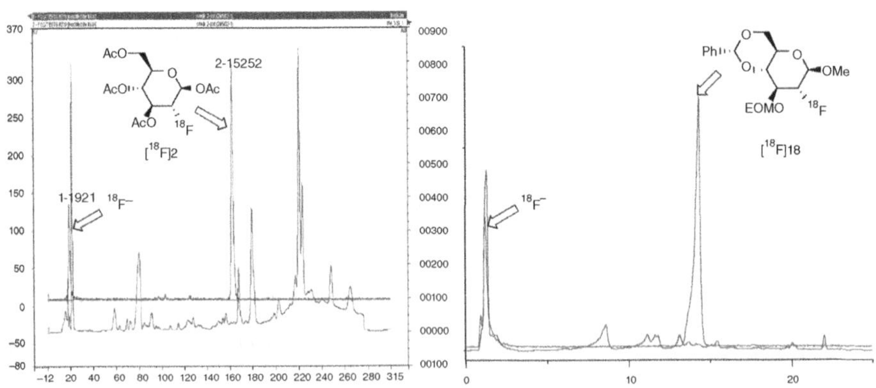

FIGURE 3 Reversed phase HPLC chromatograms (C_{18} column) of the crude protected products from conventional solution-phase FDG synthesis (left) and from the solid-phase precursor **7** (right). Gray trace, Radioactivity; Black trace, UV activity (210 nM).

solid-phase precursor **7** (right), before removal of unreacted fluoride ion and Kryptofix. From the Black UV trace (210 nM) of the conventional solution fluoridation, it is obvious that a wide range of starting material decomposition products is produced. The gray radioactive trace shows that there are only two radioactive materials present identified as fluoride running near the origin and the main radioactive peak of tetraacetoxy [^{18}F]FDG. The black UV trace of the crude product of fluoridation of the solid phase precursor **7** shows that apart from material eluting from the column near the origin there is relatively less UV active material released from the solid phase. The radioactive (gray) trace shows that there are two radioactive materials present. The peak eluting near the origin is assigned to unreacted [^{18}F]fluoride ion, and the peak near 14 min is assigned as the protected [^{18}F]FDG.

Deprotection of [^{18}F]-**8** using 6 M HCl, 120°C for 5 min provided a radioactive ion-exchange chromatogram with one main peak ([^{18}F]FDG) and one small peak (<3%) corresponding to partially deprotected [^{18}F]FDG (upper line, radioactivity trace, Fig. 4). The chromatogram from the electrochemical detector (lower line, Fig. 4) shows that the protected glucose present as a side product from the initial fluoridation reaction to prepare [^{18}F]-**8** has been deprotected and some other decomposition products formed. The formation of these by-products is a consequence of residual water present in the [^{18}F]fluoride solution and elimination due to the use of excess base in the labeling process. Comparison of the chromatograms generated using electrochemical and γ-detectors indicates that [^{18}F]FDG was produced with high specific activity. The neutralized deprotection reaction mixture was analyzed further for nonpolar impurities by reversed phase HPLC, identifying benzaldehyde as the major nonpolar impurity. Benzaldehyde was readily removed by passage of the reaction mixture through a C_{18} sep-pak cartridge (benzaldehyde retained).

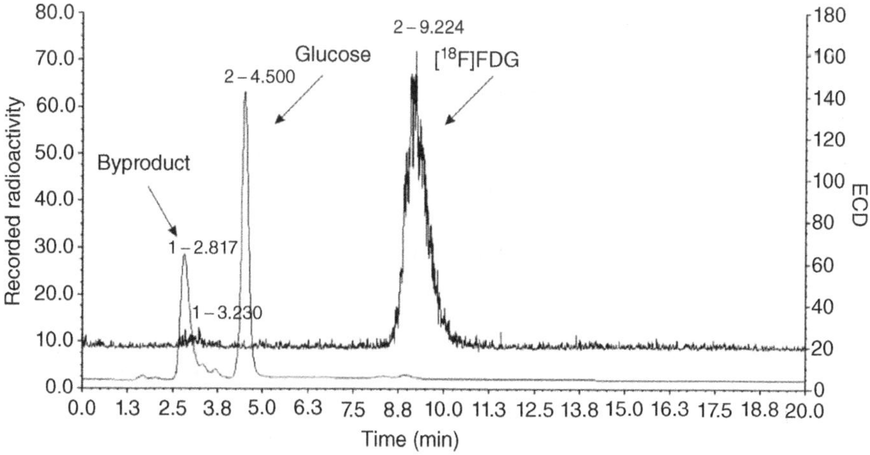

FIGURE 4 Dionex anion exchange HPLC of the reaction mixture from global deprotection yielding [^{18}F]FDG.

Sequential labeling experiments re-using the same sample of resin **7** all led to the formation of **8** with consistently high radiochemical yields (up to five cycles), indicating that the majority of the protected mannose derivative remained attached to the resin during ^{18}F labeling. This means that one batch of the resin could, in principle, be used to generate multiple doses of the radiotracer.

3 DISCUSSION

Initial studies concentrated on direct translation of the conventional solution-phase synthesis of FDG to the solid phase to produce a supported analogue of β-D-mannopyranose-1,3,4,6-tetraacetate-2-(trifluoromethanesulfonate). Reaction of this mannose triflate with stoichiometric K[^{19}F] was highly inefficient, and returned the desired fluorinated sugar in less than 5% yield along with an elimination product and β-D-glucose pentaacetate. In the conventional solution-phase radiosynthesis, this problem is overcome by employing a large excess of the mannose triflate derivative. To achieve a successful solid-phase approach, our belief was that a more chemically efficient fluoridation reaction was required. Literature precedent suggested that a 4,6-benzylidene acetal in combination with acid-labile protection at the 1- and 3-positions of D-mannopyranose would provide a superior substrate for fluoridation [31]. The ethoxymethyl group was selected for protection of the C3 alcohol as it can be removed under similar conditions used to deblock the benzylidene acetal and anomeric methoxy protecting groups. Indeed, solution fluoridation using [^{19}F]fluoride ion (KF/Kryptofix [2.2.2]) proceeded in substantially improved yield and with considerably less degradation, providing the basis for the solid-phase radiosynthesis of FDG described here.

The [^{18}F]FDG (Vector) precursor, prepared from commercially available methyl β-D-glucopyranoside in six steps, is immobilized onto aminomethylated polystyrene (Resin) through a bifunctional linker. The amide bond serves as a robust attachment to the resin and the highly activated perfluoroalkyl sulfonate functions as an exceptional leaving group. This linker allows for selective cleavage of the protected radiotracer into solution using [^{18}F] fluoride ion in high radiochemical yield and superior chemical purity, due to retention of the majority of the unreacted substrate on the resin. For the radiosynthesis using [^{18}F]fluoride ion, it is desirable to keep the reaction time as short as possible and the preferred conditions for the generation of [^{18}F]FDG required ^{18}F labeling at 86°C for 4 min in CH_3CN followed by deprotection with 6 M HCl for 5 min at 120°C. It was also found that stirring of the heterogeneous reaction mixture was important in order to achieve consistent results. Following deprotection, the ^{18}F-containing tracer is passed through disposable cartridges to remove Kryptofix [2.2.2], any residual [^{18}F]fluoride ion and benzaldehyde leaving product ready for formulation. The RLV synthesis produced an average radiochemical yield of 73% (decay corrected) and activity losses on the resin ranged between 3 and 8%, with 91–97% of the activity being collected. Comparison of UV activity traces of protected [^{18}F]FDG released from the resin precursor **7** to that prepared by the conventional solution synthesis showed that the product released from the resin contained significantly reduced levels of chemical impurities. The general approach we have described lends itself to automation or minaturization as the solid-supported precursor could easily be packaged in cartridge format or microfluidic device, suitable for the production of [^{18}F] radiopharmaceuticals. More recently, the RLV approach has been applied to the synthesis of *O*-(2-[^{18}F]fluoroethyl)-L-tyrosine ([^{18}F]FLT) [32].

4 EXPERIMENTAL SUPPLEMENT

4.1 Methyl-4,6-*O*-benzylidene-3-*O*-ethoxymethyl-β-D-glucopyranoside (2)

Mp 99–102°C; $[\alpha]_D$ −11.3 (c=0.488, $CHCl_3$); ν_{max} (neat, cm^{-1}) 3450, 2876, 1727, 1380, 1100, 1069, 1026; ^{1}H NMR (300 MHz, $CDCl_3$) δ 7.50–7.30 (5H, m), 5.53 (1H, s), 4.85 (1H, d, J=7.4 Hz), 4.78 (1H, d, J=7.4 Hz), 4.38 (1H, d, J=7.4 Hz), 4.37 (1H, dd, J=5.2, 10.3 Hz), 4.00 (1H, d, J=1.5 Hz), 3.84–3.74 (2H, m), 3.66 (1H, t, J=8.1 Hz), 3.62–3.56 (2H, m), 3.60 (3H, s), 3.58–3.40 (2H, m), 1.21 (3H, t, J=6.6 Hz); ^{13}C NMR (300 MHz, $CDCl_3$) δ 137.25, 129.28, 128.40, 126.33, 104.53, 101.77, 96.68, 82.17, 79.71, 74.09, 68.86, 66.55, 64.39, 57.67, 15.02; MS (ES$^+$) m/z 363.1 ([M+Na]$^+$); Anal. Calcd for $C_{17}H_{24}O_7$: C, 59.99; H, 7.11. Found: C, 59.83; H, 7.14.

4.2 Methyl-4,6-*O*-benzylidene-3-*O*-ethoxymethyl-β-D-mannopyranoside (3)

Mp 106–108°C; $[\alpha]_D$ −38.7 (c=0.375, $CHCl_3$); ν_{max} (neat, cm^{-1}) 3473, 2971, 2891, 1738, 1375, 1092, 1030; ^{1}H NMR (300 MHz, $CDCl_3$) δ 7.50–7.30 (5H, m), 5.56 (1H, s), 4.91 (1H, d, J=6.9 Hz), 4.83 (1H, d, J=6.9 Hz), 4.50 (1H, s), 4.34 (1H, dd, J=5.2, 10.3 Hz), 4.17 (1H, dd, J=1.5, 2.2 Hz), 4.06 (1H, t, J=9.6 Hz), 3.91 (1H, dd, J=2.9, 9.5 Hz), 3.88 (1H, t, J=10.3 Hz), 3.73–3.60 (2H, m), 3.58 (3H, s), 3.40 (1H, ddd, J=5.2, 9.6, 10.3 Hz), 2.59 (1H, d, J=1.5 Hz), 1.17 (3H, t, J=7.0 Hz); ^{13}C NMR (300 MHz, $CDCl_3$) δ 137.48, 129.15, 128.36, 126.25, 101.87, 101.50, 94.91, 77.65, 74.89, 70.56, 68.74, 67.13, 63.79, 57.50, 15.16; MS (ES$^+$) m/z 703.2 ([2M+Na]$^+$); HRMS (ES) Calcd for $C_{17}H_{24}O_7Na$: 363.1414. Found: 363.1416; Anal. Calcd for $C_{17}H_{24}O_7$: C, 59.99; H, 7.11. Found: C, 59.90; H, 7.12.

4.3 Methyl-4,6-*O*-benzylidene-3-ethoxymethyl-2-(5-iodooctafluoro-3-oxapentanesulfonate)-β-D-mannopyranoside (4)

$[\alpha]_D$ −4.0 (c=0.028, $CHCl_3$); ν_{max} (film, cm^{-1}) 2971, 2890, 1738, 1411, 1380, 1208, 1146, 1092, 1025, 915; ^{1}H NMR (300 MHz, $CDCl_3$) δ 7.50–7.30 (5H, m), 5.57 (1H, s), 5.13 (1H, d, J=2.9 Hz), 4.86 (1H, d, J=7.4 Hz), 4.78 (1H, d, J=7.4 Hz), 4.57 (1H, s), 4.34 (1H, dd, J=5.1, 10.3 Hz), 4.12 (1H, dd, J=2.9, 9.6 Hz), 3.94–3.82 (2H, m), 3.74–3.55 (2H, m), 3.56 (3H, s), 3.44 (1H, ddd, J=4.4, 9.5, 10.3 Hz), 1.13

(3H, t, $J = 6.6$ Hz); ^{13}C NMR (75 MHz, $CDCl_3$) δ 137.13, 129.28, 128.37, 126.19, 101.92, 99.16, 94.07, 83.61, 77.32, 71.04, 68.44, 67.58, 63.96, 57.50, 15.01; ^{19}F NMR (282 MHz, $CDCl_3$, ref. C_6F_6) δ 96.98, 79.97, 76.42, 48.24; MS (ES$^+$) *m/z* 769.1 ([M+Na]$^+$); Anal. Calcd for $C_{21}H_{23}O_{10}F_8IS$: C, 33.79; H, 3.11. Found: C, 33.86; H, 3.13.

4.4 Methyl-4,6-*O*-benzylidene-3-ethoxymethyl-2-(3-oxa-12,12,13,13,15,15,16,16-octafluoro-10-iodohexadecanoic acid-16-sulfonate)-β-D-mannopyranoside (5)

ν_{max} (film, cm^{-1}) 2931, 2858, 1709, 1410, 1192, 1147, 1093, 1025, 919; ^{1}H NMR (300 MHz, $CDCl_3$) δ 7.50–7.31 (5H, m), 5.58 (1H, s), 5.15 (1H, d, $J = 2.9$ Hz), 4.85 (1H, d, $J = 7.4$ Hz), 4.79 (1H, d, $J = 7.4$ Hz), 4.61 (1H, s), 4.40–4.25 (2H, m), 4.17 (1H, dd, $J = 2.9$, 10.3 Hz), 3.95–3.85 (2H, m), 3.75–3.60 (2H, m), 3.58 (3H, s), 3.51–3.42 (1H, m), 3.00–2.65 (2H, m), 2.35 (2H, t, $J = 7.4$ Hz), 1.90–1.30 (14H, m), 1.15 (3H, t, $J = 6.6$ Hz); ^{13}C NMR (75 MHz, $CDCl_3$) δ 178.22, 136.01, 128.10, 128.18, 126.46, 102.20, 99.47, 94.26, 83.84, 77.58, 71.39, 68.71, 67.90, 64.25, 57.73, 41.65, 40.69, 34.31, 29.86, 29.53, 29.49, 29.37, 28.82, 25.04, 21.48, 15.23; ^{19}F NMR (282 MHz, $CDCl_3$, ref. C_6F_6) δ 79.78, 74.37, 47.92, 43.70; MS (ES$^-$) *m/z* 929.3 ([M]$^-$); HRMS (ES$^+$) Calcd for $C_{32}H_{43}O_{12}F_8ISNa$: 953.1284. Found: 953.1266.

4.5 Methyl-4,6-*O*-benzylidene-3-ethoxymethyl-2-(3-oxa-12,12,13,13,15,15,16,16-octafluorohexadecanoic acid-16-sulfonate)-β-D-mannopyranoside (6)

ν_{max} (film, cm^{-1}) 2929, 2858, 2858, 1710, 1411, 1147, 1093, 1025, 994, 918; ^{1}H NMR (300 MHz, $CDCl_3$) δ 7.50–7.31 (5H, m), 5.58 (1H, s), 5.15 (1H, d, $J = 2.9$ Hz), 4.85 (1H, d, $J = 7.4$ Hz), 4.79 (1H, d, $J = 7.4$ Hz), 4.63 (1H, s), (1H, q, $J = 5.8$ Hz), 4.17 (1H, dd, $J = 2.9$, 10.3 Hz), 3.95–3.86 (2H, m), 3.75–3.60 (2H, m), 3.58 (3H, s), 3.51–3.42 (1H, m), 2.35 (2H, t, $J = 7.4$ Hz), 2.15–1.90 (2H, m), 1.65–1.25 (14H, m), 1.10 (3H, t, $J = 7.4$ Hz); ^{13}C NMR (75 MHz, $CDCl_3$) δ 179.34, 129.39, 128.48, 126.34, 102.08, 99.38, 94.14, 83.57, 77.46, 71.31, 68.59, 67.78, 64.11, 57.58, 34.22, 30.89, 30.67, 30.45, 29.57, 29.45, 29.36, 29.30, 24.95, 20.56, 15.09; ^{19}F NMR (282 MHz, $CDCl_3$, ref. C_6F_6) δ 79.8, 74.2, 47.9, 43.5; MS (ES$^-$) *m/z* 803.3 ([M]$^-$); HRMS (ES$^+$) Calcd for $C_{32}H_{44}O_{12}F_8SNa$: 827.2318. Found: 827.2299.

4.6 Methyl-4,6-*O*-benzylidene-3-ethoxymethyl-2-(3-oxa-12,12,13,13,15,15,16,16-octafluorohexadecanoic acid-16-sulfonate)-β-D-mannopyranoside polystyryl amide (7)

ν_{max} (on-bead, cm^{-1}) 2925, 1662, 1493, 1453, 1411, 1146, 1095, 1025; ^{19}F NMR (282 MHz, $CDCl_3$, ref. $CFCl_3$) δ −82.1, −88.1, −113.9, −118.3; Theoretical loading calcd: 0.69 mmol g^{-1}. Found (F elemental analysis) 0.59 mmol g^{-1}.

WASTE DISPOSAL INFORMATION

All hazardous chemicals and toxic materials were disposed of according to Prudent Practices in the Laboratory (Washington, DC: National Academy Press, 1995).

CHEMICAL ABSTRACTS NOMENCLATURE

Methyl-4,6-*O*-benzylidene-β-D-glucopyranoside (**1**); (14155-23-8)
Methyl-4,6-*O*-benzylidene-3-*O*-ethoxymethyl-β-D-glucopyranoside (**2**); (839730-33-5)
Methyl-4,6-*O*-benzylidene-3-*O*-ethoxymethyl-β-D-mannopyranoside (**3**); (839730-36-8)
Methyl-4,6-*O*-benzylidine-3-ethoxymethyl-2-(5-iodooctafluoro-3-oxapentane sulfonate)-β-D-mannopyranoside (**4**); (839730-37-9)
Methyl-4,6-*O*-benzylidine-3-ethoxymethyl-2-(3-oxa-12,12,13,13,15,15,16,16-octa fluoro-10-iodo-hexadecanoic acid-16-sulfonate)-β-D-mannopyranoside (**5**); (839730-38-0)
Methyl-4,6-*O*-benzylidine-3-ethoxymethyl-2-(3-oxa-12,12,13,13,15,15,16,16-octafluoro-hexadecanoic acid-16-sulfonate)-β-D-mannopyranoside (**6**); (839730-39-1)
Sodium hydride; (7646-69-7)
Chloromethylethyl ether; (3188-13-4)
Acetic anhydride; 1(08-24-7)
Sodium borohydride; (16940-66-2)
NaHMDS; sodium bis(trimethylsilyl)amide; (1070-89-9)
5-Iodooctafluoro-3-oxapentanesulfonyl fluoride; (66137-74-4)
Undec-10-enoic acid; (112-38-9)
Sodium hydrosulfite; (7775-14-6)
Amino-methylated polystyrene; Polystyrene crosslinked with 1% divinylbenzene, aminomethylated; (89551-24-6)
N,N-Diisopropylethyl amine; (7087-68-5)
Diphenylphosphoryl chloride; (2524-64-3)
Kryptofix [2.2.2]; 4,7,13,16,21,24-Hexaoxa-1,10-diazabicyclo[8.8.8]hexacosane; (23978-09-8)

REFERENCES AND NOTES

For detailed supplier information, see Appendix 1.

1. (a) M. M. Alauddin, *Am. J. Nucl. Med. Mol. Imaging.*, 2012, 2, 55; (b) H. R. Herschman, *Science*, 2003, 302, 605; (c) J. Czernin, M. E. Phelps, *Annu. Rev. Med.*, 2002, 53, 89; (d) R. Nutt, *Mol. Imaging. Biol.*, 2002, 4, 11; (e) R. G. Blasberg, J. G. Tjuvajev, *J. Clin. Invest.*, 2003, 111, 1620.
2. B. Beuthien-Baumann, K. Hamacher, F. Oberdorfer, J. Steinbach, *Carbohydr. Res.*, 2000, 327, 107.
3. T. J. Tewson, *Nucl. Med. Biol.*, 1989, 16, 533.
4. (a) L. J. Brown, N. Ma, D. R. Bouvet, S. Champion, A. M. Gibson, Y. Hu, A. Jackson, I. Khan, N. Millot, A. C. Topley, H. Wadsworth, D. Wynn, R. C. D. Brown, *Org. Biomol. Chem.*, 2009, 7,

564; (b) L. J. Brown, D. R. Bouvet, S. Champion, A. M. Gibson, Y. Hu, A. Jackson, I. Khan, N. Ma, N. Millot, H. Wadsworth, R. C. D. Brown, *Angew. Chem. Int. Ed.*, 2007, 46, 941.

5. D. R. Bundle, *J. Chem. Soc. Perkin. Trans 1.*, 1979, 2751.
6. Methyl-4,6-*O*-benzylidene-β-D-glucopyranoside was synthesized from the commercially available methyl β-D-glucopyranoside (Sigma-Aldrich, >99%)[5].
7. THF was purchased from Fisher Scientific (reagent grade, stabilized with 0.025% butylated hydroxytoluene), and distilled from sodium and benzophenone prior to use.
8. Sodium hydride (60% dispersed in mineral oil) was purchased from Sigma-Aldrich and used as received. Sodium hydride reacts violently with moisture to liberate hydrogen gas with the risk of explosion or fire; therefore, all apparatus and solvents must be rigorously dried prior to use and reactions performed under nitrogen or argon flow in a well-ventilated fume cupboard.
9. Chloromethylethyl ether (95%) was purchased from Sigma-Aldrich and used as received.
10. Methanol (99.5%) and dichloromethane (reagent grade, 99%) were purchased from Fisher Scientific and used without any purification.
11. Two further byproducts were obtained: methyl-4,6-*O*-benzylidene-2-*O*-ethoxymethyl-β-D-glucopyranoside (white solid, 260 mg, 0.76 mmol, 25%), and methyl-4,6-*O*-benzylidene-2,3-*O*-di(ethoxymethyl)-β-D-glucopyranoside (white solid, 82 mg, 0.21 mmol, 7%), data for these compounds are given below:

    Methyl-4,6-*O*-benzylidene-2-*O*-ethoxymethyl-β-D-glucopyranoside. Mp 103–105°C; $[\alpha]_D$ −20.8 (c=0.606, $CHCl_3$); ν_{max} (neat, cm^{-1}) 3437, 2975, 2879, 1453, 1390, 1089, 1046, 1025; ^{1}H NMR (300 MHz, $CDCl_3$) δ 7.51–7.30 (5H, m), 5.56 (1H, s), 4.89 (1H, d, J=6.6 Hz), 4.79 (1H, d, J=7.4 Hz), 4.38 (1H, d, J=7.4 Hz), 4.36 (1H, dd, J=5.2, 10.3 Hz), 4.02 (1H, s), 3.85–3.75 (3H, m), 3.62–3.55 (2H, m), 3.56 (3H, s), 3.46 (1H, ddd, J=4.4, 9.5, 10.3 Hz), 3.34 (1H, t, J=8.1 Hz), 1.21 (3H, t, J=6.6 Hz); ^{13}C NMR (400 MHz, $CDCl_3$) δ 137.08, 129.14, 128.26, 126.30, 103.47, 101.88, 96.67, 83.34, 80.59, 72.71, 68.70, 66.08, 64.33, 57.41, 15.01; MS (ES$^+$) *m/z* 363.3 ([M+Na]$^+$).

    Methyl-4,6-*O*-benzylidene-2,3-*O*-di(ethoxymethyl)-β-D-glucopyranoside. Mp 73–75°C; $[\alpha]_D$ −62.2 (c=0.475, $CHCl_3$); ν_{max} (neat, cm^{-1}) 2975, 2880, 1455, 1390, 1095, 1046, 1028; ^{1}H NMR (300 MHz, $CDCl_3$) δ 7.50–7.30 (5H, m), 5.52 (1H, s), 4.90 (1H, d, J=6.6 Hz), 4.89 (2H, s), 4.83 (1H, d, J=6.6 Hz), 4.36 (1H, d, J=8.1 Hz), 4.35 (1H, dd, J=4.4, 10.3 Hz), 3.87 (1H, t, J=9.2 Hz), 3.77 (1H, t, J=10.3 Hz), 3.71–3.52 (6H, m), 3.54 (3H, s), 3.43 (1H, ddd, J=5.1, 9.6, 10.3 Hz), 1.22 (3H, t, J=7.0 Hz), 1.05 (3H, t, J=6.6 Hz); ^{13}C NMR (300 MHz, $CDCl_3$) δ 137.35, 129.18, 128.35, 126.29, 104.73, 101.65, 95.97, 80.89, 78.13, 76.89, 68.95, 66.23, 63.98, 63.75, 57.39, 15.13, 14.91; MS (ES$^+$) *m/z* 421.4 ([M+Na]$^+$).
12. Dimethylsulfoxide (anhydrous, ≥99.5%) and acetic anhydride (Reagentplus, ≥99.0%) were purchased from Sigma-Aldrich and used as received.
13. Sodium borohydride (granular, ≥98%) was purchased from Sigma-Aldrich and used as received. This reagent reacts violently with moisture to liberate hydrogen gas with the risk of fire; therefore, all apparatus and solvents must be rigorously dried prior to use.
14. In addition, some mixed fractions were obtained containing some of compound **3** (2.74 g) and pure **2** (1.44 g, 4.23 mmol, 10%).
15. NaHMDS (sodium bis(trimethylsilyl)amide) 1.0 M solution in THF was purchased from Sigma-Aldrich and used without further purification.
16. Iodooctafluoro-3-oxapentanesulfonyl fluoride was purchased from Apollo Scientific Ltd, and used as received.
17. Undec-10-enoic acid (≥95%), $Na_2S_2O_4$ (85%) were purchased from Sigma-Aldrich and acetonitrile (HPLC grade) from Fisher Scientific, used without further purification.
18. Zinc metal powder (99.998%, 100 mesh) was purchased from Sigma-Aldrich and used as received.
19. The acidic conditions described require careful control of reaction time and temperature to prevent loss of protecting groups.
20. Aminomethylated polystyrene resin HL (50–100 mesh, 1% DVB cross-linked) was obtained from NovaBioChem (now EMD Millipore) loading 1.5 mmol g^{-1}. In this work, the beads were not washed before use. However, it may be advisable to use a Soxhlet continuous extraction apparatus with a suitable solvent such as CH_2Cl_2 to clean up the beads prior to use. If a washing step is carried

out, beads should be carefully dried at 30–50°C using a vacuum oven. Aminomethylated polystyrene resins (1% DVB) with mesh sizes 110–200 and 200–400 gave similar results.

21. *N,N*-Disopropylethylamine (anhydrous, 99%) and diphenylphosphoryl chloride (99%) were purchased from Sigma-Aldrich. All were used as received.
22. Polystyrene resins may suffer mechanical damage during stirring with a magnetic stirrer bar, and should only be stirred at slow speeds. The use of large, heavy stirrer bars should be avoided. In most situations, agitation by other methods such as gentle shaking is preferable.
23. Weights of isolated resins are only provided as a guide here, and are not used to estimate reaction yields due to mechanical losses encountered during manipulations of the resin. Loadings were obtained by F elemental analysis.
24. Three lines were attached to allow evaporation, nitrogen flow, and addition of reagents (Fig. 2). A custom-made glassy carbon reactor was used in the manual labelling experiments described here. A diagram is provided. Glass Wheaton reaction vials have also been used for labelling experiments with other RLVs. High activity studies (5–6 Ci) were performed on the TRACERlab$_{MX\text{-}FDG}$ system (Coincidence Box). Although this was set up for solution phase chemistry, incorporation yields remained satisfactory (68–77% incorporation).
25. Kryptofix [2.2.2] (4,7,13,16,21,24-hexaoxa-1,10-diazabicyclo[8.8.8]hexa cosane) was purchased from Sigma-Aldrich and used without further purification.
26. Stirring of the RLV was found to be import in order to obtain consistent, and high levels of incorporation.
27. C_{18}Sep-pak cartridges were purchased from Waters, 55–105 μm particle size, and were used to adsorb analytes of from the aqueous solutions.
28. The silica sep-pak cartridges were purchased from Waters and preconditioned with Et_2O prior to use.
29. Reverse-phase HPLC was carried out using a Gilson system (gamma and UV detection) equiped with a μ-Bondapak column (3.9 × 300 mm), eluting with H_2O/CH_3CN (gradient 19:1 → 1:19) at 2 ml/min.
30. Ion-exchange HPLC was carried out using a Dionex system (gamma and electrochemical detectorε) equipped with a Carbo Pac PA10 column (4 × 250 mm) with a Carbo Pac PA10 guard (4 × 50 mm), eluting with NaOH (50–52%)/H_2O (2 ml : 600 ml, isocratic) at 1 ml min^{-1}.
31. (a) S. Levy, E. Livni, D. Elmaleh, W. Curatolo, *J. Chem. Soc. Chem. Commun.*, 1982, 17, 972; (b) B. Doboszewski, G. W. Hay, W. A. Szarek, *Can. J. Chem.*, 1986, 65, 412.
32. A. C. Topley, V. Isoni, T. A. Logothetis, D. Wynn, H. Wadsworth, A. M. R. Gibson, I. Khan, N. J. Wells, C. Perrio, R. C. D. Brown, *Chem. Eur. J.*, 2013, 19, 1720.

CHAPTER 18

PRODUCTION AND REACTION OF [^{11}C]CARBON DISULFIDE FOR THE SYNTHESIS OF [^{11}C]DITHIOCARBAMATES

PHILIP W. MILLER

Department of Chemistry, Imperial College London, South Kensington, London, UK

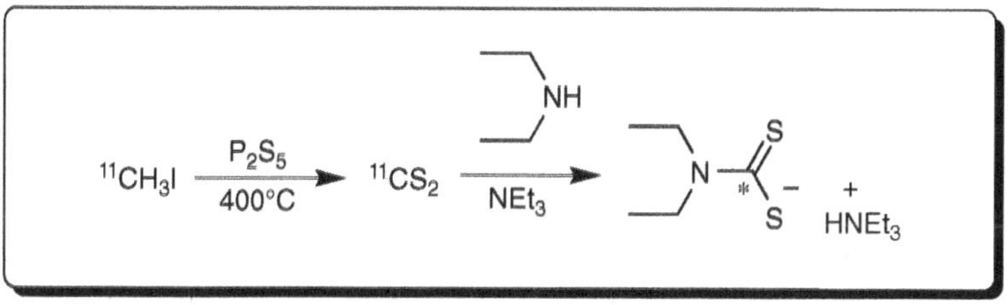

1 INTRODUCTION

The short 20.4 min half-life of the commonly used C11 positron-emitting radioisotope precludes long and complex multistep syntheses [1, 2]. The labeling chemistry of C11 is therefore dominated by simple one-step, one-pot reaction procedures with small and reactive C11 precursor molecules. The most widely used C11 precursor molecules include $^{11}CH_3I$, $^{11}CH_3OTf$, $^{11}CO_2$, $^{11}CH_4$, ^{11}CO, $^{11}COCl_2$ and $^{11}CH_2O$ since they can be produced either directly within the cyclotron target or within minutes post end of bombardment (EOB) via fast reaction processes. $^{11}CH_3I$ is the most commonly used precursor for rapid one-step labeling of amines, alcohols, and thiols, and for Pd-mediated C–C bond-forming reactions. It is now commonly used for the clinical production of a range of C11 PET tracers. Although $^{11}CH_3I$ is now the reliable workhorse of C11 radiolabeling, its chemistry is not suitable for labeling many molecules in all the desired atomic positions. It is restricted to labeling on the periphery of the tracer molecule in what may be considered more metabolically susceptible positions. Hence, there is still a need a for other precursor molecules capable of labeling in the core structural positions of tracer molecules and for labeling in other more exotic atomic positions. ^{11}CO [3–5] and $^{11}CO_2$ [6–9]

Radiochemical Syntheses, Volume 2: Further Radiopharmaceuticals for Positron Emission Tomography and New Strategies for Their Production, First Edition. Edited by Peter J. H. Scott.

are currently attracting considerable interest for C11 in carbonyl positions within the skeletal core structure of tracer molecules. Although academically very interesting, this chemistry is currently at a development stage and has still to be widely adopted for the clinical production of common PET tracers. C11 carbon disulfide [10], similar to ^{11}CO or $^{11}CO_2$, is at an early stage of development and could be used for the C11 labeling in a variety of atomic positions and for generating different C11 functional groups.

Carbon disulfide is a commonly used laboratory solvent and reagent that is produced on an industrial scale. It is commonly used for the production of xanthates, cellulose, and other sulfur-containing chemicals. It has also found a number of applications in the synthesis of fine chemicals, mainly for the synthesis of thioureas, dithiocarbamates, isothiocyanates, and thioesters [11–13]. In contrast to the isoelectronic carbon dioxide, which is a gas at room temperature, carbon disulfide is a liquid (bp=46°C) and is considered to be more reactive owing to the weaker C=S bonds. CS_2 behaves as an electrophile that is susceptible to nucleophilic attack from amines, alcoholates, and carbanions. It is this reactivity, ease of handling (as liquid rather than a gas), and diverse range of chemistry that makes the $^{11}CS_2$ equivalent a potentially useful precursor for C11 PET radiolabeling. Herein, the rapid and high yielding route to $^{11}CS_2$ from $^{11}CH_3I$ is described together with the synthesis and radio-HPLC characterization of simple dithiocarbamate molecules.

2 PROCEDURES

2.1 Synthesis Procedures

CAUTION: All radiochemical syntheses must be carried out using appropriate equipment in a facility authorized for the use of radioactive materials. Personal protective equipment must be worn, and all local radiation safety laws followed.

2.2 Production of $^{11}CH_3I$

[^{11}C]carbon dioxide was produced using a Siemens Eclipse HP cyclotron by 11 MeV proton bombardment of a target containing nitrogen and 1% oxygen. In a typical irradiation, a 5 μA proton beam for 5 min was used. At EOB $^{11}CO_2$ was transferred with a helium sweep from the target at a flow rate of 200 ml/min into an automated Siemens $^{11}CH_3I$ GPC Explora system. $^{11}CO_2$ was converted to $^{11}CH_4$ via a high-temperature (400°C) gas-phase reaction over a nickel catalyst. $^{11}CH_4$ was cryogenically trapped at −196°C using a liquid nitrogen trap. On release, $^{11}CH_4$ was converted to $^{11}CH_3I$ using the high-temperature (700°C) gas-phase iodination reaction.

2.3 Production of $^{11}CS_2$

$^{11}CH_3I$ was transferred from the Siemens system to a hot cell under a flow of helium (10 sccm) directly into a reactor column packed with a mixture of P_2S_5 and sand. The glass Omnifit column (10 mm diameter, 150 mm length, part no. 56003) contained a

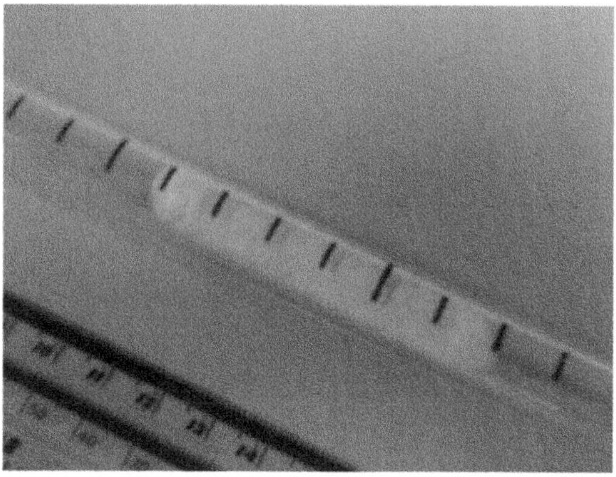

FIGURE 1 Omnifit column (10 mm diameter, 150 mm length) packed with P_2S_5 and sand mixture plugged at either end with glass wool.

1:1.5 mixture of P_2S_5 and chromatography sand (2.4 and 3.6 g), plugged at either end with glass wool (Fig. 1) and placed in a Carbolite horizontal single zone tube furnace oven (model MTF10/25/130). The oven was heated to 400°C for 10 min prior to the reaction. The outlet of the P_2S_5 column was connected to a receptor vial containing acetonitrile, which trapped the $^{11}CS_2$ at room temperature. A bag was attached to the vial outlet in order to contain any excess radioactive gases that were not trapped in solution. The conversion process can be monitored by placing pin diodes at column oven and the trapping vial. An initial maximum of radioactivity is observed at the column oven that quickly depletes as the $^{11}CH_3I$ is converted to $^{11}CS_2$ and flows to the trapping vial. A typical production of $^{11}CS_2$ was completed less than 8 min from the end of $^{11}CH_3I$ production with 150–200 MBq of activity obtained under the cyclotron bombardment conditions stated earlier (5 μA for 5 min). The total processing time from EOB to complete $^{11}CS_2$ trapping was approximately 20 min. Radio-HPLC analysis of the acetonitrile trapping solution confirmed the formation of $^{11}CS_2$ by coinjection of a cold CS_2 reference sample. Typically, radiochemical yields were quantitative (Fig. 2) however, small amounts of $^{11}CH_3I$ were occasionally present (<5%). The same P_2S_5/sand column could be used for multiple productions on the same day but give poor conversions if used again the following day. Conversions of $^{11}CH_3I$ to $^{11}CS_2$ were also depleted if the oven was not given sufficient time to warm up or if gas flow rates were greater than 10 sccm (Fig. 3).

2.4 Radio-HPLC Conditions

A 25 μl aliquot was removed from the reaction vials and injected for UV-vis and radio-HPLC analysis (solvent 50:50 acetonitrile/deionized water; flow, 1.5 ml min^{-1}; column Phenomenex SphereClone ODS, 150 × 4.6 mm). The radio-HPLC system consisted of an Agilent 1100 series with quaternary pump and diode array UV detector connected in series to a NaI radiodetector of in-house design.

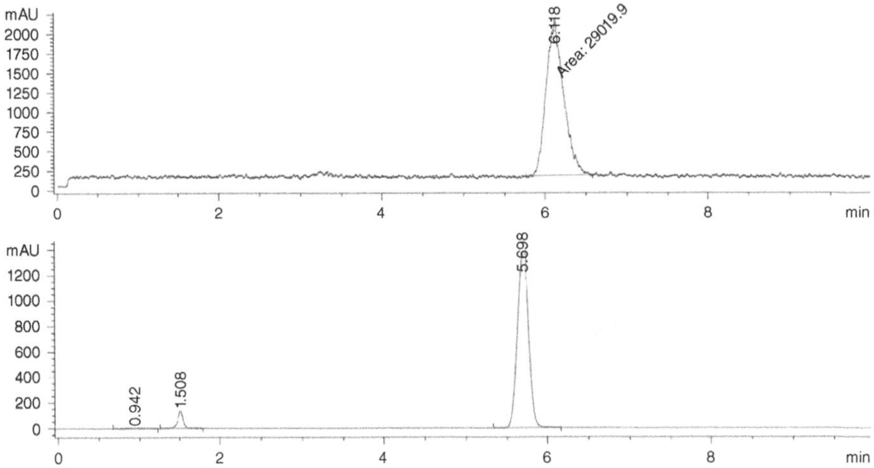

FIGURE 2 Top: radio-HPLC trace of $^{11}CS_2$. Bottom: coinjection UV trace of CS_2 (210 nm). Note: there is an approximate time delay of 0.4 min between the radio and UV peaks.

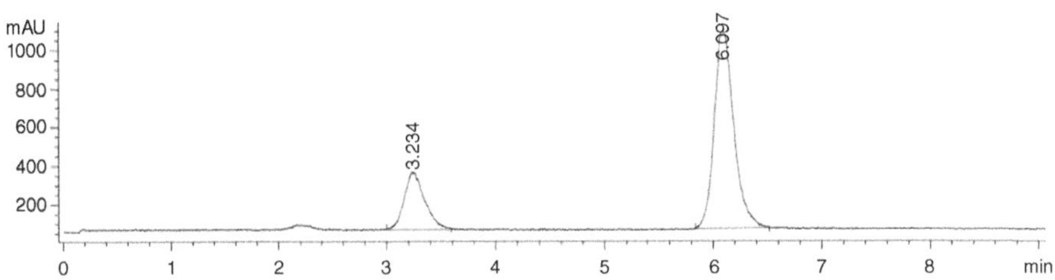

FIGURE 3 Radio-HPLC trace showing partial conversion of $^{11}CH_3I$ (3.2 min) to $^{11}CS_2$ (6.1 min).

2.5 Dithiocarbamate Formation

In a typical reaction, an aliquot of $^{11}CS_2$ solution (200 µl) was added to a solution of diethylamine (10 µl) and triethylamine (10 µl) in acetonitrile (300 µl), and the reaction allowed to stand at room temperature for 5 min. The reaction with aniline was allowed to stand at room temperature for 15 min to ensure complete conversion. The identities of the labeled [^{11}C]dithiocarbamate salts were confirmed using radio-HPLC analysis by coinjection of authentic reference samples or by further reaction to methylated target molecules.

2.6 Alkylation Reactions

The [^{11}C]dithiocarbamate salts could be reacted with methyl iodide to form the corresponding methylated dithiocarbamates. In a typical reaction 5 µl of methyl iodide was added to the [^{11}C]dithiocarbamate solution allowed to stand at room temperature for 5 min before radio-HPLC analysis. There was an instantaneous color change from light greenish blue to clear colorless on addition of the methyl iodide.

3 DISCUSSION

The routine use of automated synthesis units for $^{11}CH_3I$ production has resulted in its widespread availability and ease of access. $^{11}CH_3I$ is therefore not only an important C11 labeling reagent but also a viable precursor for the synthesis of $^{11}CS_2$ and other labeling reagents [14]. The conversion of $^{11}CH_3I$ to $^{11}CS_2$ by the method described earlier is both rapid and efficient and is achieved using easily accessible, low-cost commercially available equipment and chemicals. Typical conversions are greater than RCY 95% under optimized conditions of temperature (400°C) and gas flow rates (10 sccm). Trapping in acetonitrile provides a convenient method of handling $^{11}CS_2$; aliquots can be easily removed and added to nucleophilic reagents. As a reagent, $^{11}CS_2$ is a reactive electrophile that will rapidly react (<5 min) with primary and secondary alkyl amines to form the corresponding dithiocarbamate salts (Figs. 4 and 5). Reaction of $^{11}CS_2$ with the less nucleophilic amine, aniline (Figs. 6 and 7), requires a longer reaction time of 15 min for good conversion to the dithiocarbamate salt. The use of a strong base to deprotonate the

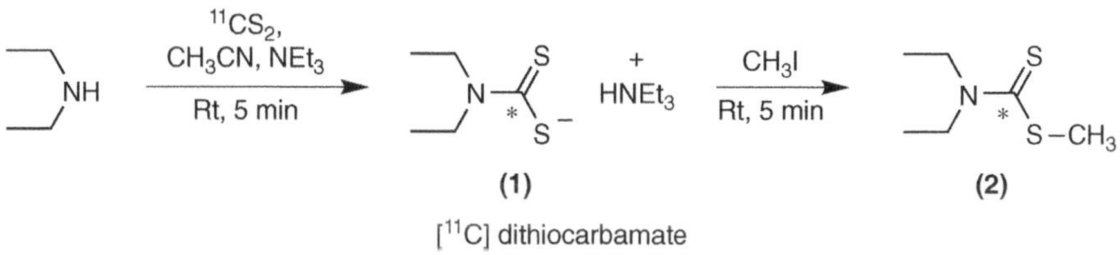

FIGURE 4 Reaction of $^{11}CS_2$ with diethylamine to form the [^{11}C]dithiocarbamate salt (**1**) followed by methylation to give (**2**).

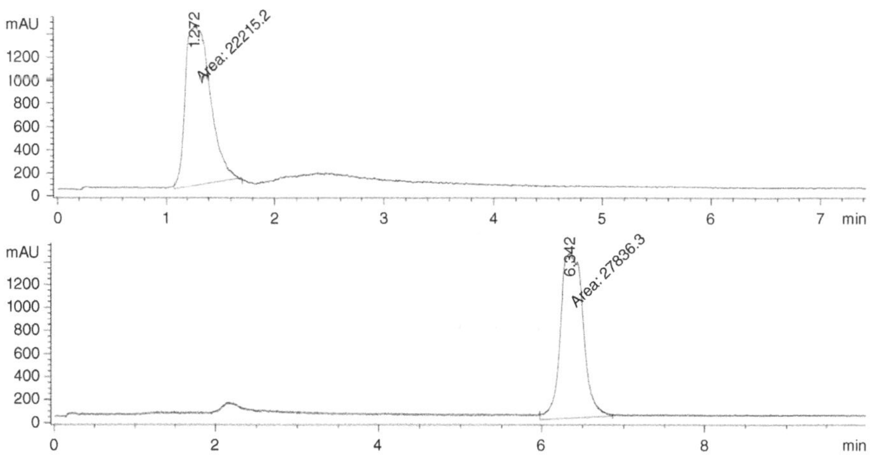

FIGURE 5 Top: radio-HPLC trace of [^{11}C]triethylammonium diethylcarbamodithioate (**1**). Note: there is no longer an $^{11}CS_2$ peak at 6.1 min. Bottom: radio-HPLC trace of [^{11}C]methyl *N,N*-diethylcarbamodithioate (**2**).

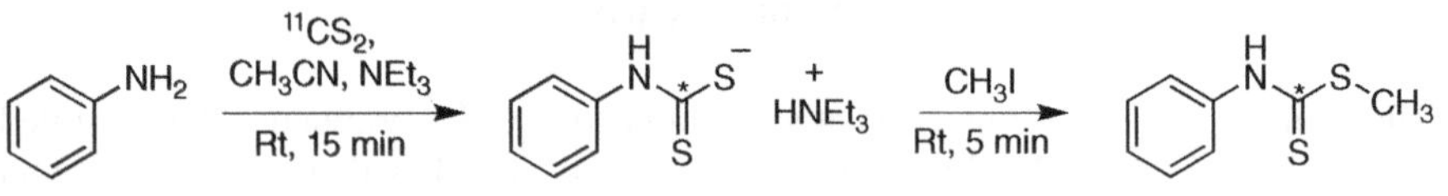

(3) (4)

[^{11}C] dithiocarbamate

FIGURE 6 Reaction of $^{11}CS_2$ with aniline to form the [^{11}C]dithiocarbamate salt (**3**) followed by methylation to give (**4**).

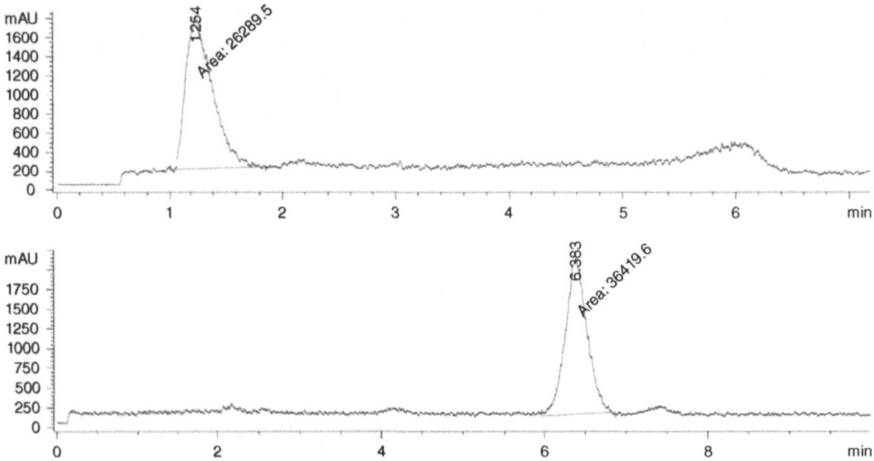

FIGURE 7 Top: radio-HPLC trace of [^{11}C]triethylammonium phenylcarbamodithioate (**3**). Note: there is still some residual $^{11}CS_2$ at 6.1 min. Bottom: radio-HPLC trace of [^{11}C]methyl *N*-phenylcarbamodithioate (**4**).

aniline may improve its reaction with $^{11}CS_2$. The labeled dithiocarbamate salts can be rapidly and cleanly converted to the neutral alkylated species, in one pot, simply by adding methyl iodide; other alkyl halides may be expected to react equally well. It is anticipated that $^{11}CS_2$ may become a useful precursor for the synthesis of sulfur-containing C11 tracer compounds and for labeling in more challenging atomic positions within the core skeletal structure such as heterocyclic rings.

ACKNOWLEDGMENTS

PWM gratefully acknowledges Imanova Ltd. for access to radioisotope production facilities and for use of radioanalytical equipment. The Royal Society is acknowledged for supporting the purchase of equipment and consumables necessary for $^{11}CS_2$ production (RG110449).

WASTE DISPOSAL INFORMATION

All hazardous chemicals and toxic materials were disposed of according to Prudent Practices in the Laboratory (Washington, D.C.: National Academy Press, 1995).

CHEMICAL ABSTRACTS NOMENCLATURE (REGISTRY NUMBER)

Aniline (62-53-3)
Acetonitrile (75-05-8)
Carbon disulfide (75-15-0)
Diethylamine (109-89-7)
Methyl iodide (74-88-4)
Methyl *N,N*-diethylcarbamodithioate (686-07-7)
Methyl *N*-phenylcarbamodithioate (701-73-5)
Phosphorus pentasulfide (1314-80-3)
Triethylamine (121-44-8)
Triethylammonium diethylcarbamodithioate (2391-78-8)
Triethylammonium phenylcarbamodithioate (43009-16-1)

REFERENCES AND NOTES

For detailed supplier information, see Appendix 1.

1. P. W. Miller, N. J. Long, R. Vilar, A. D. Gee, *Angew. Chem. Int. Ed.*, 2008, 47, 8998.
2. P. J. H. Scott, *Angew. Chem. Int. Ed.*, 2009, 48, 6001.
3. B. Langstom, O. Itsenko, O. Rahman, *J. Label. Compd. Radiopharm.*, 2007, 50, 794.
4. S. Kealey, P. W. Miller, N. J. Long, C. Plisson, L. Martarello, A. D. Gee, *Chem. Commun.*, 2009, 25 3696.
5. J. Eriksson, J. van den Hoek, A. D. Windhorst, *J. Label. Compd. Radiopharm.*, 2012, 55, 223.
6. A. A. Wilson, A. Garcia, S. Houle, N. Vasdev, *Org. Biomol. Chem.*, 2010, 8, 428.
7. A. A. Wilson, A. Garcia, S. Houle, O. Sadovski, N. Vasdev, *Chem.-Eur. J.*, 2011, 17, 259.
8. J. M. Hooker, A. T. Reibel, S. M. Hill, M. J. Schueller, J. S. Fowler, *Angew. Chem. Int. Ed.*, 2009, 48, 3482.
9. P. J. Riss, S. Y. Lu, S. Telu, F. I. Aigbirhio, V. W. Pike, *Angew. Chem. Int. Ed.*, 2012, 51, 2698.
10. P. W. Miller, D. Bender, *Chem.-Eur. J.*, 2012, 18, 433.
11. A. D. Dunn, W. D. Rudorf, *Carbon Disulfide in Organic Chemistry*, Chichester, West Sussex: Ellis Horwood Ltd, 1989.
12. N. Azizi, F. Aryanasab, M. R. Saidi, *Org. Lett.*, 2006, 8, 5275.
13. W. D. Rudorf, *J. Sulfur. Chem.*, 2007, 28, 295.
14. J. M. Hooker, M. Schoenberger, H. Schieferstein, J. S. Fowler, *Angew. Chem. Int. Ed.*, 2008, 47, 5989.

CHAPTER 19

ONE-POT, DIRECT INCORPORATION OF [$^{11}CO_2$] INTO CARBAMATES

Christian K. Moseley and Jacob M. Hooker

Athinoula A. Martinos Center for Biomedical Imaging, Department of Radiology, Massachusetts General Hospital, Harvard Medical School, Charlestown, Massachusetts, USA

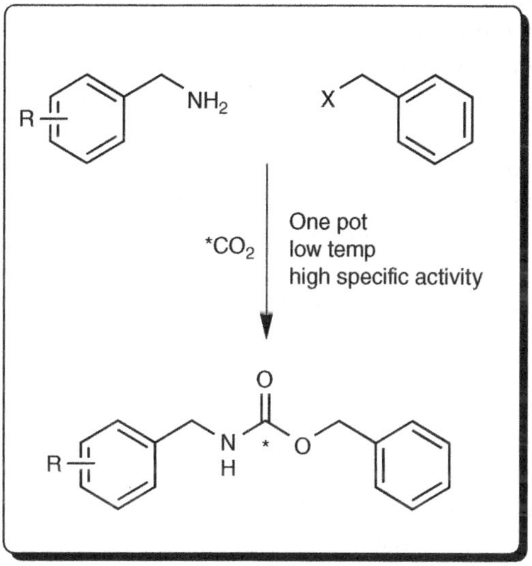

1 INTRODUCTION

The overall and expanding usefulness of positron emission tomography (PET) as an imaging modality is dependent upon the number and variety of radiotracers that can be synthesized and subsequently used to investigate human physiology and disease. Given the prevalence of carbon in bioactive compounds, carbon-11 is a particularly useful positron-emitting isotope for use in radiochemistry and PET imaging. However, the short half-life of carbon-11 ($t_{1/2}$ = 20.38 min) often hinders

Radiochemical Syntheses, Volume 2: Further Radiopharmaceuticals for Positron Emission Tomography and New Strategies for Their Production, First Edition. Edited by Peter J. H. Scott.

the use of conventional synthetic chemistry techniques in radiosynthesis. The overall synthesis and purification time after cyclotron bombardment needs to be short (<40 min) because it has a dramatic impact on the total amount of radioactivity that can be used for PET imaging of the final product. As decay occurs, both the specific activity and the radiochemical yield (end-of-synthesis yield, not decay corrected) are reduced. Chemical strategies that work in traditional (nonradioactive) transformations must be adapted to operate under the time and mass (high specific activity) constraints of carbon-11 in order to be used for the synthesis of PET radiotracers.

Carbon-11 is produced from a cyclotron in the form of $^{11}CO_2$ or $^{11}CH_4$ in sufficient quantity and with high specific activity for PET imaging. Therefore (with only a few very rare exceptions), all reagents used in carbon-11 radiochemistry are derived from one of these two bombardment products. Many syntheses require the conversion of carbon to a more reactive reagent (termed a carbon-11 synthon), such as methyl iodide, $^{11}CH_3I$, or methyl triflate, $^{11}CH_3OTf$. The canonical strategy for carbon-11 incorporation has in fact been through the use of these reagents in methylation reactions. Other synthon examples including [^{11}C]phosgene or [^{11}C] carbon monoxide can be used to access higher oxidation state moieties such as carboxylic acids, amides, ureas, carbamates, and their associated derivatives [1, 2]. Although these synthons are quite useful from a purely synthetic perspective, the synthon synthesis process itself is time consuming and often inefficient from engineering constraints (e.g., reagent trapping efficiency). This often reduces the radiochemical yield by approximately half and can have relatively low overall chemical conversion. Coupled together, a loss of more than half of the initial radioactivity during synthon formation is very common [3]. Moreover, synthon preparation often requires additional automated equipment, which hinders the widespread adaptation of new chemical reactions in clinical (nonchemistry research) settings. Although new carbon-11 synthons are needed and have high value themselves, development of new techniques for the direct incorporation of $^{11}CO_2$ has certain advantages to synthon use: (i) reduced overall synthesis time, (ii) no losses through chemical conversion/manipulation, and (iii) no need for additional equipment.

Compounds to *fix* carbon-11 CO_2 have existed for decades and include organometallic reagents such as Grignard reagents and organolithiums, as well as silanamines [4–8]. However, organometallic and organosilicon reagents suffer from problems that limit the complexity of synthesizable functional groups, reproducibility, and automation [9]. Simpler methods have since been devised to fix $^{11}CO_2$ in solution by merely bubbling $^{11}CO_2$ through a reaction mixture containing a *fixation base*. This efficient and direct use of $^{11}CO_2$ has broadened the spectrum of carbon-11-labeled molecules available to radiochemists by increasing access to functional groups and decreasing technical hurdles (Fig. 1).

One functional group in particular, the carbamate, is significant for its use in the modular synthesis of organic compounds and its chemically and metabolically stable properties. The development of a one-pot method to directly incorporate $^{11}CO_2$ into a carbamate moiety has simplified the synthetic process for many biological molecules. To test the validity of $^{11}CO_2$ fixation in a compound with known

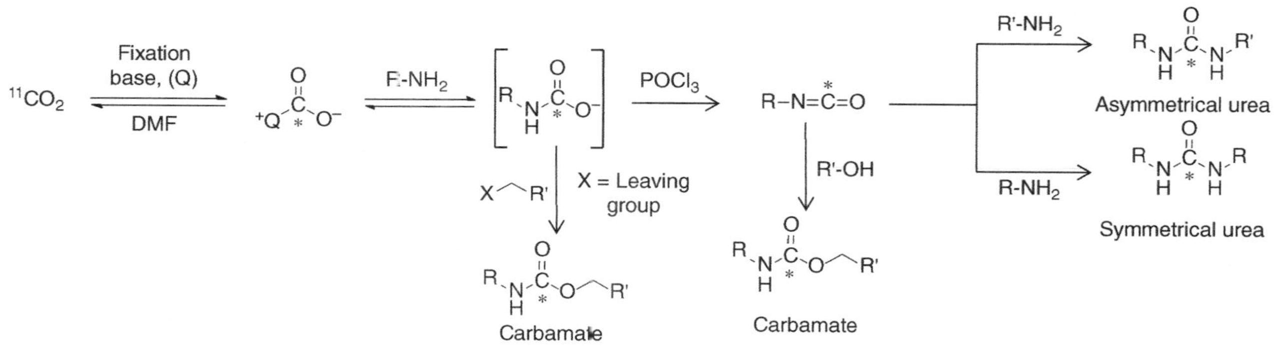

FIGURE 1 [^{11}C]CO_2 Fixation applications. ^{11}C-*carbonyl* carbamates can be synthesized by direct [^{11}C]CO_2 fixation methods or through the subsequent use of intermediate carbamate salts. The process is versatile, also allowing access to asymmetric and symmetric ^{11}C-*carbonyl* ureas.

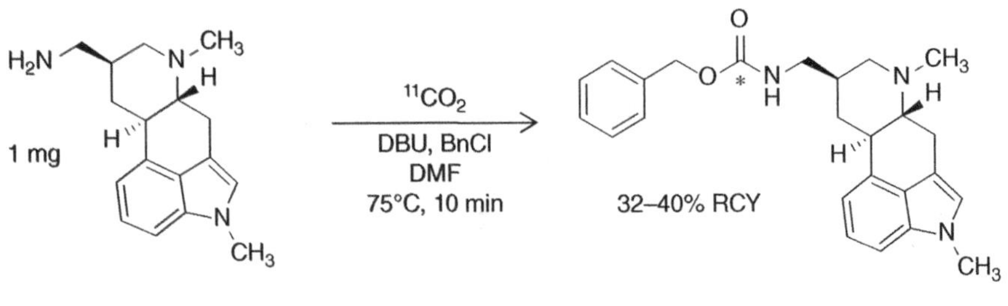

FIGURE 2 Synthesis of [^{11}C]metergoline.

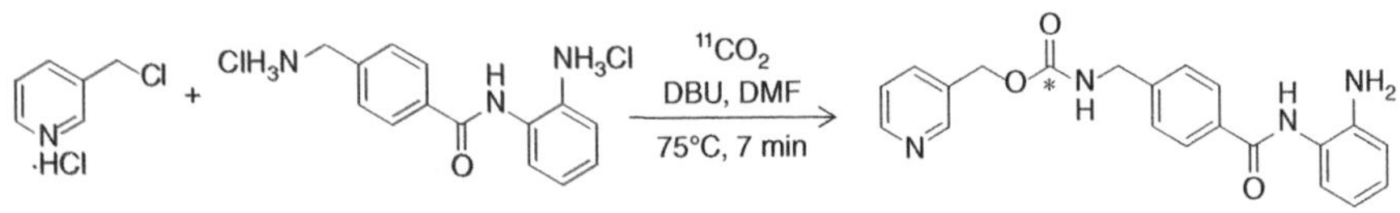

FIGURE 3 Synthesis of [^{11}C]MS-275. It is worth noting that the more basic amine is participating in the reaction. The aniline-labeled product is not observed.

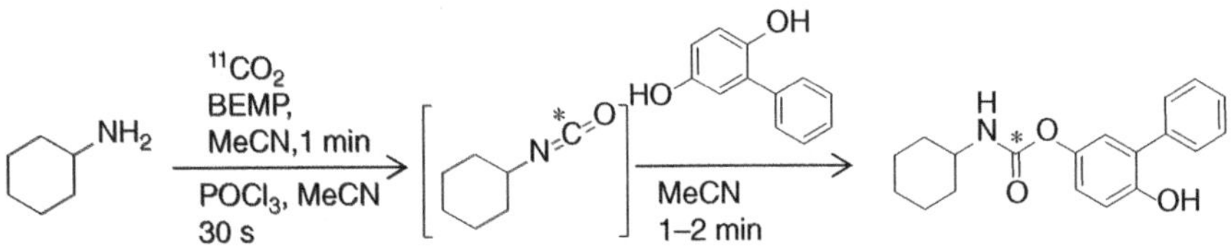

FIGURE 4 Synthesis of [^{11}C]CURB.

pharmacological activity, the method was applied to the carbamate-containing molecules metergoline [10] (Fig. 2) and MS-275 [11] (Fig. 3). $^{11}CO_2$ fixation is a versatile process, and through the rapid transformation of $^{11}CO_2$ into [^{11}C]isocyanate as seen in the synthesis of [^{11}C]CURB (Fig. 4), carbon can also be indirectly introduced into carbamates [12]. The successful synthesis of metergoline as well as several other carbamate-containing compounds highlights the utility of one-pot, direct $^{11}CO_2$ fixation in radiosynthesis.

2 PROCEDURES

CAUTION: All radiochemical syntheses must be carried out using appropriate equipment in a facility authorized for the use of radioactive materials. Personal protective equipment must be worn, and all local radiation safety laws followed.

2.1 Production of $^{11}CO_2$

[^{11}C]Carbon dioxide was generated by a cyclotron by the nuclear reaction, ^{14}N(p,a)^{11}C, using a nitrogen/oxygen (1000 ppm) target. A CO_2 trap is often used to purify the target gas as both oxygen and by-product nitric oxides may be present and may interfere with reactions. Target gas purification is typically done in one of two ways. Cryogenic purification condenses the target gas into a container cooled by liquid nitrogen. This concentrates the $^{11}CO_2$ sample while noncondensable gases are allowed to escape and impurities such as NO_x are removed by chemical traps [13]. Another method utilizes solid-phase immobilization to remove contaminants. A solid material such as molecular sieves sequesters the CO_2 as impurities are separated; the CO_2 is then thermally liberated from the trap [14]. For our screening experiments, the radioactivity was transferred in a large volume of N_2 (~1 l) through a custom-built, automated molecular sieve trap (Alltech 4 Å, 50–80 mesh crushed mol. sieves, 330 mg packed in a 1/4 in. i.d. SS tube). At the end of bombardment (EOB), the target gas was delivered to the radiochemistry synthesis lab using approximately 1 l of N_2 (at ~300 ml/min when the $^{11}CO_2$ trap and release system was used). For direct trapping of purified $^{11}CO_2$ from the target into reaction solutions (300 µl or less), the gas flow was reduced to a predetermined rate using a needle valve.

2.2 General Synthesis of $^{11}CO_2$ Carbamates

Separate solutions (300 mM each) of the amine, 1,8-diazabicyclo[5.4.0]undec-7-ene (DBU), and chloride were prepared in dimethylformamide (DMF) that had been previously sparged with helium (He) gas to remove any dissolved CO_2 from the atmosphere. Aliquots (100 µl) of each solution were combined in a reaction vessel, which was then sealed with a septum and screw cap. The resulting 300 µl solution (0.1 M in each reagent) was sparged with He gas for at least 2–3 min prior to EOB. For high specific activity syntheses, sparging the combined solution with high-purity gas for 10–15 min was typically sufficient to remove carbon dioxide from atmospheric sources. [^{11}C]Carbon dioxide was released in a stream of He (~50 ml/min) from an automated trap and release system (see details in the preceding text) and introduced to the reaction solution under constant flow at room temperature. The amount of $^{11}CO_2$ trapped was monitored and recorded, and the capture step continued until the radioactivity curve reached a maximum, approximately 2 min. The outlet gas of the reaction vessel was passed through a trapping vessel of either 1 M sodium hydroxide (NaOH) (aq.) or solid soda lime and the flow through radioactivity was determined. This amount (the radioactivity that passed through the reaction solution but was not trapped) was typically less than 5% of the total. After removing the inlet and outlet needles, the sealed vessel was transferred into a heating bath. Most reactions were carried out for 10 min at 75°C.

2.3 Selection of Reagents and Conditions

We screened six compounds as catalysts for carbamate formation: DBU, 1,5-diazabicyclo[4.3.0]non-5-ene (DBN), 4-(dimethylamino)pyridine (DMAP), 1,4-diazabicyclo[2.2.2]octane (DABCO), triethylamine, and morpholine. We investigated the reaction between benzyl bromide and 4-fluorobenzylamine in several solvents and found that in the absence of catalyst, no background reaction occurred. Under all conditions we examined, the amidine DBU was the only catalyst where product formation was observed in less than 10 min. At 20 min, the product was observed for reactions with DBN. Each of the other potential catalysts was ineffective. In cold model reactions with DBU, DMF was superior to dimethyl sulfoxide (DMSO) and acetonitrile (MeCN). In fact, the desired reaction did not occur at all in MeCN. We repeated this analysis with $^{11}CO_2$ and found that DMF was best, but DMSO was nearly as effective, and MeCN provided radiochemical yields that were roughly half of with DMF. The optimum temperature for carbamate formation fell between 50 and 100°C. Very little, if any, product was observed at temperatures lower than 50°C with solutions of DBU in DMF [10].

A phosphazene compound, 2-tert-butylimino-2-diethylamino-1,3-dimethyl-perhydro-1,3,2-diazaphosphorine (BEMP), has also shown utility as an effective CO_2 fixation agent and successfully traps CO_2 in solutions of DMF or MeCN. Reactions were conducted at ambient temperature and produced high radiochemical yield of carbamates and ureas [12, 15].

A variety of amines may be used as nucleophiles for $^{11}CO_2$ incorporation, including aliphatic primary and secondary amines. Less nucleophilic amines, including anilines, tend to result in poor yields. Appropriate precursor amines for $^{11}CO_2$ fixation can be prepared from benzyl carbamate-containing molecules through deprotection of the desired molecule with H_2 and palladium on carbon (Pd/C). Reduction in the concentration of the amine has minimal effect on yield in comparison to reducing the amount of trapping and alkylating agents.

Regarding the leaving group of alkyl electrophiles—of the substrates we analyzed in model reactions, benzyl bromide was as effective as benzyl chloride (BnCl) in terms of carbamate formation. However, we noted substantial differences in the two leaving groups using the trace concentrations of $^{11}CO_2$. We speculate that this is due to competitive alkylation pathways observed with the bromides, including alkylation of amines and DBU. Consistent with this idea, we were never able to form the desired carbamate with alkyl iodides such as methyl iodide [10]. However, carbamate formation with carbon-11 and alkyl iodides can be accomplished with BEMP.

2.4 Synthesis of [^{11}C]Metergoline

Carbon-11 labeling was accomplished using a modified procedure of the screening conditions outlined earlier. Briefly, 1 mg of precursor dissolved in 100 µl of DMF was combined with 100 µl of a solution of DBU (300 mM in DMF) and then treated with a 100 µl aliquot of benzyl chloride (300 mM in DMF) in a cone-bottom 5 ml reaction vessel equipped with a septum and screw cap. The resulting solution was sparged with helium for 15 min. At the EOB, the target gas from the cyclotron was passed through the solution and was captured. (The expansion of the target gas through the reaction solution required ~2.5 min.) The vessel was then heated to 75°C for 7.5 min, at which time 100 µl of trifluoroacetic acid was added (an excess relative to DBU and amine). The volatile activity was removed by sparging the solution with He (~50 ml/min, this required ~0.5 min). The solution was diluted with 1.0 ml of water and purified by high-performance liquid chromatography (HPLC). Radiochemical yield (32 ± 2%, n = 3) was calculated as detailed in Section 3.1 [10].

2.5 Synthesis of [^{11}C]MS-275 (Entinostat)

Carbon-11 labeling was accomplished using a modified procedure of the general conditions outlined earlier. Atmospheric carbon dioxide from separate reagent solutions (DBU in DMF (1.0 M) and 3-picolyl chloride hydrochloride (0.30 M)) was removed by sparging the solutions with a stream of ultrapure helium gas for 30 min. An aliquot (100 µl) of each DMF solution was added into a V-shaped vial with a septum-sealed cap containing 4-(aminomethyl)-*N*-(2-aminophenyl)benzamide·2HCl (2.0 mg). The combined contents were mixed by vortex and then sparged with helium gas for at least 10 min. At the EOB, the target gas from the cyclotron was trapped and purified on molecular sieves. Subsequent heating of the sieves to 250°C released a bolus of $^{11}CO_2$ which required a lower volume of He carrier gas (flowing at ~75 ml/min) and optimized delivery to the vessel. Delivery continued until radioactivity peaked in the vessel. Trapping efficiencies for both steps were typically greater than 95%. After introduction of $^{11}CO_2$, the vessel was sealed and heated to 75°C for 7 min, at which time the solution was removed from heat and sparged. The reaction mixture was acidified with formic acid (0.1 ml) to ensure that the total residual [$^{11}CO_2$]carbon dioxide was removed, and then sparged again. The mixture was diluted with HPLC eluent and purified by HPLC. The radiochemical yield was 25% (decay corrected to EOB), and the specific activity for the imaging studies was 2.7–6.2 Ci/µmol (calculated at EOB). Typical radiosynthesis and purification time was 50 min [11].

2.6 Synthesis of [^{11}C]CURB

An alternative labeling strategy for [^{11}C]carbamates includes the production of an isocyanate intermediate. Despite the simplicity of a one-pot reaction at ambient temperature and pressure, this reaction requires careful attention to stoichiometry. Excess amine will react with the generated [^{11}C]isocyanate, forming symmetric ureas. Excess phosphoryl trichloride ($POCl_3$) will reduce this problem; however, it will also consume the nucleophile. When the stoichiometry is controlled, the carbamate salt can be dehydrated to [^{11}C]isocyanate and transformed into asymmetrical ureas (when reacted with amines) or carbamates (when reacted with alcohols or phenols). In brief $^{11}CO_2$, flowing at 10 ml/min in a stream of N_2 was bubbled into a 1 ml conical vial through a solution containing anhydrous MeCN (100 µl), cyclohexylamine (0.1 mg), and BEMP (5 µl, 17.3 mmol) as a fixation agent. 100 µl of $POCl_3$ (0.2% v/v) in MeCN was added, followed by the quenching agent 2-phenyl-1,4-dihydroquinone (2 mg) in 100 µl MeCN, to form the desired carbamate. Heating the reaction was unnecessary to form product. The mixture was diluted with HPLC eluent and purified by HPLC. Radiochemical conversion yields were determined by radio-HPLC analysis. Purification conditions were as follows: Phenomenex C18 Luna(2) 10 µ, 250 × 10 mm eluted with 70% methanol (MeOH)/30% H_2O containing 1% formic acid at 7 ml/min. The product was eluted at 9–10 min. [^{11}C]CURB was obtained 27 min after the end of radionuclide production while specific activities were (2500 ± 500) mCi/µmol [12].

3 QUALITY CONTROL PROCEDURES

CAUTION: All radiopharmaceuticals produced for clinical use must have local regulatory approval before use in humans. Trained personnel must carry out quality control (QC) procedures, and each dose must meet all established QC criteria before release to the clinic. Many methods exist to demonstrate the quality of a radiopharmaceutical for clinical use. Consult local regulations for appropriate QC procedures.

3.1 Determination of Radiochemical Yield

In a radiochemical synthesis, only a fraction of the total mass of radioactive material will undergo a reaction to form products, and of those products, only a portion will be the desired final product. For example, some amount of the total $^{11}CO_2$ bubbled into the DBU and precursor solution will be lost to the atmosphere or will not undergo a reaction. The radiochemical conversion in this instance is the total amount of radioactive $^{11}CO_2$ consumed by the reaction and incorporated into a molecule. To find this conversion value, the reaction solutions were placed in contact with miniature radiation detectors (Si PIN diode detectors (various sizes, Carroll & Ramsey Associates, Berkeley, CA) equipped with a triple-channel amplifier (model 101-HDC-3; Carroll & Ramsey Associates, Berkeley, CA) and computer interfaced using NI data acquisition card and LabVIEW (National Instruments, Austin, TX)). The reaction vessels were equipped with an inlet and outlet line using needles through the septa cap. After quenching the solution, the residual $^{11}CO_2$ trapped in solution was then removed from solution by sparging with a stream of He gas. When the radioactivity in the solution became constant (ignoring decay), an aliquot of the solution was analyzed by radio-thin-layer chromatography (TLC) and/or radio-HPLC.

The radiochemical conversion is instrumental to identifying the radiochemical yield. Following the previous example, the radiochemical yield is equal to the product of the radiochemical conversion (amount of $^{11}CO_2$ involved in the reaction divided by the total $^{11}CO_2$ introduced, decay corrected) and the percent yield of the desired compound. We considered the percentage of radioactivity coincident with reference compound to be the percentage yield of the final product (see Fig. 5).

3.2 Radiochemical Identity and Purity

Analytical HPLC was performed using a Knauer HPLC system (Sonntek, Inc., Woodcliff Lake, NJ, United States) with a model K-5000 pump, a model 87 variable wavelength monitor, and NaI radioactivity detector. Radiochemical identity was determined through the observation of a coincident radioactivity signal accompanying the UV signal of a coinjected reference compound.

Radiochemical purity was also determined by TLC using Macherey–Nagel POLYGRAM SIL G/UV254 plastic-backed TLC plates and measuring radioactivity distribution with Bioscan system 200 imaging scanner (Bioscan, Inc., Washington, D.C.). In almost all cases, radio-TLC was the more conservative

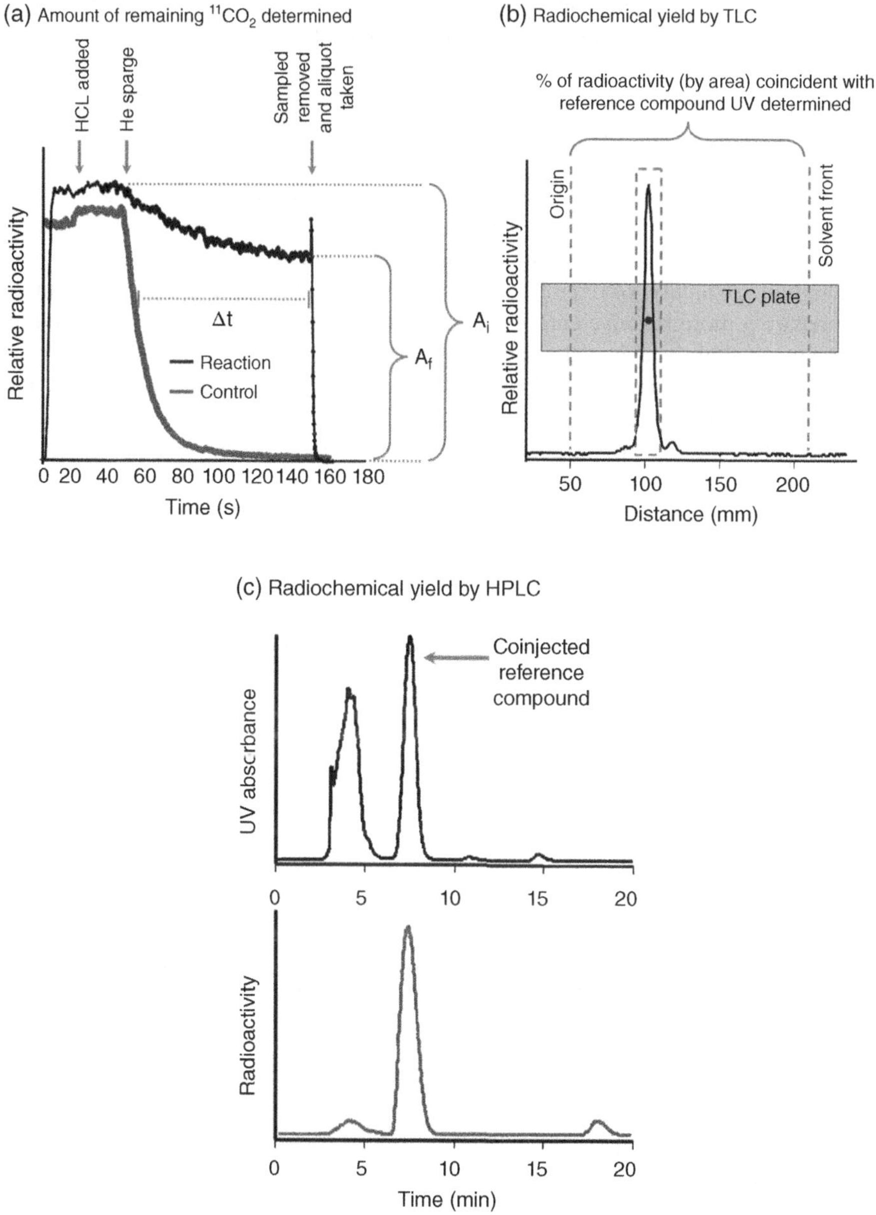

FIGURE 5 Radiochemical yield calculation. (a) To determine the amount of volatile radioactivity in the sealed reaction vessel (primarily $^{11}CO_2$) at the end reactions, each solution was sparged with He after acidification. A representative data set is presented (black trace). In validation experiments, >99% of $^{11}CO_2$ trapped in a solution of DBU was released using this method (gray trace). The relative radioactivity remaining (A_f) was decay corrected (Δt) and divided by the initial radioactivity (A_i). (b and c) Radioactivity remaining in the sample was analyzed by radio-TLC and/or radio-HPLC to determine radiochemical purity. *Reported radiochemical yields are the product of the quotient from (a) and the % yield from (b).*

estimate of radiochemical yield (although the two methods were within 5% of one another), and therefore, this value was used to report yields.

4 DISCUSSION

Desirable radiosyntheses require reaction schemes with minimal steps and equipment to make radiotracers available to a wide range of radiopharmacies. The simplicity of direct $^{11}CO_2$ fixation meets this need by increasing radiochemical yield through the reduction of overall synthesis time and additional equipment required to prepare a more reactive carbon-11 reagent. Mild reaction conditions including ambient or slightly elevated temperatures between 50 and 100°C, for less than 10 min, provided sufficient product yields. The reaction is also tolerant to moisture; the role of water has been investigated with NMR spectroscopy in reactions containing DBU and CO_2. In the absence of water, neither neat DBU nor DBU in DMF has a significant (observable) interaction with CO_2 even when using excess CO_2 under high pressure. The introduction of up to 10 mg of water only minimally decreased yields, suggesting that dry solvents are not a requirement for successful $^{11}CO_2$ fixation [16]. Direct incorporation of $^{11}CO_2$ negates the need for expensive equipment designed solely to produce a more reactive form of radiolabeled carbon. This highly robust, versatile, and rapid one-pot technique meets the criteria for radiosynthesis using short-lived isotopes and expands the library of accessible radiotracers.

WASTE DISPOSAL INFORMATION

All hazardous chemicals and toxic materials were disposed of according to Prudent Practices in the Laboratory (Washington, DC: National Academy Press, 1995).

CHEMICAL ABSTRACTS NOMENCLATURE (REGISTRY NUMBER)

Acetonitrile (75-05-8)
DBU (6674-22-2)
DMF (68-12-2)
Sodium hydroxide (1310-73-2)
DBN (3001-72-7)
DMAP (1122-58-3)
DABCO (280-57-9)
Triethylamine (121-44-8)
Morpholine (110-91-8)
Benzyl bromide (100-39-0)
4-Fluorobenzylamine (140-75-0)
DMSO (67-68-5)

BEMP (98015-45-3)
Benzyl chloride (100-44-7)
Trifluoroacetic acid (76-05-1)
3-Picolyl chloride hydrochloride (6959-48-4)
Phosphoryl trichloride (10025-87-3)
Cyclohexylamine (108-91-8)
2-Phenyl-1,4-dihydroquinone (1079-21-6)
Methanol (67-56-1)
Formic acid (64-18-6)

REFERENCES AND NOTES

For detailed supplier information, see Appendix 1.

NOTE

1. Unless otherwise noted, chemicals were purchased from Sigma-Aldrich (United States) and used without additional purification. *N,N*-Dimethylformamide (DMF, 99.8%, water <50 ppm, extra dry) was purchased from Acros Organics. Radioactive syntheses with carbon-11 were performed in 3.0–5.0 ml vials with reduced volume conical bottoms (Wilmad, LG-4395) using silicon septa.

REFERENCES

For detailed supplier information, see Appendix 1.

1. D. Roeda, F. Dollé, *Curr. Top. Med. Chem.*, 2010, 10, 1680.
2. P. W. Miller, N. J. Long, R. Vilar, A. D. Gee, *Angew. Chem. Int. Ed.*, 2008, 47, 8998.
3. P. J. H. Scott, *Angew. Chem. Int. Ed.*, 2009, 48, 6001.
4. J. M. Buchanan, A. B. Hastings, F. B. Nesbett, *J. Biol. Chem.*, 1949, 180, 435.
5. C. Prenant, J. Sastre, C. Crouzel, A. Syrota, J. Labelled *J. Label. Compd. Radiopharm.*, 1987, 24, 227.
6. A. R. Studenov, M. S. Berridge, D. V. Soloviev, M. Matarrese, S. Todde, *Nucl. Med. Biol.*, 1999, 26, 431.
7. M. van der Meij, N. I. Carruthers, J. D. M. Herscheid, J. A. Jablonowski, J. E. Leysen, A. D. Windhorst, *J. Label. Compd. Radiopharm.*, 2003, 46, 1075.
8. S. Ram, R. E. Ehrenkaufer, D. M. Jewett, *Int. J. Rad. Appl. Instrum. A.*, 1986, 37, 391.
9. S. Ram and R. E. Ehrenkaufer, *Int. J. Rad. Appl. Instrum. B.*, 1988, 15, 345.
10. J. M. Hooker, A. T. Reibel, S. M. Hill, M. J. Schueller, J. S. Fowler, *Angew. Chem. Int. Ed.*, 2009, 48, 3482.
11. J. M. Hooker, S. W. Kim, D. Alexoff, Y. Xu, C. Shea, A. Reid, N. Volkow, J. S. Fowler, *ACS Chem. Neurosci.*, 2009, 1, 65.
12. A. A. Wilson, A. Garcia, S. Houle, O. Sadovski, N. Vasdev, *Chem. Eur. J.*, 2011, 17, 259.
13. T. J. Tewson, W. Banks, M. Franceschini, J. Hoffpauir, *Int. J. Rad. Appl. Instrum. A.*, 1989, 40, 765.
14. B. H. Mock, M. T. Vavrek, G. K. Mulholland, *Nucl. Med. Biol.*, 1995, 22, 667.
15. A. A. Wilson, A. Garcia, S. Houle, N. Vasdev, *Org. Biomol. Chem.*, 2010, 8, 428.
16. D. J. Heldebrant, P. G. Jessop, C. A. Thomas, C. A. Eckert, C. L. Liotta, *J. Org. Chem.*, 2005, 70, 5335.

CHAPTER 20

RADIOSYNTHESIS OF [^{11}C]CARBOXAMIDES VIA REACTION OF [^{11}C]CO_2 CAPTURED BY A Cu(I)-BASED CATALYST SYSTEM WITH BORONIC ACID ESTER PRECURSORS

PATRICK J. RISS[1,2], SHUIYU LU[3], SANJAY TELU[3], FRANKLIN I. AIGBIRHIO[1], AND VICTOR W. PIKE[3]

[1] *Wolfson Brain Imaging Centre, Addenbrooke's Hospital, University of Cambridge, Cambridge, UK*
[2] *Kjemisk Institutt, Universitetet i Oslo, Oslo, Norway*
[3] *Molecular Imaging Branch, National Institute of Mental Health, Bethesda, Maryland, USA*

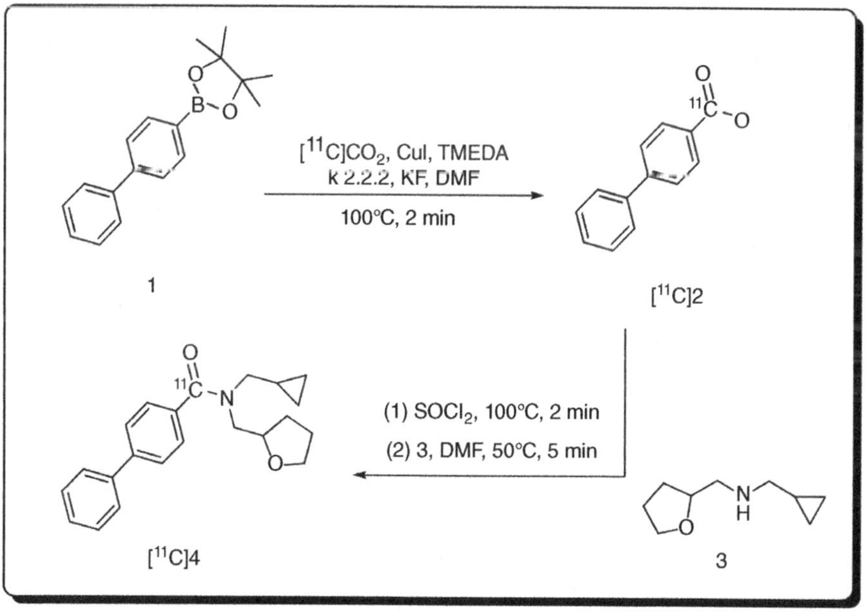

Radiochemical Syntheses, Volume 2: Further Radiopharmaceuticals for Positron Emission Tomography and New Strategies for Their Production, First Edition. Edited by Peter J. H. Scott.

1 INTRODUCTION

Carbon-11 ($t_{1/2}$=20.4 min) is routinely accessible in high activity and in high specific radioactivity from the cyclotron-promoted $^{14}N(p,\alpha)^{11}C$ nuclear reaction [1]. The chemical form of the carbon-11 is determined by the inclusion of hydrogen or oxygen in the nitrogen gas target leading to the formation of [^{11}C]methane or [^{11}C]carbon dioxide, respectively. Direct use of either of these primary precursors for radiolabeling is quite constricted due to their limited reactivities. Instead, conversion is necessary into ^{11}C-labeling agents with more versatile reactivity, for example, [^{11}C]CH_3I, [^{11}C]CH_3OTf, or [^{11}C]CO [2, 3]. Such chemical transformations, however, come at the expense of time and, consequently, the specific radioactivity of the produced radiotracers. In this regard, direct installation of carbon-11 into radiotracers is more desirable so that rapid and simple radiosyntheses can be achieved [4, 5].

So far, the use of [^{11}C]CO_2 in radiotracer synthesis has been confined largely to the ^{11}C-carboxylation of organometallic reagents to produce ^{11}C-labeled carboxylic acids [4]. These methods require diligent control of reaction conditions, and are limited in terms of the structural and functional group diversities of products that may be obtained. More recently, protocols for the synthesis of ^{11}C-labeled carbamates and ureas have been reported [5].

We have developed an efficient method for producing no-carrier-added [^{11}C] carboxylic acids with high functional group tolerance and relatively undemanding reaction conditions using boronic acid esters as labeling precursors (Fig. 1) [6]. Cyclotron-produced [^{11}C]CO_2 is first captured with a Cu(I)-based catalyst system and then reacted with a boronic acid ester. This method allows one to produce [^{11}C]carboxylic acids, efficiently in the presence of other functional groups; these may be onward converted into [^{11}C]esters and [^{11}C]amides *in situ*. Boronic acids and esters are widely available or can be easily prepared. Herein, we use the radiosynthesis of a prospective oxytocin receptor ligand, [^{11}C]*N*-(cyclopropylmethyl)-*N*-((tetrahydrofuran-2-yl)methyl)-[1,1′-biphenyl]-4-carboxamide ([^{11}C]**4**), as an example to demonstrate the protocol's utility and efficiency.

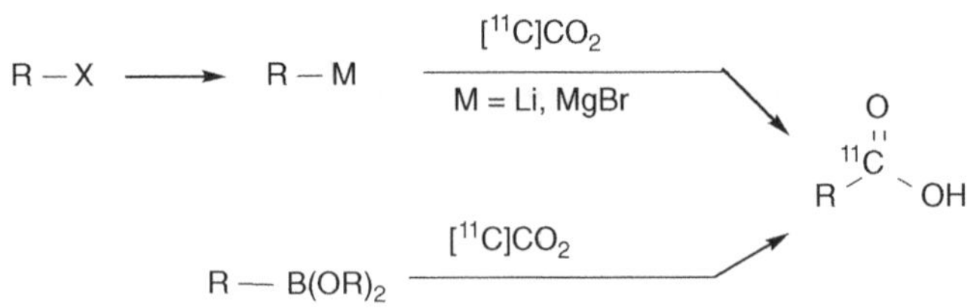

FIGURE 1 Radiosyntheses of [^{11}C]carboxylic acids from [^{11}C]CO_2 with classical organometallic reagents or, as described here, novel boronic acid esters.

2 PROCEDURES

2.1 Synthesis Procedures

CAUTION: All radiochemical syntheses must be carried out using appropriate equipment in a facility authorized for the use of radioactive materials. Personal protective equipment must be worn and all local radiation safety laws followed.

2.1.1 Materials

2-([1,1′-Biphenyl]-4-yl)-4,4,5,5-tetramethyl-1,3,2-dioxaborolane (**1**) was purchased from TCI America (Portland, OR). Chemicals and solvents were purchased in highest available purity from Burdick & Jackson (Muskegon, MI), Sigma-Aldrich (St. Louis, MO), TCI America (Portland, OR), or W. R. Grace & Co. (Columbia, MD). 1-Cyclopropyl-*N*-((tetrahydrofuran-2-yl)methyl)methanamine (**3**) was prepared as previously described [7]. All purchased chemicals were used without further purification. Radioactivity was measured using an Atomlab 300 dose calibrator (Biodex Medical Systems), with corrections for background radioactivity and decay.

Radiochemistry

Production of [^{11}C]Carbon Dioxide Carbon-11 was produced on a PETtrace cyclotron (GE Medical Systems, Milwaukee, WI) using the $^{14}N(p,\alpha)^{11}C$ nuclear reaction. An aluminum target (internal volume 75 ml) containing a compressed mixture of nitrogen gas containing oxygen (0.5%) (initially 160 psi, 1.1×10^6 Pa, 11 bar) was irradiated with a 10 µA proton beam (16.5 → 3 MeV) for 10 min. At the end of the irradiation, gas was released from the target and the $^{11}CO_2$ (~400 mCi; ~14.8 GBq) was trapped on a molecular sieve 13× (80/100; 300 mg; W. R. Grace) column.

Radiosyntheses of [^{11}C]2 and [^{11}C]4 A radiosynthesis module (Tracerlab FX C Pro; GE Medical Systems), housed within a lead-shielded hot-cell, was used for all radiochemistry procedures. The apparatus was modified to bypass the part constructed for gas-phase iodination of [^{11}C]methane due to its redundancy in this work (Fig. 2).

The reaction vessel was charged with TMEDA (*N,N,N′,N′*-tetramethylethylenediamine; 0.5 mmol), CuI (0.3 µmol), kryptofix 2.2.2 (1 µmol), KF (1 µmol) and 2-([1,1′-biphenyl]-4-yl)-4,4,5,5-tetramethyl-1,3,2-dioxaborolane (**1**, 24 µM) in DMF (400 µl). [^{11}C]CO_2 was released from the molecular sieve into the reaction mixture by heating the column to 330°C while purging the column with He carrier gas (15 ml/min). Release of [^{11}C]CO_2 and accumulation of radioactivity in the reaction mixture were monitored using two PIN diode detectors (1-in.) that were connected to a B-HC-2000 module (Bioscan, Inc.; Washington, DC). Figure 3 illustrates representative [^{11}C]CO_2 release and trapping profiles.

When [^{11}C]carbon dioxide release had ended, the reactor was sealed and heated at 90°C for 5 min, and then rapidly cooled to RT using a gas stream jet generated from liquid nitrogen. Aqueous formic acid (0.1 M, 10 ml) was added to the reaction

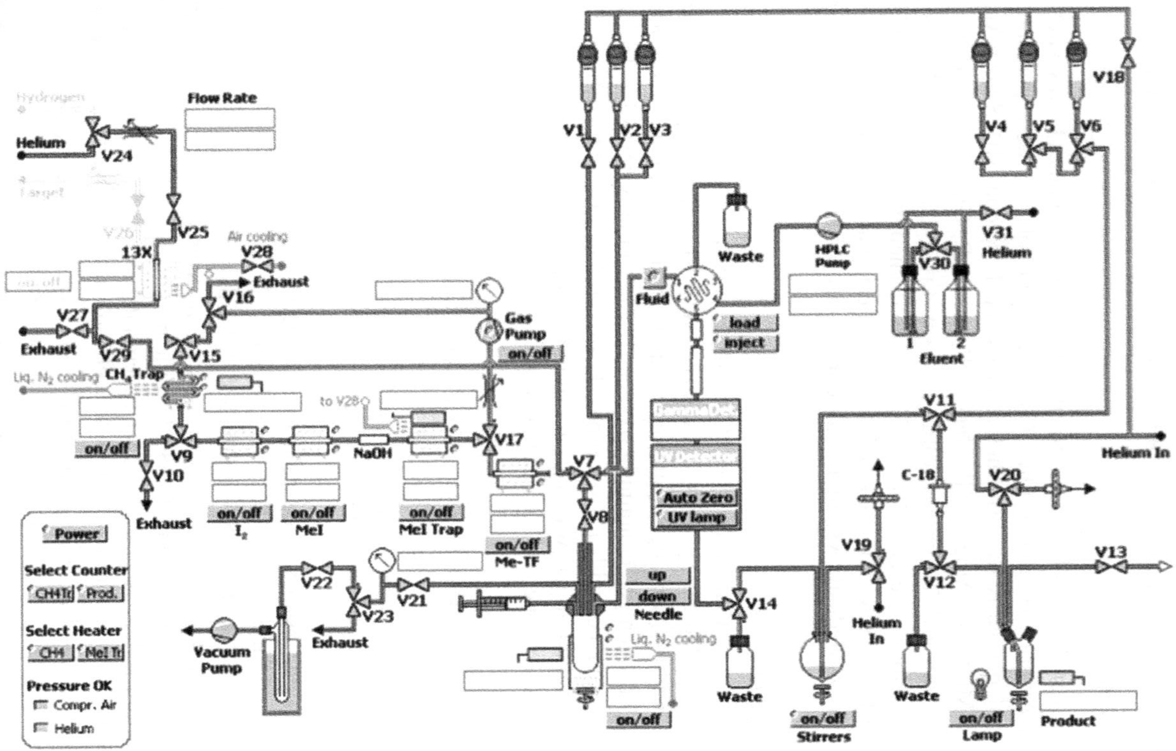

FIGURE 2 Configuration of [^{11}C]CO_2 gas flow on the Tracerlab FX C Pro radiosynthesis module. The recirculation setup is bypassed via a direct connection of valves V7 and V29.

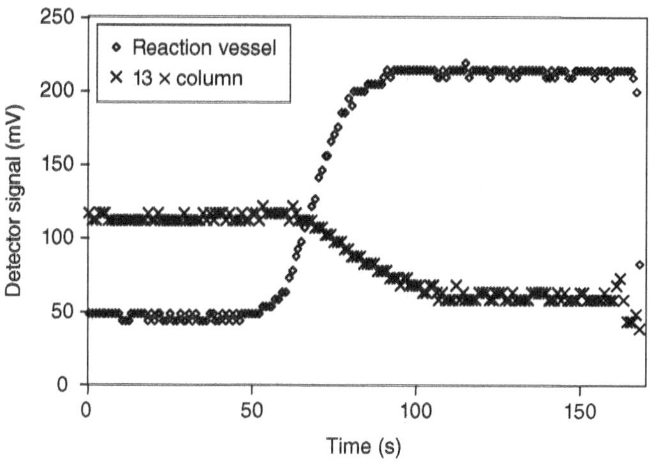

FIGURE 3 Exemplified profiles for [^{11}C]CO_2 release from the molecular sieves 13× column (×) and trapping in the reaction vessel (◊).

mixture and the diluted homogeneous solution was passed through a C18 plus SPE cartridge (Waters). The cartridge was dried in a stream of He and [^{11}C]**2** was eluted with DMF. $SOCl_2$ (25 mmol) in DMF (0.5 ml) was added to [^{11}C]**2**, and the mixture was heated to 100°C for 2 min. Subsequently, the reaction mixture was cooled to 50°C and amine **3** (20 mg, 130 μmol) [7] was added. After 5 min at 50°C, the mixture was diluted with water and injected onto a Luna RP18(2) column (10 μm, 250 mm × 10 mm, Phenomenex, Torrance, CA) eluted with MeCN-25 mM $HCOONH_4$ (60: 40 v/v) at 4.75 ml/min (Fig. 4). The product [^{11}C]**4** was collected (t_R = 14.3 min), concentrated in vacuo and the residue formulated in saline–EtOH (9:1 v/v; 10 ml). The radiochemical purity of [^{11}C]**4** was greater than 98% and the decay-corrected radiochemical yield was 20%.

2.2 Quality Control Procedures

CAUTION: All radiopharmaceuticals produced for clinical use must have local regulatory approval before use in humans. Trained personnel must carry out QC procedures, and each dose must meet all established QC criteria before release to the clinic.

Mass spectrometry was performed on a LCQ DECA/AS2000 LC-MS instrument (Thermo Fisher Scientific, San Jose, CA). Analytical HPLC was performed on an apparatus comprised of a System Gold 126 pump and a System Gold 166 variable wavelength detector, using 32 Karat software for data acquisition and analysis (Beckman Coulter). A Luna RP18(2) column (10 μm; 100 Å; 250 × 4.6 mm; Phenomenex) was eluted with MeCN-25 mM aqueous $HCOONH_4$ (60:40 v/v) at 1.75 ml/min. Eluate was monitored for UV absorbance at 254 nm and for radioactivity. A representative analytical chromatogram of [^{11}C]**4** is shown in Fig. 4.

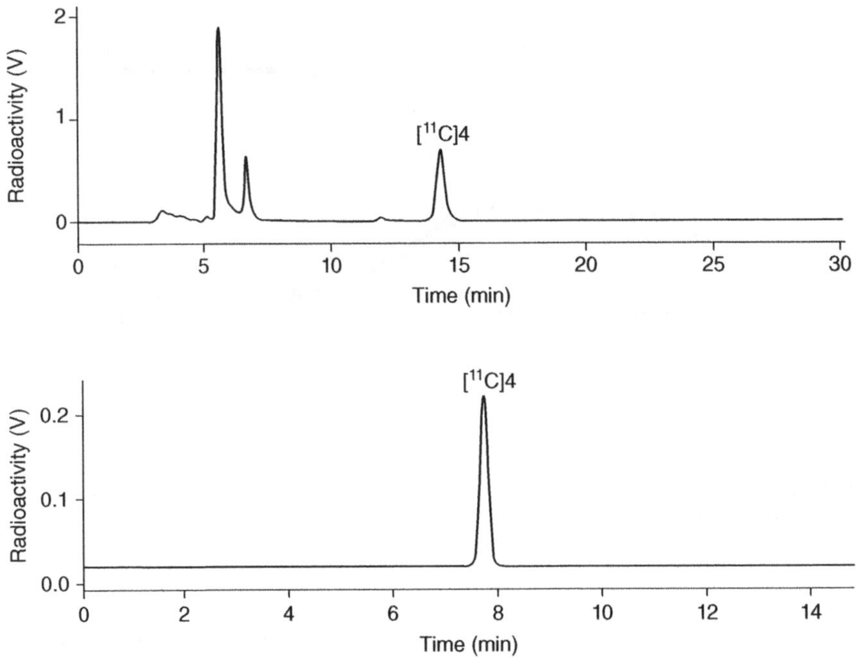

FIGURE 4 Example of chromatograms from the reverse-phase HPLC purification of reaction [^{11}C]**4** from reaction mixture (top) and the analysis of purified and formulated [^{11}C]**4** (below). HPLC conditions are described in the text.

2.2.1 *Measurements of Specific Radioactivity*

The specific radioactivity of the final [^{11}C]**4** product was determined by radio-HPLC. Thus, a calibration curve was prepared from six individual dilutions of **4** in order to establish the correlation between carrier mass of **4** in the injected sample and integrated UV absorbance signal area. A sample of formulated [^{11}C]**4** was then analyzed by HPLC. Specific activity (mCi/μmol; MBq/μmol) was then calculated as the ratio of [^{11}C]**4** radioactivity in the sample (mCi; MBq) to the amount of carrier **4** (μmol), corrected for physical decay to the end of synthesis. The specific radioactivity of [^{11}C]**4** was 1.5 Ci/μmol at 43 min from radionuclide production. The radiochemical purity of [^{11}C]**4** was greater than 98%.

3 DISCUSSION

In no-carrier-added ^{11}C-radiochemistry, carboxylation reactions are carried out with sub-stoichiometric amounts of CO_2 at very low partial pressure. In accordance with Henry's law, the amount of $^{11}CO_2$ in solution is very low even within a pressurized reaction vessel. Most of the dissolved CO_2 will be lost to the gas phase upon heating unless it is held in solution whilst still preserving its reactivity.

Strongly basic imines, such as 1,8-diazabicycloundec-7-ene (DBU) or phosphazen bases are known to retain $[^{11}C]CO_2$ in solution [5]. In this method, we devised a catalyst system comprising TMEDA and CuI that can efficiently trap $^{11}CO_2$ in solution and effectively promote the carboxylation reaction. This Cu(I)-based catalyst system was found to be more efficient than other Cu(I)-based copper–bisoxazoline and copper *N*-heterocyclic carbene (Cu-NHC) systems [8].

Using this catalyst system, we were able to efficiently prepare a variety of $[^{11}C]$ carboxylic acids within only 10 min from the end of radionuclide production. Other substituents, such as halogen, –CN, $–NO_2$, $–CONR_2$, –CHO and to some extent, –OH, were tolerated in the first-step reaction. No stringent precautions in terms of exclusion of air, a source of carrier CO_2, were necessary to obtain the product with high specific radioactivity because the precharged reactor was under constant helium gas purge before delivery of the radioactivity.

We extended this methodology to the labeling of a model amide, a prospective oxytocin receptor radioligand, $[^{11}C]$**4** [7]. After compound **1** was reacted with $[^{11}C]CO_2$, the solvent, TMEDA and unreacted $^{11}CO_2$ were removed by solid phase extraction. To ensure that the ^{11}C-labeled carboxylic acid was fully free of moisture and ready for reaction in the next step, the nitrogen gas flow must be carefully optimized during the removal of solvent and excess base using SPE cartridge. The residue containing $[^{11}C]$**2** was redissolved in DMF and $SOCl_2$ was added. Conversion of the acid into amide $[^{11}C]$**4** was accomplished by the addition of amine **3** at 50°C, followed by heating for 5 min. The product was obtained in useful RCY, and within two half-lives from end of radionuclide production.

In conclusions, the present method provides a concise procedure for CuI-mediated carboxylation of boronic acid esters in the presence of TMEDA, KF and K 2.2.2 and subsequent conversion into an amide. The obtained $[^{11}C]$carboxamide is of high radiochemical purity and high specific radioactivity. Given the high abundance of carboxyl groups and their derivatives in PET radiotracers, and the availability of $[^{11}C]CO_2$ at most PET centers, we expect this novel methodology to be widely adapted for PET radioligand development.

ACKNOWLEDGMENTS

PJR was supported by a Medical Research Council (UK) postdoctoral fellowship as a visiting worker at the National Institutes of Health. ST, SL, and VWP were supported by the Intramural Research Program of the National Institutes of Health (NIMH).

WASTE DISPOSAL INFORMATION

All hazardous chemicals and toxic materials were disposed of according to Prudent Practices in the Laboratory (Washington, DC: National Academy Press, 1995).

CHEMICAL ABSTRACTS NOMENCLATURE

Acetonitrile [75-05-8]
Ammonia [7664-41-7]
Ammonium formate [540-69-2]
2-(4-Biphenylyl)-4,4,5,5-tetramethyl-1,3,2-dioxaborolane [144432-80-4]
Copper iodide [7681-65-4]
Dimethylformamide (DMF) [68-12-2]
Formic acid [64-18-6]
4,7,13,16,21,24-hexaoxa-1,10-diazabicyclo[8.8.8]hexacosan (kryptofix 2.2.2) [23978-09-8]
Potassium carbonate [584-08-7]
Potassium fluoride [7789-23-3]
Thionyl chloride ($SOCl_2$) [7719-09-7]
N,N,N′,N′-tetramethylethylenediamine (TMEDA) [110-18-9]

REFERENCES AND NOTES

For detailed supplier information, see Appendix 1.

NOTES

1. A glass column (10 mm × 250 mm) packed with ascarite was attached to the vent line in order to prevent the release of [^{11}C]CO_2 that was not trapped in the reaction mixture.
2. Trapping efficiency was calculated as the percentage of radioactivity retained in the reaction mixture over the total amount of [^{11}C]CO_2 released from the molecular sieve 13× column.

REFERENCES

1. (a) G. Stocklin, V. W. Pike (Eds), *Radiopharmaceuticals for Positron Emission Tomography—Methodological Aspects*. Kluwer Academic Publishers, Dordrecht, 1993; (b) M. J. Welch, C. S. Redvanly (Eds), *Handbook of Radiopharmaceuticals—Radiochemistry and Applications*. John Wiley and Sons, Chichester, 2003.
2. (a) G. Antoni, T. Kihlberg, B. Långström, ^{11}C: Labeling chemistry and labeled compounds in *Handbook of Nuclear Chemistry*, Vol. 4 (A. Vértes, S. Nagy, Z. Klencsár (Eds)), Kluwer Academic Publishers, Amsterdam, 2003. p. 119–157; (b) P. W. Miller, N. J. Long, R. Vilar, A. D. Gee, *Angew. Chem. Intl. Ed.* 2008, 47, 8998; (c) M. Allard, E. Fouquet, D. James, M. Szlosek-Pinaud, *Curr. Med. Chem.* 2008, 15, 235; (d) P. J. H. Scott, *Angew. Chem. Intl. Ed.* 2009, 48, 6001; (e) Z. Tu, R. H. Mach, *Curr. Topics Med. Chem.* 2010, 10, 1060.
3. (a) G. J. Meyer, S. L. Waters, H. H. Coenen, A. Luxen, B. Mazière, B. Långström, *Eur. J. Nucl. Med.* 1995, 22, 1420; (b) B. Långström, O. Itsenko, O. Rahman, *J. Label. Compd. Radiopharm.* 2007, 50, 794.
4. (a) V. W. Pike, M. N. Eakins, R. M. Allan, A. P. Selwyn, *J. Radioanal. Chem.* 1981, 64, 291; (b) V. W. Pike, M. N. Eakins, R. M. Allan, A. P. Selwyn, *Int. J. Appl. Radiat. Isot.* 1982, 33, 505; (c) E. Livni, D. R. Elmaleh, S. Levy, G. L. Brownell, W. H. Strauss, *J. Nucl. Med.* 1982, 23, 169;

(d) M. van der Meij, N. I. Carruthers, J. D. M. Herscheid, J. A. Jablonowski, J. E. Leysen, A. D. Windhorst, *J. Label. Compd. Radiopharm.* 2003, 46, 1075.

5. (a) J. M. Hooker, A. T. Reibel, S. M. Hill, M. J. Schueller, J. S. Fowler, *Angew. Chem. Intl. Ed.* 2009, 48, 3482; (b) A. A. Wilson, A. Garcia, S. Houle, O. Sadovski, N. Vasdev, *Chem. Eur. J.* 2011, 17, 259.
6. P. J. Riss, S. Lu, S. Telu, F. I. Aigbirhio, V. W. Pike, *Angew. Chem. Intl. Ed.* 2012, 51, 2698.
7. B. R. Bellenie, N. P. Barton, A. J. Emmons, J. P. Heer, C. Salvagno, *Bioorg. Med. Chem. Lett.* 2009, 19, 990.
8. (a) J. Takaya, S. Tadami, K. Ukai, N. Iwasawa, *Org. Lett.* 2008, 10, 2697; (b) T. Ohishi, M. Nishiura, Z. Hou, *Angew. Chem. Intl. Ed.* 2008, 47, 5792; (c) T. Ohishi, L. Zhang, M. Nishiura, Z. Hou, *Angew. Chem. Intl. Ed.* 2011, 50, 8114.

CHAPTER 21

ETHANOL AS A SOLVENT FOR CARBON-11 RADIOCHEMISTRY

XIA SHAO, MARIA V. FAWAZ, KEUNSAM JANG, AND PETER J. H. SCOTT

Department of Radiology, University of Michigan School of Medicine, Ann Arbor, Michigan, USA

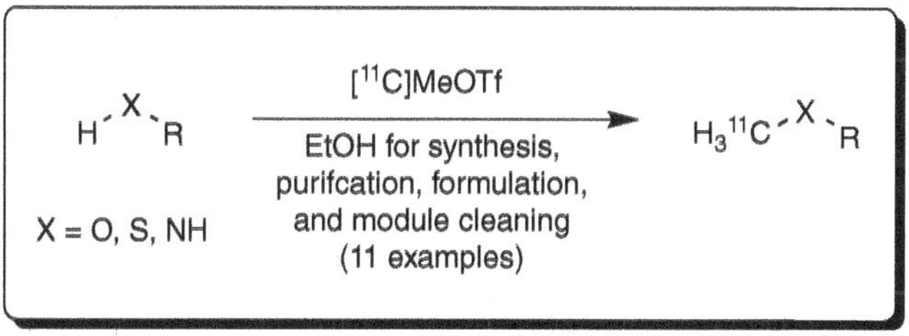

1 INTRODUCTION

Green chemistry is the development of products (or processes) that minimize or eliminate the use, generation, and disposal of hazardous chemical substances [1]. In our laboratory, we have been applying the principles of green chemistry to radiochemical syntheses, recently demonstrating that carbon-11 radiochemistry is compatible with ethanol and water [2]. To date, eleven carbon-11 labeled radiopharmaceuticals have been prepared using ethanol as the only organic solvent throughout the entire manufacturing process (synthesis, purification, formulation, and synthesis module cleaning), and all radiopharmaceutical doses prepared using ethanol radiochemistry are suitable for human administration (Fig. 1, Table 1).

The development of ethanol-based radiochemistry has the potential of simplifying workflow in PET radiopharmaceutical synthesis facilities by changing residual solvent analysis from a daily quality control (QC) test of multiple different organic solvents to a single GC analytical method for ethanol that might, with regulatory approval, be relegated to a quarterly or annual test. In addition, the removal of all other organic solvents from the process simplifies laboratory operations as it eliminates the need to purchase inventory, handle, and properly dispose of other hazardous solvents. The methods in development can be used as stand-alone syntheses in

Radiochemical Syntheses, Volume 2: Further Radiopharmaceuticals for Positron Emission Tomography and New Strategies for Their Production, First Edition. Edited by Peter J. H. Scott.

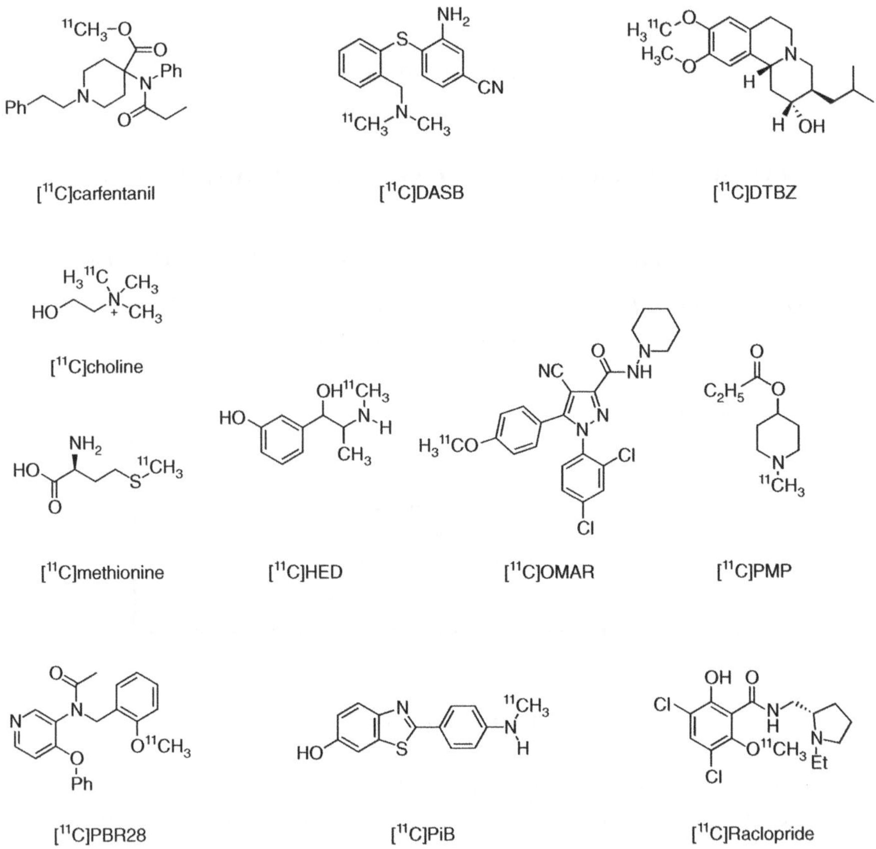

FIGURE 1 Radiotracers amenable to synthesis and purification using only ethanol.

their own right, or individual components (e.g., synthesis or purification) can be used in conjunction with the other synthetic approaches outlined in this book, as well as the previous volume in the series.

2 SYNTHESIS PROCEDURES

CAUTION: All radiochemical syntheses must be carried out using appropriate equipment in a facility authorized for the use of radioactive materials. Personal protective equipment must be worn and all local radiation safety laws followed.

2.1 Production of $[^{11}C]CO_2$

$[^{11}C]CO_2$ was produced by General Electric (GE) PETTrace cyclotron. The target [3] was loaded with $[^{14}N]N_2$ gas [4] and bombarded with a proton beam (40 µA beam for 30 min) to generate approximately 3Ci of $[^{11}C]CO_2$ by the $^{14}N(p,\alpha)^{11}C$ nuclear reaction.

TABLE 1 Synthesis and Quality Control Data

Product	*n*	Method	Precursor	Purification	RCY[a]	RCP[b]	SA[c]	pH
Carfentanil	2	[^{11}C]MeOTf/reactor	TBA salt	Sep-pak	6.8%	>95%	>30,000	5.5
Choline	10	[^{11}C]MeI/sep-pak	Free base	Sep-pak	7.3%	>95%	ND[d]	5.5
DASB	13	[^{11}C]MeOTf/loop	Free base	HPLC	3.0%	>95%	10,900	6.0
(+)DTBZ	12	[^{11}C]MeOTf/reactor	TBA salt	HPLC	6.2%	>95%	31,400	5.5
HED	31	[^{11}C]MeOTf/reactor	Free base	HPLC	5.0%	>95%	10,100	6.0
Methionine	3	[^{11}C]MeI/reactor	Thiolactone + NaOH	Evap.	4.0%	>95%	ND[d]	6.0
OMAR	3	[^{11}C]MeOTf/loop	TBA salt	HPLC + sep-pak	4.1%	>95%	4900	6.0
PBR28	8	[^{11}C]MeOTf/loop	TBA salt	HPLC + sep-pak	2.8%	>95%	5700	6.0
PiB	3	[^{11}C]MeOTf/loop	Free base	HPLC + sep-pak	2.3%	>95%	11,000	5.5
PMP	26	[^{11}C]MeOTf/reactor	HCl salt + $NaHCO_3$	HPLC	4.4%	>95%	Mass < LOD[e]	6.0
Raclopride	144	[^{11}C]MeOTf/loop	TBA salt	HPLC	3.6%	>95%	20,750	6.0

[a] RCY = non-decay-corrected radiochemical yield at EOS based on 3Ci of starting [^{11}C]CO_2.
[b] RCP = Radiochemical purity.
[c] SA = Specific activity in Ci/mmol determined at EOS.
[d] ND = Not determined as there are no specific activity limits established for endogenous ligands.
[e] LOD = 25 μg/ml limit of detection for [^{11}C]PMP.

2.2 Conversion of $[^{11}C]CO_2$ to $[^{11}C]MeI$ and $[^{11}C]MeOTf$

$[^{11}C]CO_2$ was delivered from the target by nitrogen gas pressure to a column filled with a mixture of molecular sieves (0.3 g) [5] and Shimalite-Ni catalyst (0.2 g) [6] at room temperature. The column was sealed under H_2, and then heated to 350°C for 20 s to reduce the $[^{11}C]CO_2$ to $[^{11}C]CH_4$. $[^{11}C]CH_4$ was passed through a phosphorus pentoxide desiccant [7] column and trapped in a carbosphere [8] column at −78°C. Gaseous $[^{11}C]CH_4$ then entered a circulation loop, which included a gas pump, iodine column at 100°C [9], a reactor tube at 720°C, two adjacent Ascarite II columns [10], and a porapak column [11] at room temperature. The gaseous mixture was then circulated for 5 min, while $[^{11}C]CH_3I$ accumulated on the porapak column. $[^{11}C]CH_3I$ was then converted to $[^{11}C]CH_3OTf$ (if required) by passing it through a silver triflate-Graphpac column [12], heated to 190°C.

2.3 Preparation of TBA Salts of Precursors

The desmethyl precursor (10 mg) was dissolved in ethanol (100 µl) and water (50 µl) in a bullet vial by shaking with a vortex mixer for 30 s. Tetrabutylammonium hydroxide (100 µl of a 1 M solution in methanol) was added and the vial was mixed with a vortex mixer for another 30 s. The resulting solution was diluted with water (6 ml) and passed through a C18 extraction disk (preconditioned with ethanol (5 ml) and water (10 ml)). The disk was washed with additional water (10 ml) and dried under a nitrogen stream. The product was then eluted into a new vial with ethanol (2 ml), and the resulting eluent was dispensed into 10 bullet vials (1 mg precursor in 200 µl ethanol per vial). The vials were placed in a vacuum dessicator and evaporated to dryness overnight under vacuum. Vials were stored in the refrigerator and used within 90 days of production.

2.4 General Procedure for Reactor Syntheses

The TRACERlab $FX_{C\text{-}Pro}$ synthesis module was configured for reactor syntheses with HPLC or SPE purification as previously described [2, 13]. The precursor (~1.0 mg) was dissolved in ethanol (100 µl) and loaded directly into the synthesis module reactor. In the event that a reformulation was involved, additional setup was as follows: vial 4: sterile water for injection, USP (7 ml) [14]; vial 5: ethanol (0.5 ml); vial 6: 0.9% NaCl for injection, USP (9.5 ml) [15]; round-bottomed dilution flask: Milli-Q water (20–50 ml). $[^{11}C]MeOTf$ or $[^{11}C]MeI$ was bubbled through the precursor solution in the reactor at 15 ml/min for 3–5 min. Following reaction, the mixture was purified by semi-preparative HPLC (DTBZ, HED, and PMP), SPE (carfentanil) or by simple evaporation of the solvent (methionine). Final formulations were then passed through a 0.22-µm filter into a sterile dose vial and submitted for QC testing.

2.5 General Procedure for Loop Chemistry

The TRACERlab $FX_{C\text{-}Pro}$ synthesis module was configured for loop chemistry as previously described [2, 13]. The appropriate precursor (1.0 mg) was dissolved in ethanol (100 µl) and loaded into the 2 ml steel HPLC loop and conditioned with nitrogen gas for 20 s at 10 ml/min. In the event that a reformulation was involved, additional setup was as follows: vial 4: sterile water for injection, USP (7 ml) [14]; val 5: ethanol (0.5 ml); vial 6: 0.9% NaCl for injection, USP (9.5 ml) [15]; round-bottomed dilution flask: Milli-Q water (20–50 ml). [^{11}C]MeOTf was passed through the HPLC loop at 15–40 ml/min for 3–5 min. Following reaction, the mixture was purified by semi-preparative HPLC. The product peak was collected either directly for injectable mobile phases, or reformulated into ethanolic saline. The final formulation was then passed through a 0.22-μm filter into a sterile dose vial and submitted for QC testing.

2.6 General Procedure for Sep-pak Syntheses

The TRACERlab $FX_{C\text{-}Pro}$ synthesis module was configured for sep-pak syntheses as previously described [13]. Precursors were dissolved in ethanol (20 μl) and loaded onto a sep-pak. The sep-pak was then installed on the synthesis module and [^{11}C]MeI, produced according to the general procedure outlined in Section 3.2, was passed through the sep-pak at 20 ml/min flow rate for 3 min. After this time, the sep-pak was washed and then the radiopharmaceutical was eluted and formulated as described in the following text.

2.7 Purification and Formulation

2.7.1 [^{11}C]Carfentanil

[^{11}C]Carfentanil was prepared according to the general procedure for reactor chemistry described in Section 2.4. [^{11}C]MeOTf was bubbled through the precursor solution at 15 ml/min at rt for 3 min. After production, 1% ammonium hydroxide (1 ml) was added to the reaction vessel. The crude reaction mixture was then conveyed via helium gas to the dilution flask containing 5 ml of 1% ammonium hydroxide. This mixture was passed through a 3M Empore C2 extraction disk where the [^{11}C]carfentanil was trapped. 3 ml of 20% ethanol followed by 7 ml Milli-Q water were then passed through to remove impurities from the disk. The disk was then dried by passing helium gas through for 1.0 min and the [^{11}C]carfentanil was eluted off with ethanol (0.5 ml) and diluted with sterile water for injection (9.5 ml) [16]. The formulated product was then passed through a 0.22 μm filter into a sterile dose vial, and the product was submitted for QC testing. Typical end-of-synthesis (EOS) yields were 6.8% (n=2).

2.7.2 [^{11}C]Choline

[^{11}C]Choline was synthesized by [^{11}C]methylation of *N,N*-dimethylaminoethanol (DMAE), immobilized on a sep-pak, with [^{11}C]methyl iodide according to the general procedure for sep-pak chemistry described in Section 2.6. 40 μl of DMAE was dissolved into 20 μl of ethanol and loaded onto a CM sep-pak, and [^{11}C]methyl iodide was subsequently passed through the CM sep-pak at 20 ml/min flow rate for 3 min. The sep-pak was washed with 5 ml of ethanol and 20 ml of Milli-Q water. The product was eluted with 0.5 ml of USP saline, into the product vial and diluted with sterile water for injection, USP. The final dose was then transferred into a sterile dose vial through a 0.22 μm sterile filter and submitted for QC testing. Typical EOS yields were 7.3% (n=10).

2.7.3 [11C]DASB

[^{11}C]DASB was prepared according to the general procedure for loop chemistry described in Section 2.5. [^{11}C]MeOTf was passed through the HPLC loop at 40 ml/min for 3 min. Following reaction, the mixture was purified by semi-preparative HPLC (column: Phenomenex Luna CN, 150 × 10 mm, mobile phase: 5 mM NaOAc in 80% ethanol, pH: 5.0, flow rate: 4 ml/min, typical trace: Fig. 2). The product peak (t_R ~ 7 min) was collected into 50 ml of water. The solution was then passed through a C18 sep-pak, and washed with 7 ml sterile water. The product was then eluted with 0.5 ml of USP ethanol followed by 9.5 ml of USP saline. This final formulation was then passed through a 0.22-μ filter into a sterile dose vial and submitted for QC testing. Typical EOS yields were 3.0% (n = 13).

2.7.4 [11C]Dihydrotetrabenazine ([11C]DTBZ)

[^{11}C]DTBZ was prepared according to the general procedure for reactor chemistry described in Section 2.4. [^{11}C]CH_3OTf was bubbled into the reaction vessel at 15 ml/min rate for 3 min followed by 2 min of stirring. The reaction mixture was diluted with 1 ml of HPLC mobile phase, and purified by semi-preparative HPLC (column: Phenomenex Luna C8(2) 10 × 100 mm; mobile phase: 20 mM NaH_2PO_4 85/15 water/ethanol; flow rate: 3 ml/min; typical trace: Fig. 3). The product fraction (~3 ml) was collected (t_R ~ 10–11 min) into the product collection vial where it was simultaneously diluted with isotonic saline (7 ml). The resulting formulation was passed through a 0.22 μm sterilizing filter into a sterile dose vial and submitted for QC testing. Typical EOS yields were 6.2% (n = 6).

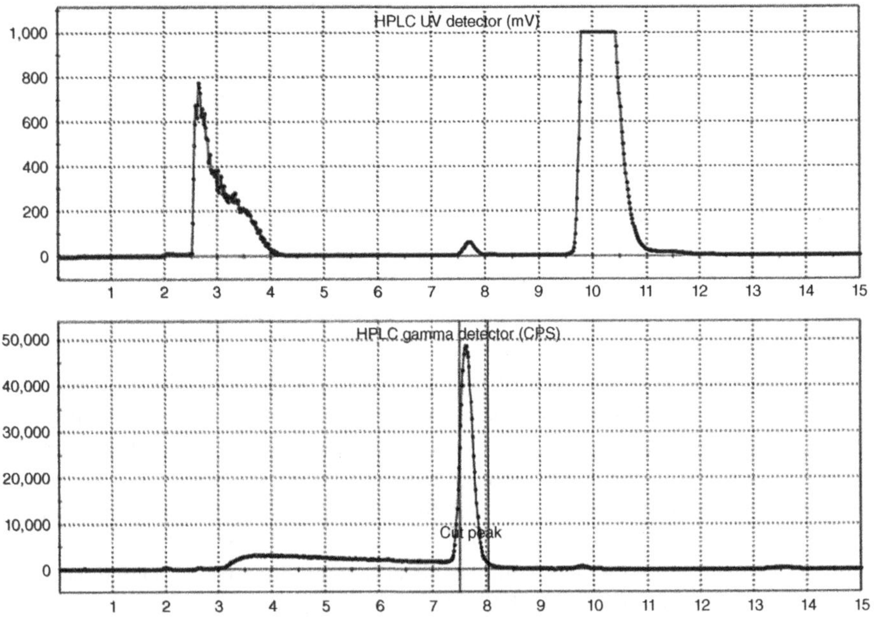

FIGURE 2 Typical ethanolic semi-preparative HPLC trace for [^{11}C]DASB.

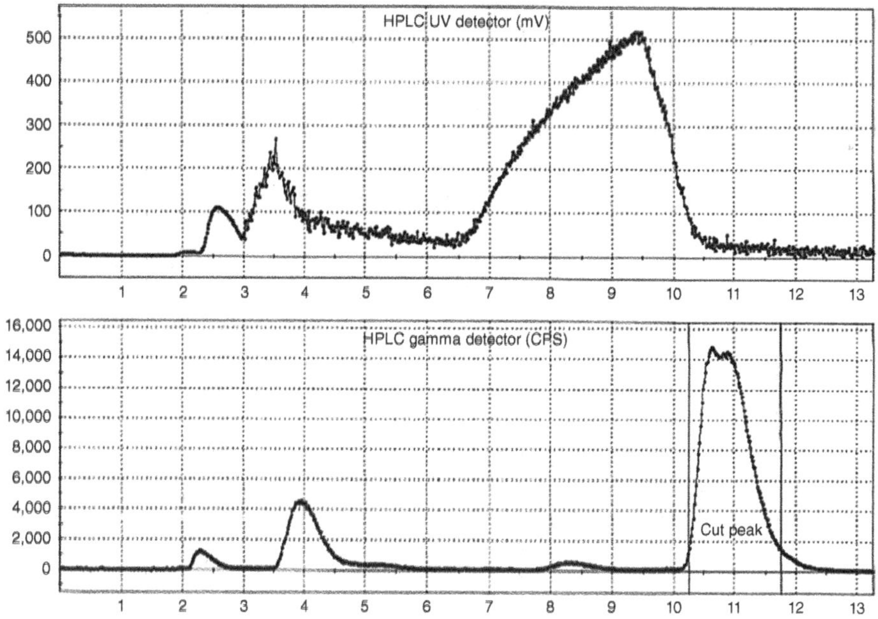

FIGURE 3 Typical ethanolic semi-preparative HPLC trace for [^{11}C]DTBZ.

2.7.5 [^{11}C]Hydroxyephedrine ([^{11}C]HED)

[^{11}C]HED was prepared according to the general procedure for reactor chemistry described in Section 2.4. [^{11}C]CH_3OTf was bubbled into the reaction vessel containing an ethanolic solution of metarminol (0.5 mg in 100 µl) at 15 ml/min for 3 min. The reaction mixture was then diluted with 1 ml of HPLC mobile phase and purified by semi-preparative HPLC (column: Phenomenex Luna C8(2) 10×150 mm; mobile phase: 1 ml of 20 mM NH_4OAc in 99/1 water/ethanol; flow rate: 4 ml/min; typical trace: Fig. 4). The product fraction (~8 ml) was collected (t_R ~ 12 min) into the product collection vial where it was simultaneously diluted with isotonic saline (2 ml). The formulated product was passed through a 0.22 µm sterilizing filter into a sterile dose vial and submitted for QC testing. Typical EOS yields were 5.0% (n=31).

2.7.6 [^{11}C]Methionine

[^{11}C]Methionine was prepared according to the general procedure for reactor chemistry described in Section 2.4. [^{11}C]CH_3I was made according to the general procedure outlined earlier and then bubbled through a solution of L-homocysteine thiolactone (2.0 mg) in sodium hydroxide (0.1 M, 0.5 ml) at 15 ml/min for 3 min. After reaction, the residual [^{11}C]CH_3I was evaporated at 55°C under a helium gas stream for 5 min. The product was then diluted with 0.9% sodium chloride for injection, USP (9.0 ml) and 45 mM sodium phosphates, USP (0.5 ml). The final formulated product was then passed through a 0.22 µm filter into a sterile dose vial, and submitted for QC testing. Typical EOS yields were 4.0% (n=3).

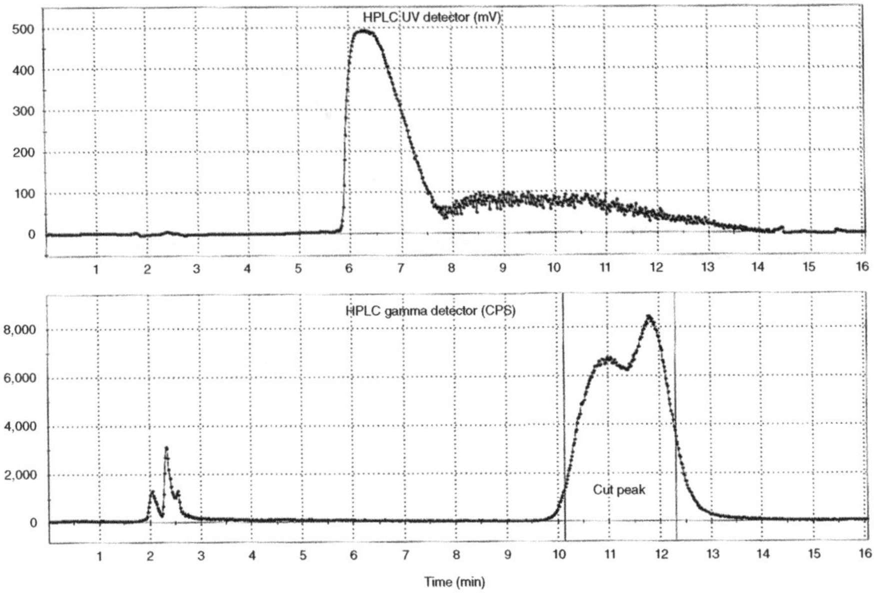

FIGURE 4 Typical ethanolic semi-preparative HPLC trace for [^{11}C]HED.

2.7.7 [^{11}C]OMAR

[^{11}C]OMAR was prepared according to the general procedure for loop chemistry described in Section 2.5. [^{11}C]MeOTf was passed through the HPLC loop at 40 ml/min for 3 min. Following reaction, the mixture was purified by semi-preparative HPLC (column: Phenomenex Luna C-5, 100×10 mm, mobile phase: 20 mM NH_4OAc in 45% ethanol, flow rate: 3 ml/min, typical trace: Fig. 5). The product peak (t_R ~ 15 min) was collected into 50 ml of water. The solution was then passed through a C18 sep-pak, and washed with 7 ml sterile water. The product was then eluted with 0.5 ml of USP ethanol followed by 9.5 ml of USP saline. This final formulation was then passed through a 0.22-μ filter into a sterile dose vial and submitted for QC testing. Typical EOS yields were 4.1% (n=3).

2.7.8 [^{11}C]PBR28

[^{11}C]PBR28 was prepared according to the general procedure for loop chemistry described in Section 2.5. [^{11}C]MeOTf was passed through the HPLC loop at 15 ml/min for 5 min. Following reaction, the mixture was purified by semi-preparative HPLC (column: Phenomenex Luna C-5, 100×10 mm, mobile phase: 20 mM NH_4OAc in 40% ethanol, flow rate: 3 ml/min, typical trace: Fig. 6). The product peak (t_R ~ 10 min) was collected into 50 ml of water. The solution was then passed through a C18 sep-pak, and washed with 7 ml sterile water. The product was then eluted with 0.5 ml of USP ethanol followed by 9.5 ml of USP saline. This final formulation was then passed through a 0.22-μ filter into a sterile dose vial and submitted for QC testing. Typical EOS yields were 2.8% (n=8).

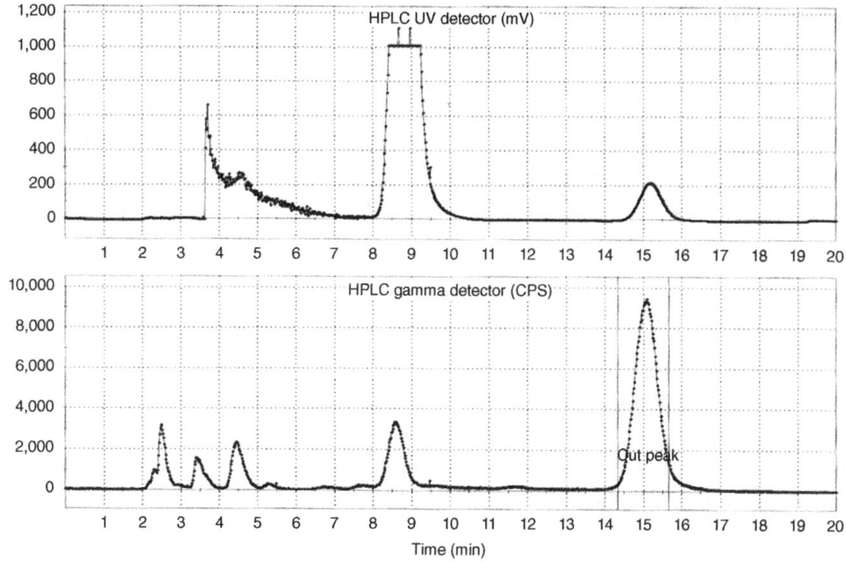

FIGURE 5 Typical ethanolic semi-preparative HPLC trace for [^{11}C]OMAR.

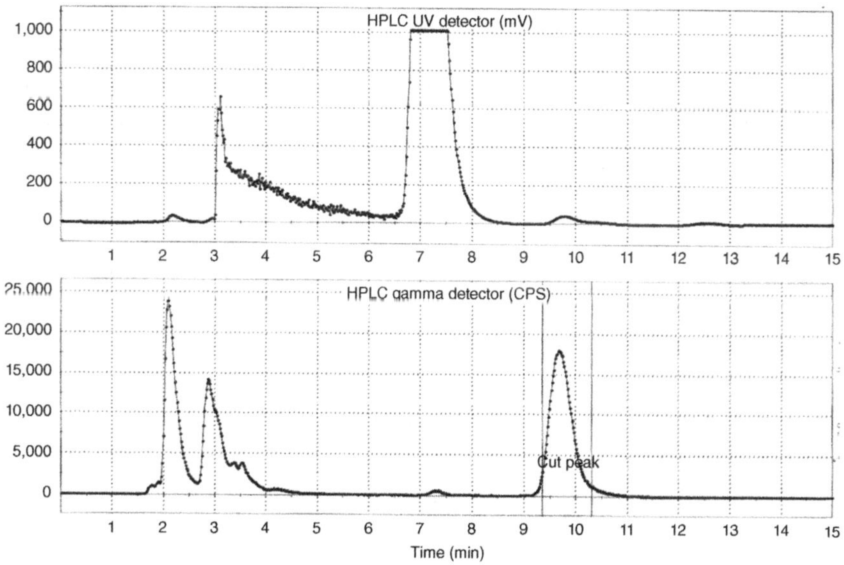

FIGURE 6 Typical ethanolic semi-preparative HPLC trace for [^{11}C]PBR28.

2.7.9 [^{11}C]Pittsburgh Compound B ([^{11}C]PiB)

[^{11}C]PBR28 was prepared according to the general procedure for loop chemistry described in Section 2.5. [^{11}C]MeOTf was passed through the HPLC loop at 15 ml/min for 5 min. Following reaction, the mixture was purified by semi-preparative

HPLC (column: Phenomenex Jupiter 5 μ C4, 150 × 10 mm, mobile phase: 20 mM NH_4OAc in 30% ethanol, flow rate: 4 ml/min, typical trace Fig. 7). The product peak (t_R ~ 8–9 min) was collected into 25 ml of water. The solution was then passed through a C18 sep-pak, and washed with 3 ml of sterile water. The product was eluted with 0.5 ml of USP ethanol followed by 9.5 ml of USP saline. This final formulated product was filtered through a 0.22-μm sterile filter into a sterile dose vial and submitted for QC testing. Typical EOS yields were 2.3% (n = 3).

2.7.10 [^{11}C]PMP

[^{11}C]PMP was prepared according to the general procedure for reactor chemistry described in Section 2.4. [^{11}C]MeOTf was prepared using the general procedure outlined earlier and bubbled through the [^{11}C]PMP precursor solution (0.5 mg was dissolved in ethanol (100 μl), and 0.1 M $NaHCO_3$ (30 μl) was added) at 15 ml/min for 3 min. The reaction mixture was then diluted with 1 ml of HPLC mobile phase, and purified by semi-preparative HPLC (column: Phenomenex Luna C18, 150 × 10 mm, mobile phase: 50 mM NH_4OAc in 5% EtOH, flow rate: 4 ml/min, typical trace: Fig. 8). The product peak was collected (t_R ~ 12–14 min) for 2 min (8 ml), diluted with USP saline (2 ml) to provide a final ethanol concentration less than 5%, and passed through a 0.22 μm sterilizing filter into a sterile dose vial. The product was then submitted for QC. Typical EOS yields were 3.7% (n = 6).

2.7.11 [^{11}C]Raclopride

[^{11}C]Raclopride was prepared according to the general procedure for loop chemistry described in Section 2.5. [^{11}C]MeOTf was passed through the HPLC loop at 40 ml/min for 3 min. Following reaction, the mixture was purified by semi-preparative

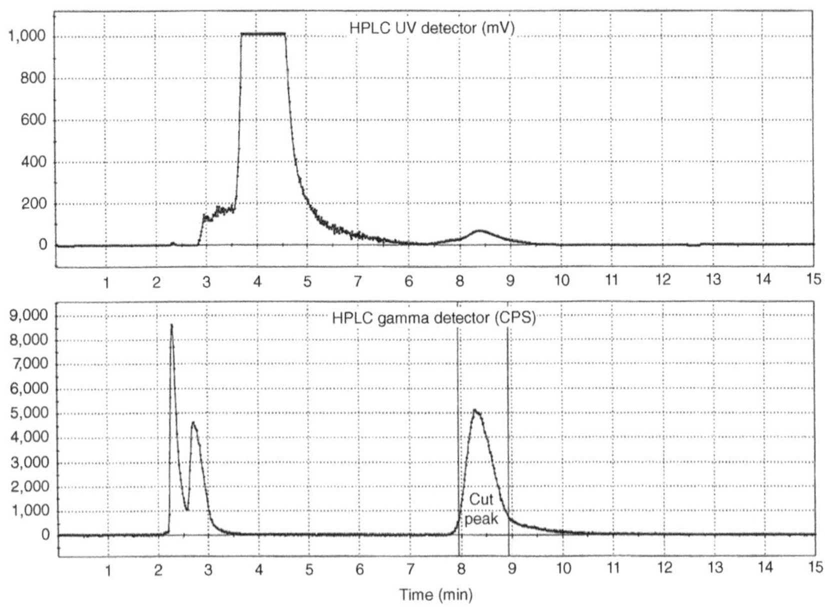

FIGURE 7 Typical ethanolic semi-preparative HPLC trace for [^{11}C]PiB.

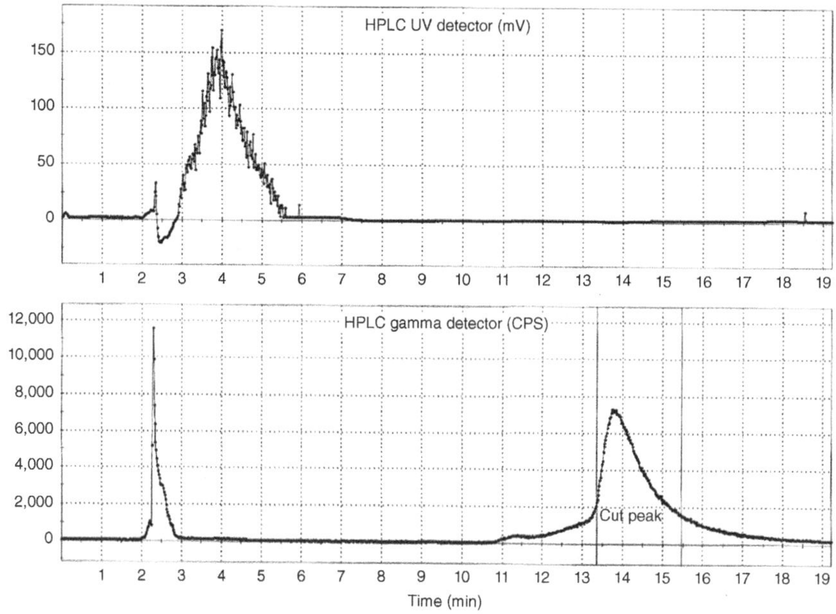

FIGURE 8 Typical ethanolic semi-preparative HPLC trace for [^{11}C]PMP.

HPLC (column: Phenomenex Luna NH_2, 250×10mm, mobile phase: 20mM NH_4OAc in 10% ethanol, flow rate: 3ml/min, typical trace: Fig. 9). The radioactive peak was collected (t_R ~ 10min) and diluted with 5.5ml USP saline. This aqueous solution was filtered through a 0.22-μm sterile filter into a sterile dose vial and submitted for QC testing. Typical EOS yields were 3.6% (n=144).

2.8 Synthesis Module Cleaning

2.8.1 Daily Clean Cycle

The module is reset to its normal configuration. Between daily runs, the vials, reaction vessel, and HPLC loop are washed and dried. The vials are loaded as follows: vial 1: water (1ml); vial 2: ethanol (1ml); vial 3: ethanol (3ml); vial 4: ethanol (10ml); vial 5: ethanol (10ml); vial 6: 70% ethanol (10ml). The water from vial 1 is initially passed through the reactor and the HPLC loop to waste, followed by the ethanol. Helium is then blown through for 5min to dry the HPLC loop. The reaction vessel is then heated to 80°C and placed under vacuum for 10min to dry it. The reformulation components are also then cleaned with ethanol and dried by using helium flow. The daily clean cycle takes approximately 25min.

2.8.2 Weekly Clean Cycle

The module is reset to its normal configuration. The first "clean" cycle washes everything with water. This is followed by a "clean/disinfect" cycle with 70% ethanol, and finally, a "dry" cycle using only ethanol. During the "clean" and "clean/disinfect" programs, water and 70% ethanol are washed from the vials through the reactor and the HPLC loop to waste, respectively. The dry cycle washes ethanol from the

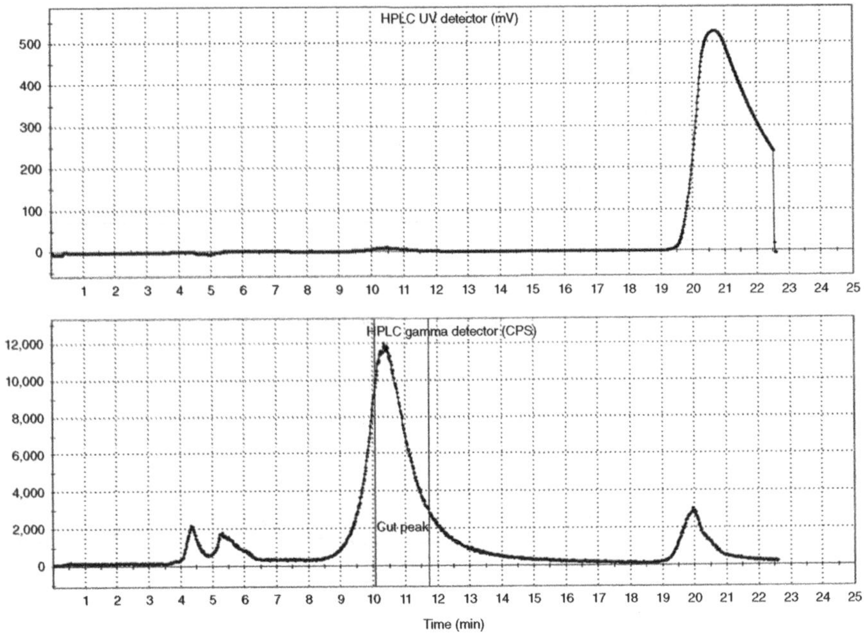

FIGURE 9 Typical ethanolic semi-preparative HPLC trace for [^{11}C]Raclopride.

vials through the reaction vessel and HPLC loop to waste, before heating the reaction vessel to 80°C and placing it under vacuum for 10 min. During the "dry" cycle, the reformulation components are cleaned with ethanol and dried by using helium flow.

3 QC PROCEDURES

CAUTION: All radiochemicals produced for clinical use must have local regulatory approval (e.g., FDA, EMEA, MHRA, PFSB, etc.) prior to human use. QC procedures must be carried out by trained personnel and each dose must meet all established QC criteria before release to the clinic.

QC procedures of each radiotracer, based upon the current requirements for radiopharmaceuticals laid out in the US Pharmacopeia [17], are summarized below. Complete QC data for batches produced using the methods disclosed herein are summarized in Table 1. Each dose met all of the established QC criteria.

3.1 Chemical Purity and Radiochemical Purity/Identity

Chemical and radiochemical purity/identity are analyzed using an HPLC equipped with a radioactivity detector and a UV detector using the methods outlined below. Radiochemical purity for doses must be greater than 95%, and identity is confirmed by comparing the retention time of the radiolabeled product with that of the corresponding unlabeled reference standard.

3.1.1 [^{11}C]Carfentanil

Column: Phenomenex Luna C8(2) 5 μ, 100 × 2.0 mm; mobile phase: 35% MeOH: 65% 20 mM NH_4OAc; pH 4.5; flow rate: 0.8 ml/min; oven: 40°C; UV: 218 nm; t_R = 3.0–4.0 min (Fig. 10).

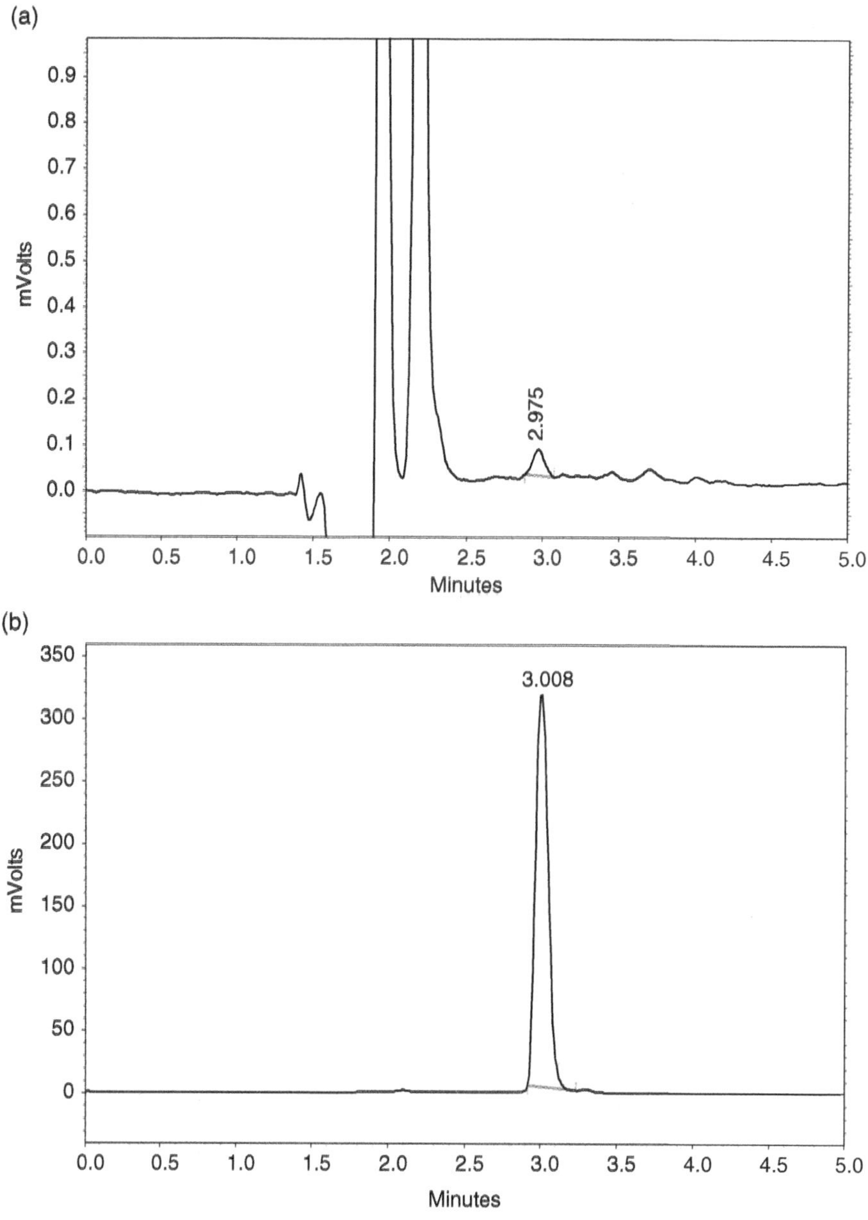

FIGURE 10 Carfentanil Analytical HPLC Trace: (a) UV, (b) Radioactivity.

3.1.2 [^{11}C]Choline

Column: Waters IC-Pak Cation M/D Column, 150×3.9 mm; mobile phase: 5 mM aqueous hydrochloric acid, flow rate: 1.0 ml/min; conductivity detector: negative polarity; t_R=7.5–8.5 min (Fig. 11).

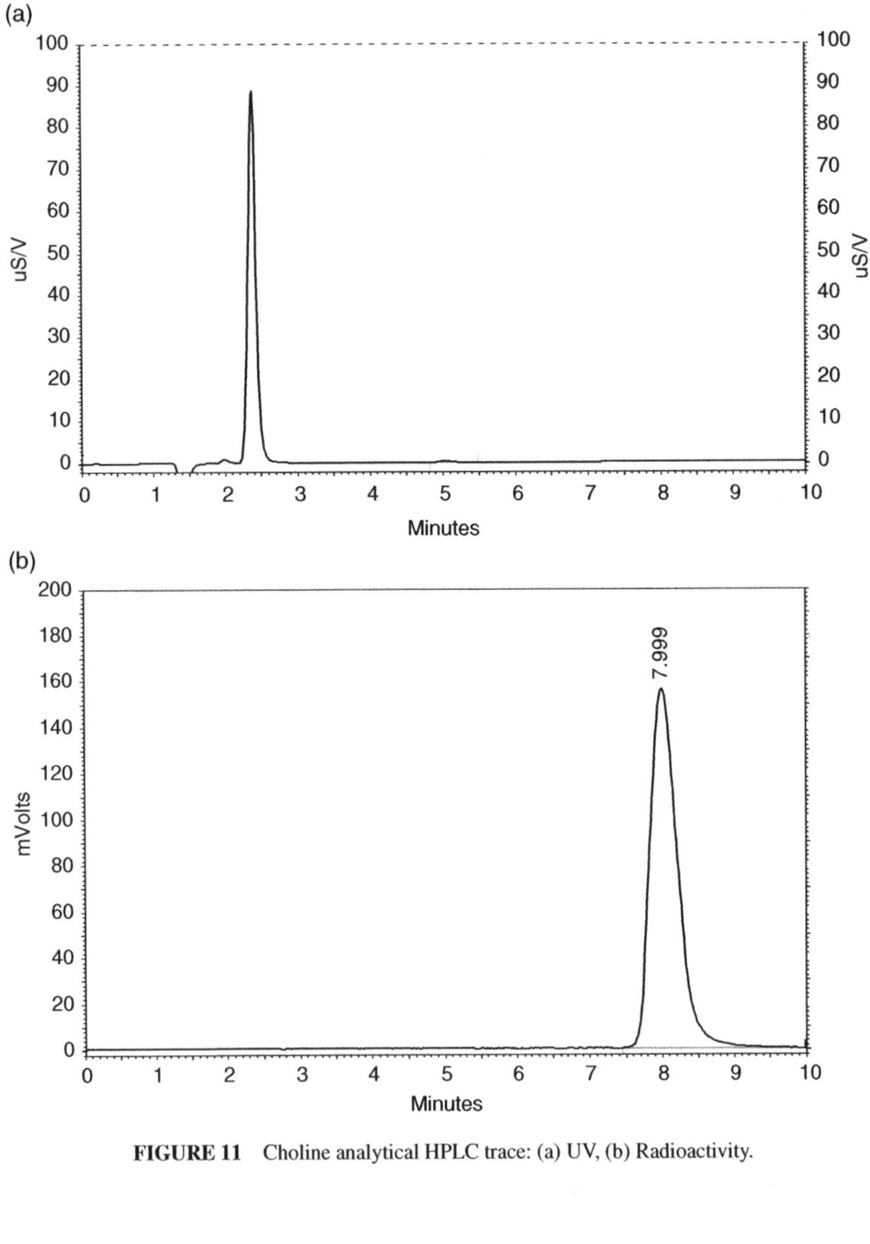

FIGURE 11 Choline analytical HPLC trace: (a) UV, (b) Radioactivity.

3.1.3 [^{11}C]DASB

Column: Phenomenex Luna C18 5 μ, 100×2.0 mm; mobile phase: 35% MeOH: 65% 20 mM NH_4OAc; pH: 4.5; flow rate: 0.3 ml/min; oven: 30°C; UV: 254 nm; t_R=5.0–6.0 min (Fig. 12).

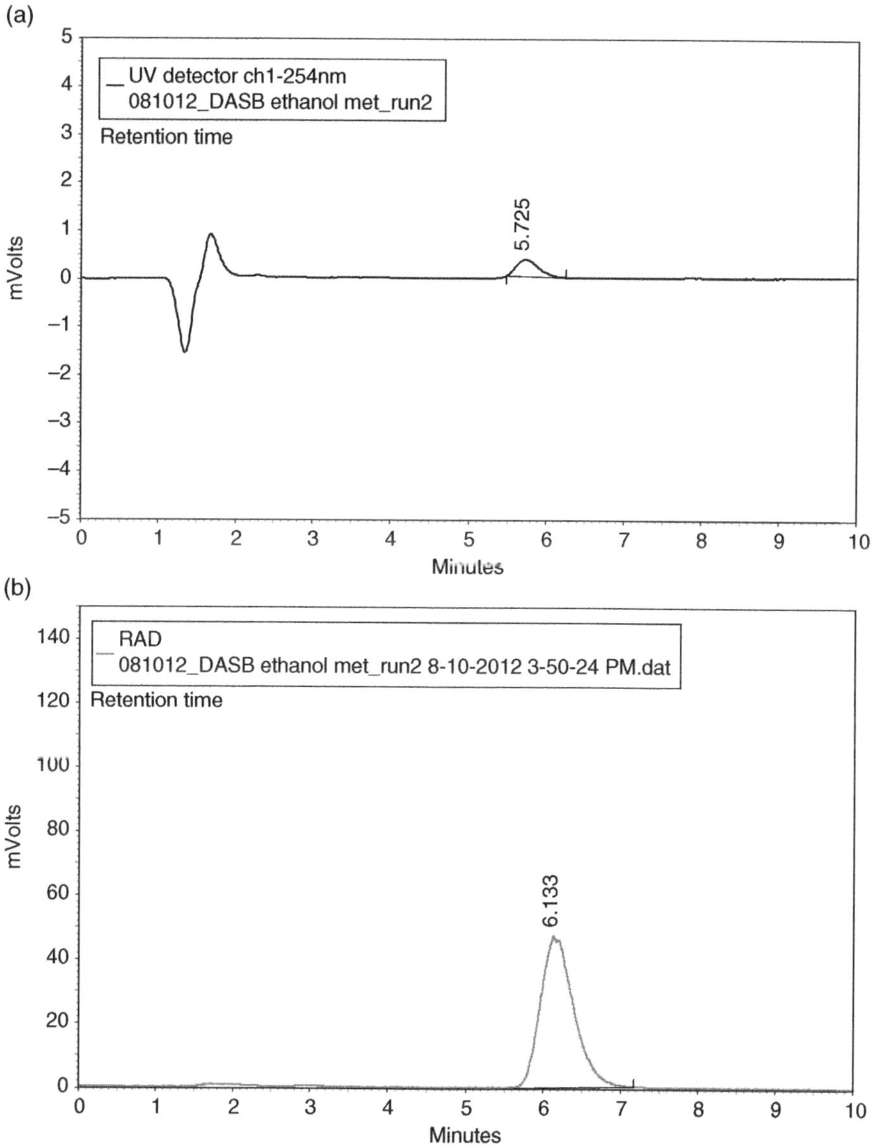

FIGURE 12 DASB analytical HPLC trace: (a) UV, (b) Radioactivity.

3.1.4 [¹¹C]DTBZ

Column: Phenomenex Luna C18 5 μ, 100 × 2.0 mm column; mobile phase: 25:75 methanol:water and 20 mM NH_4OAc; pH 4.5; flow rate: 0.6 ml/min; oven: 40°C; UV: 240 nm; t_R = 3.0–4.0 min (Fig. 13).

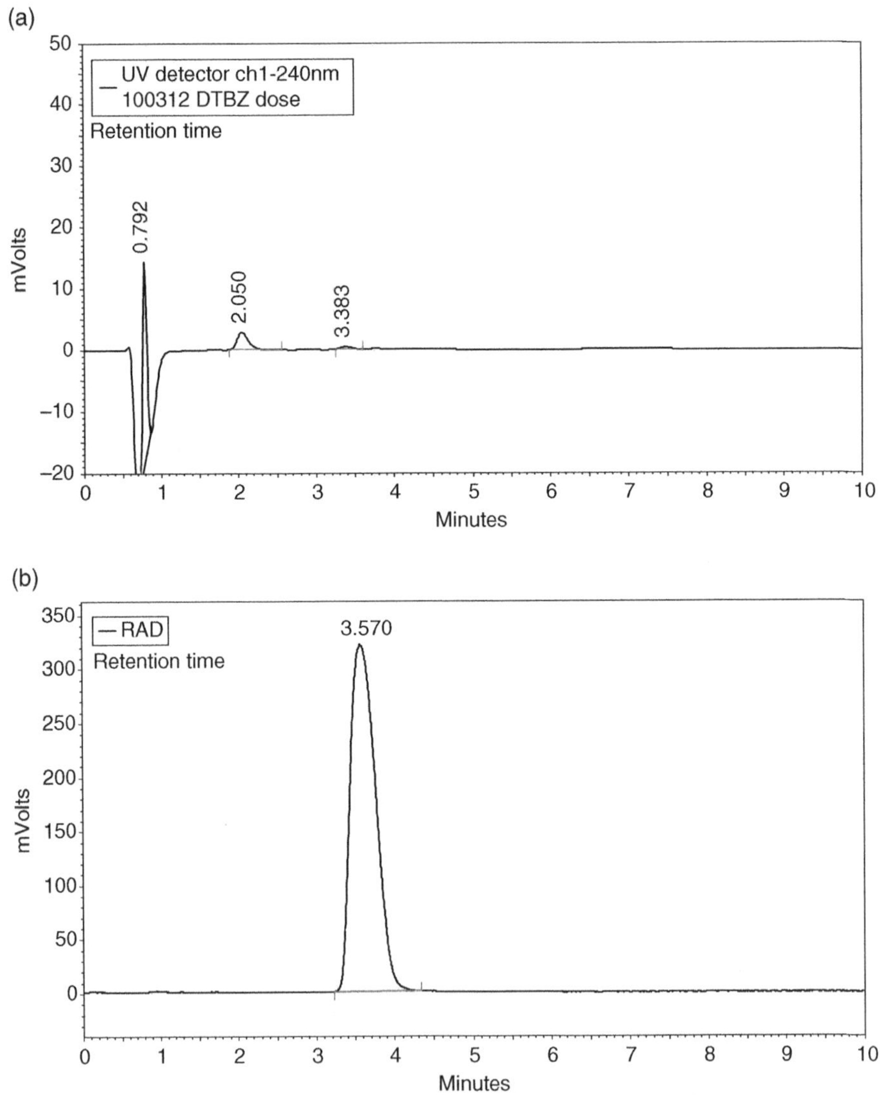

FIGURE 13 DTBZ analytical HPLC trace: (a) UV, (b) Radioactivity.

3.1.5 [^{11}C]HED

Column: Phenomenex Luna C18(2) 3 μ, 150 × 4.6 mm; mobile phase: 3.7% MeCN/20 mM NH_4OAc; pH 4.5; flow rate: 1.0 ml/min; oven: 30°C; UV: 280 nm; t_R = 6.5–7.5 min (Fig. 14).

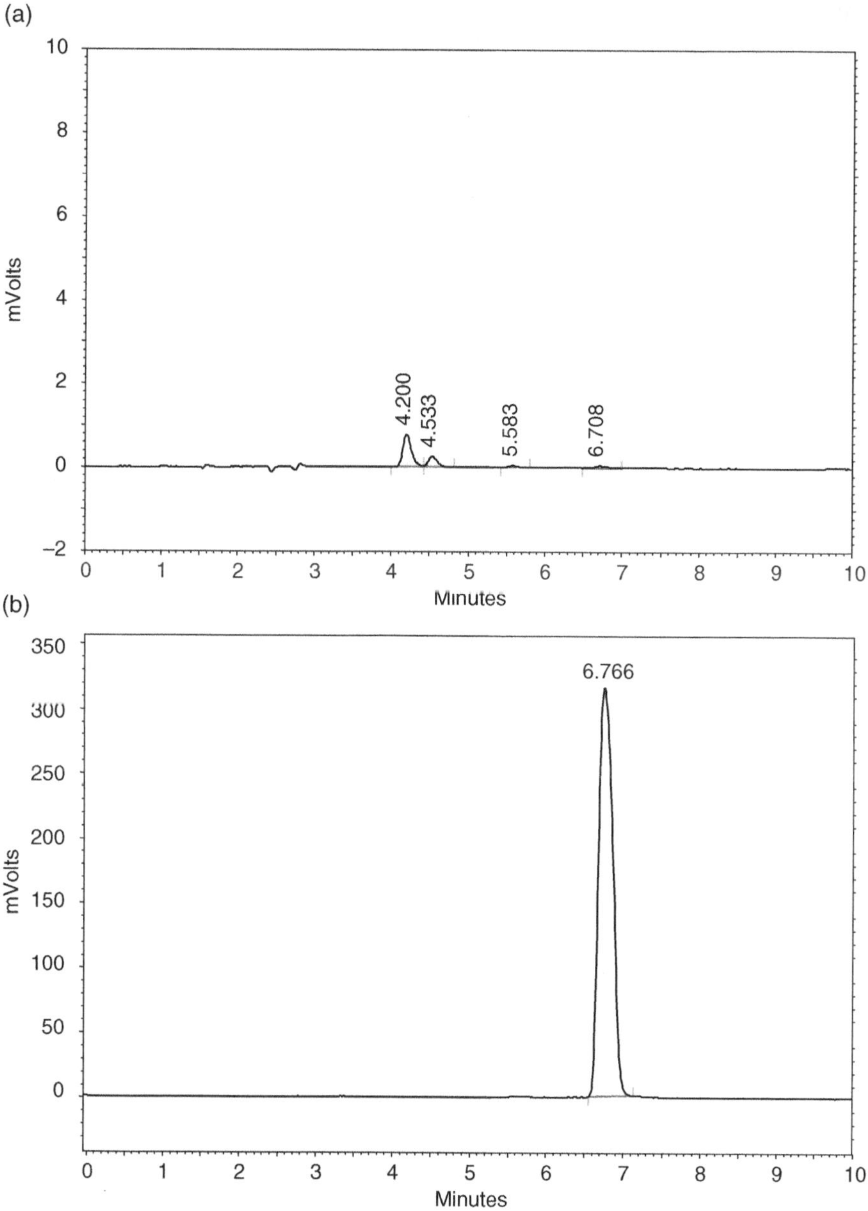

FIGURE 14 HED analytical HPLC trace: (a) UV, (b) Radioactivity.

3.1.6 [^{11}C]Methionine

Column: Phenomenex Luna C18(2) 5 µ, 150×4.6 mm; mobile phase: 8 mM $Cu(OAc)_2$: 17 mM L-proline: 30 mM NaOAc; flow rate: 1.5 ml/min; oven: 40°C; UV: 304 nm; t_R = 10.0–11.0 min (Fig. 15).

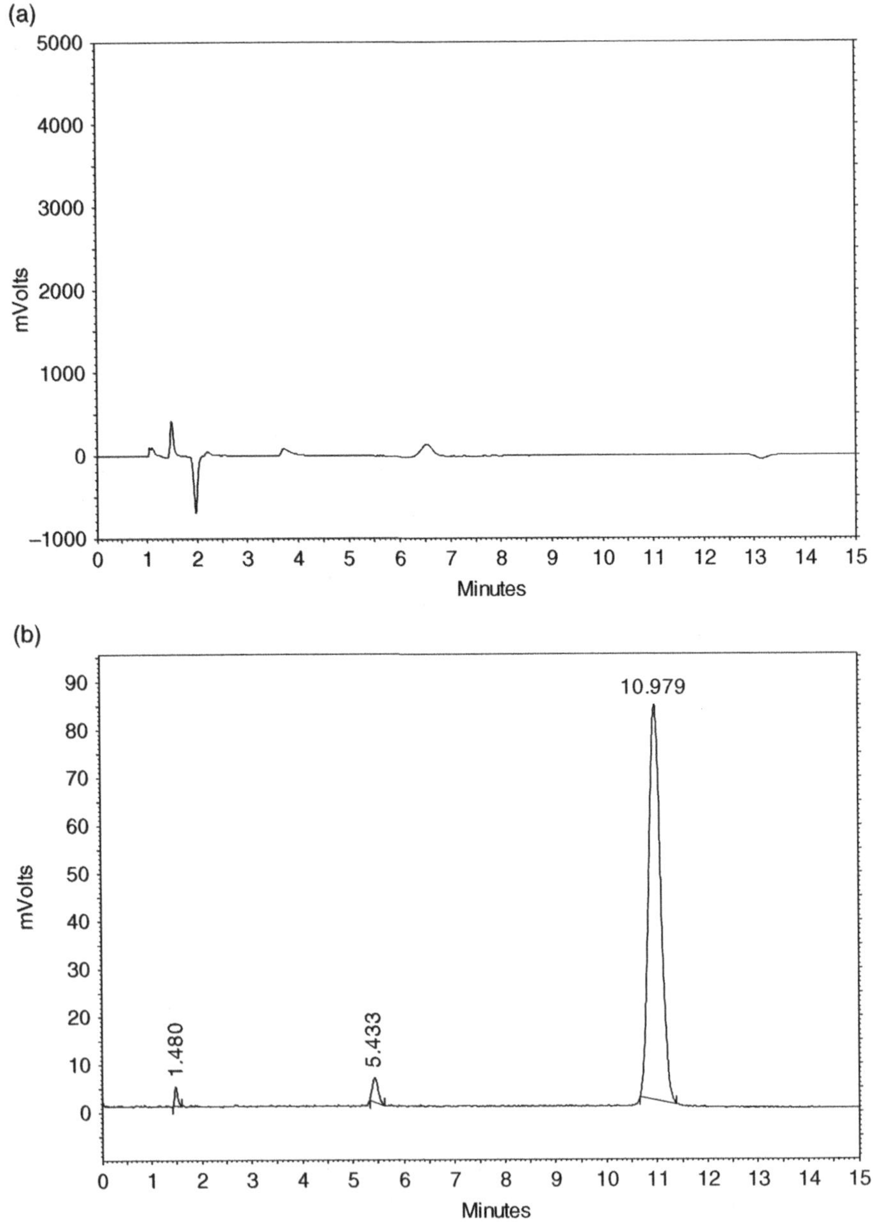

FIGURE 15 Methionine analytical HPLC trace: (a) UV, (b) Radioactivity.

3.1.7 [^{11}C]OMAR

Column: Phenomenex Jupiter C18 5 μ, 250 × 4.6 mm; mobile phase: 60% MeCN: 20% MeOH: 20% 20 mM NaH_2PO_4; flow rate: 1.5 ml/min; oven: 30°C; UV: 254 nm; t_R = 4.0–5.0 min (Fig. 16).

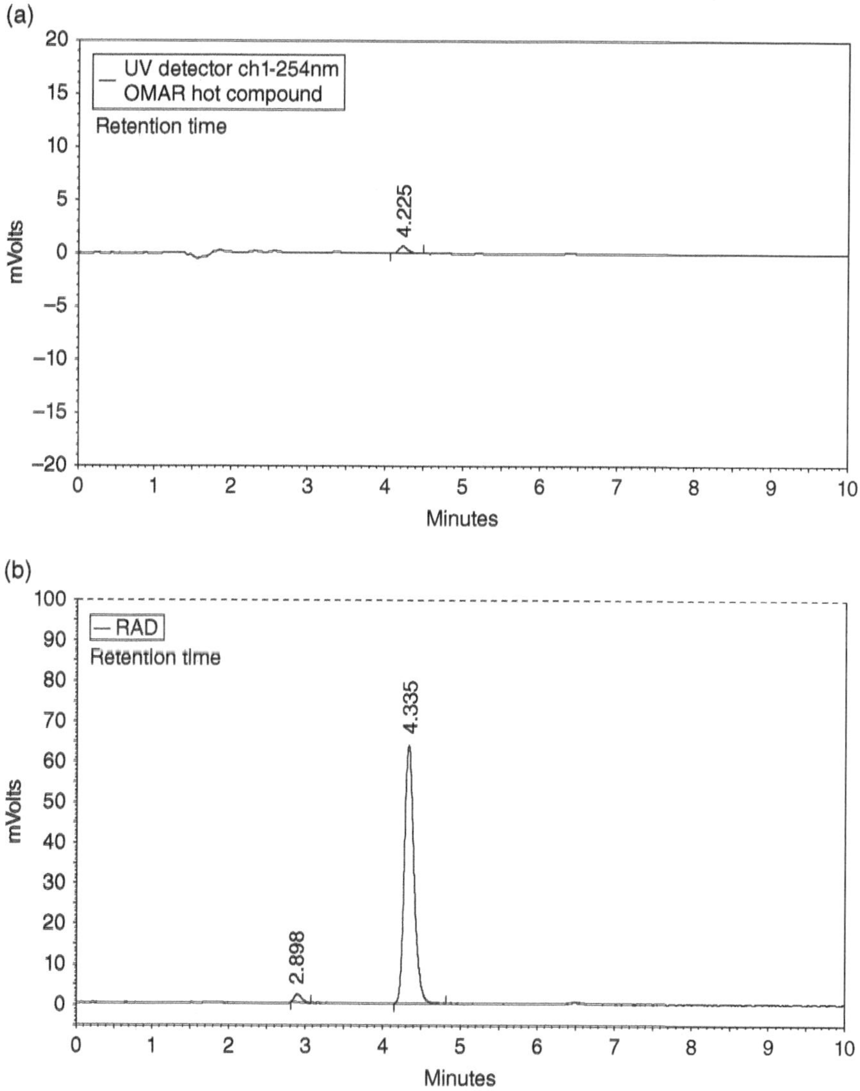

FIGURE 16 OMAR analytical HPLC trace: (a) UV, (b) Radioactivity.

3.1.8 [11C]PBR28

Column: Phenomenex Luna C18(2) 5 μ, 150×4.6 mm; mobile phase: 45% MeCN: 10 mM ammonium formate; flow rate: 3.0 ml/min; oven: 40°C; UV: 220 nm; t_R=7.5–8.5 min (Fig. 17).

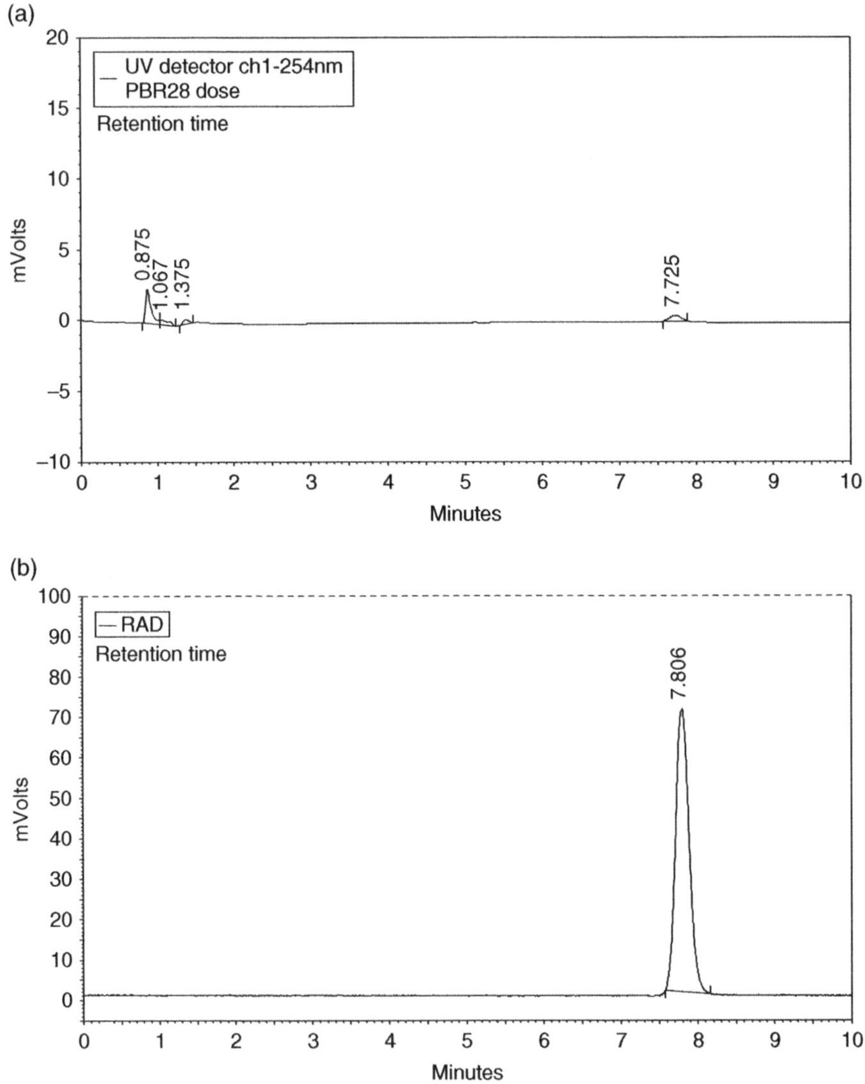

FIGURE 17 PBR28 analytical HPLC trace: (a) UV, (b) Radioactivity.

3.1.9 [^{11}C]PiB

Column: Phenomenex Luna C18(2) 5 m, 100 × 2.0 mm; mobile phase: 45% MeOH: 55% 20 mM NH_4OAc; pH: 4.5; flow rate: 0.6 mL / min; oven: 40°C; UV: 350 nm; t_R = 3.5–4.5 min (Fig. 18).

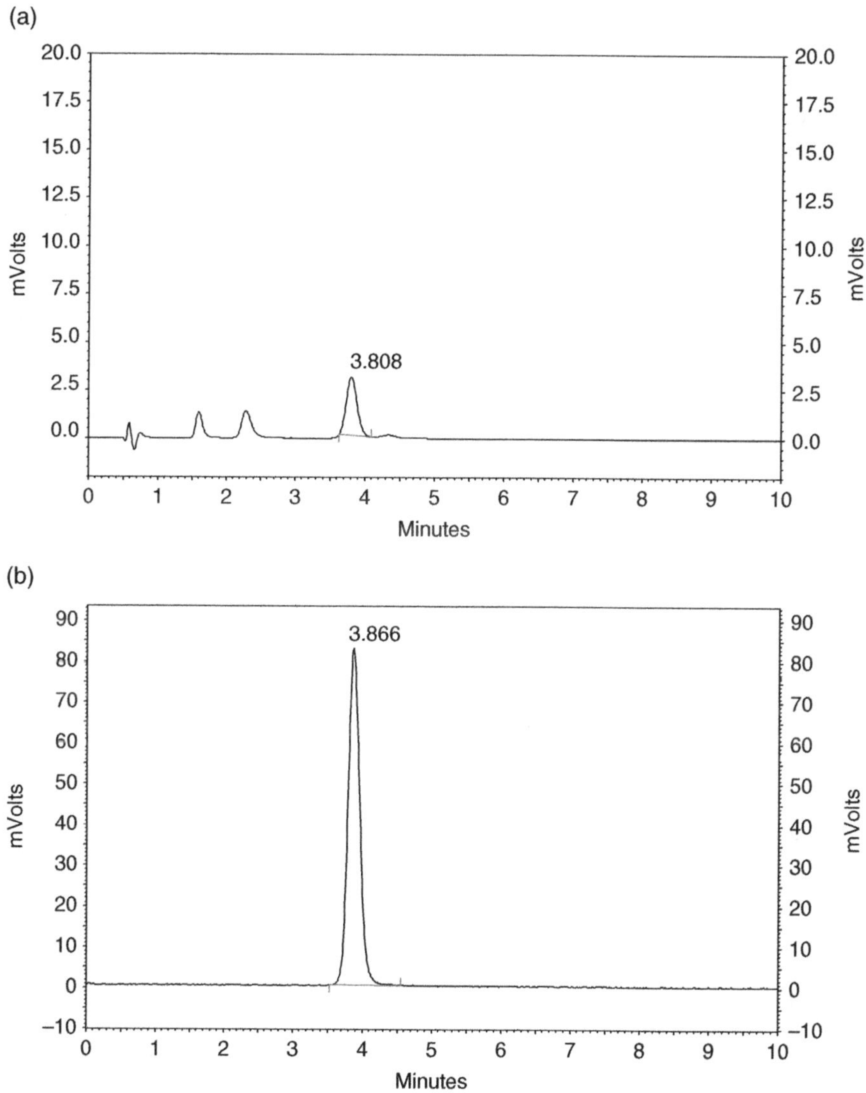

FIGURE 18 PiB Analytical HPLC Trace: (a) UV, (b) Radioactivity.

3.1.10 [^{11}C]PMP

Column: Phenomenex Luna C18(2) 5 μ, 150 × 2.0 mm; mobile phase: 5% MeOH: 95% 20 mM NH_4OAc; pH 4.5; flow rate: 0.35 ml/min; UV: 220 nm; t_R = 5.5–6.5 min (Fig. 19).

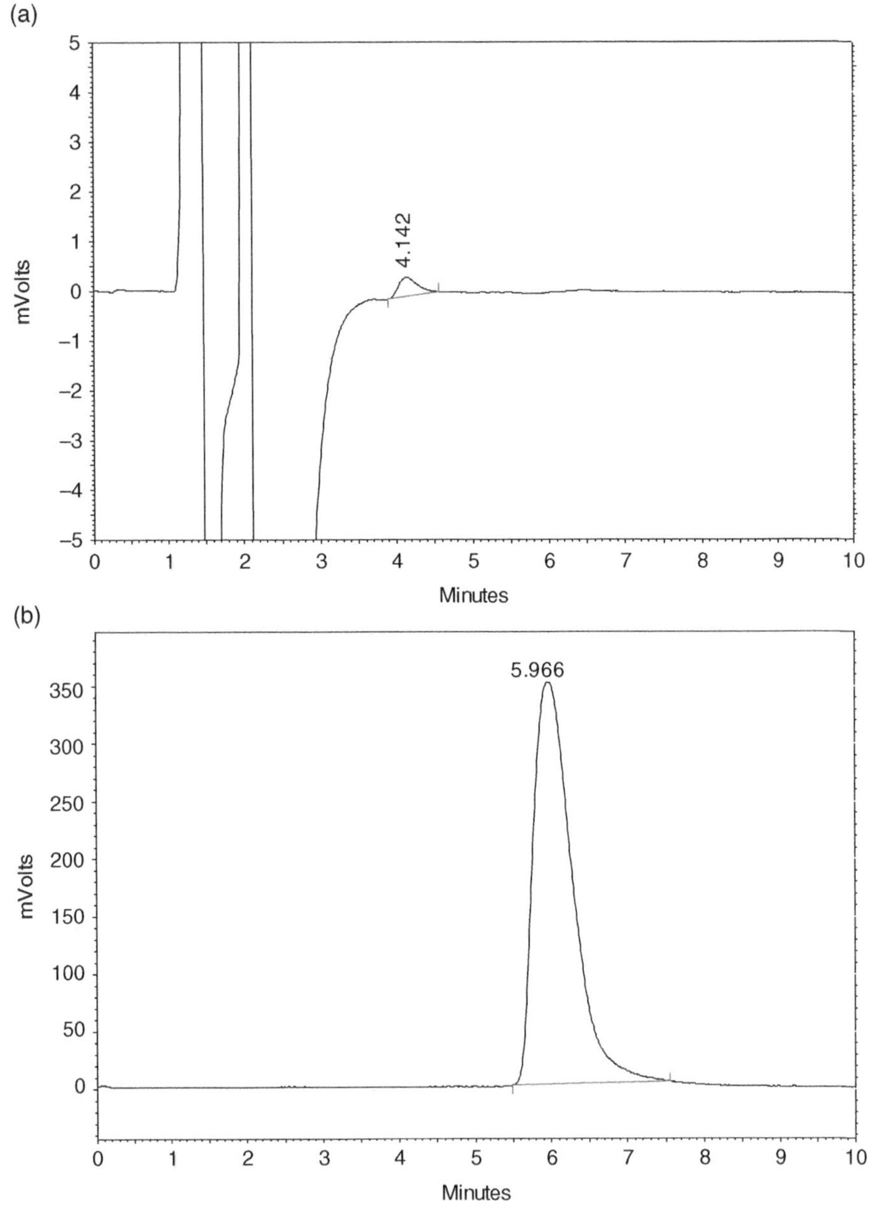

FIGURE 19 PMP Analytical HPLC Trace: (a) UV, (b) Radioactivity.

3.1.11 [^{11}C]Raclopride

Column: Phenomenex Luna C18 5 μ, 100×2.0 mm; mobile phase: 35% MeOH: 65% 20 mM NH_4OAc; pH: 4.5; flow rate: 0.8 ml/min; oven: 40°C; UV: 218 nm; t_R = 3.0–4.0 min (Fig. 20).

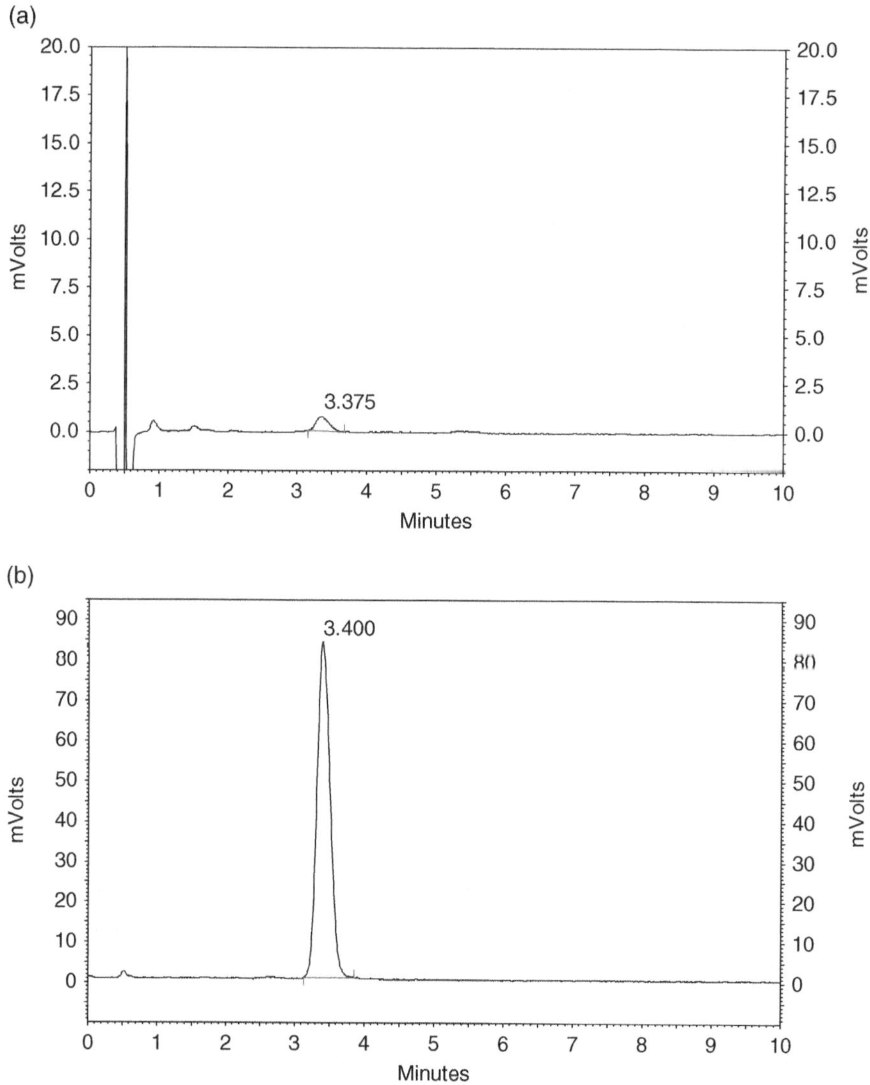

FIGURE 20 Raclopride Analytical HPLC Trace: (a) UV, (b) Radioactivity.

3.2 Visual Inspection

QC samples were examined behind a PET L-block and had to be clear, colorless, and free of any particulate matter.

3.3 Dose pH

Dose pH was determined by applying small amounts of the QC sample to colorpHast® pH 2.0–9.0 nonbleeding pH-indicator strips [18] and determined by visual comparison to the scale provided. Dose pH had to be 4.5–7.5.

3.4 Radionuclidic Identity

Radionuclidic identity is confirmed by measuring the half-life of radiopharmaceutical doses and comparing it to the known half-life of carbon-11 (20.4 min). Activities are measured using a Capintec dose calibrator [19] and half-life is calculated using equation (1). Calculated half-life must be 18.4–22.4 min and all doses met this criterion.

$$T_{1/2} = -\ln 2(\text{Time differennce}/(\text{In}\,(\text{ending activity}/\text{starting activity}))) \quad (1)$$

3.5 Sterile Filter Integrity Test

The sterile filter with needle which was used to convey the dose into the dose vial was connected to a nitrogen supply via a regulator. The filter was immersed in water and the pressure was gradually increased while observing for a steady stream of bubbles emanating from the needle. If this happened at a pressure greater than the filter acceptance level, then the filter was considered to have been intact and thus passed.

3.6 Endotoxin Analysis

The endotoxin content of doses was analyzed by a Charles River Laboratories EndoSafeTM Portable Testing System [20] as per the US Pharmacopoeia [21]. Doses had to contain less than or equal to 175 endotoxin units (EUs) which were adjusted to 17.5 EU/ml as our formulation volume was 10 ml.

3.7 Sterility Testing

Samples of radiopharmaceutical doses were added to tubed fluid thioglycolate media (FTM) and to tubed tryptic soy broth media (TSB). FTM was used to test for anaerobes, aerobes, and microaerophiles; TSB was used to test for fastidious and nonfastidious organisms [22]. These media tubes and positive and negative controls were incubated at 30–35°C (FTM) and 22°C (TSB) for 14 days according to current USP guidelines [23]. Tubes were visually inspected on the third, eight, and fourteenth day after inoculation and were compared to positive and negative controls. Positive controls had to show growth (turbidity) and the test doses and negative controls had to have no growth after 14 days in order to pass the test for sterility. Positive controls could be established separately as growth promotion tests according to USP guidelines [24].

WASTE DISPOSAL INFORMATION

All hazardous chemicals and toxic materials were disposed of according to Prudent Practices in the Laboratory (Washington, DC: National Academy Press, 1995).

CHEMICAL ABSTRACTS NOMENCLATURE (REGISTRY NUMBER)

Acetonitrile (75-05-8)
(1*R*,2*S*)-3-[-2-amino-1-hydroxy-propyl]phenol (Metaraminol, 54-49-9)
Ammonium acetate (631-61-8)
Ammonium formate (540-69-2)
Copper(II) acetate (142-71-2)
2-(Dimethylamino)ethanol (108-01-0)
Ethanol (64-17-5)
Methanol (67-56-1)
Monosodium phosphate (7558-80-7)
L-Proline (147-85-3)
Sodium acetate (127-09-3)
Tetrabutylammonium hydroxide (2052-49-5)

REFERENCES AND NOTES

For detailed supplier information, see Appendix 1.

1. (a) P. T. Anastas, J. C. Warner, *Green Chemistry Theory and Practice*, 1998, Oxford University Press, New York; (b) I. T. Horváth, P. T. Anastas, *Chem. Rev.*, 2007, 107, 2169; (c) C. J. Li, B. M. Trost, *Proc. Natl. Acad. Sci. U. S. A.*, 2008, 105, 13197; (d) R. A. Sheldon, *Chem. Soc. Rev.*, 2012, 41, 1437; (e) W. Zhang, B. W. Cue, *Green Techniques for Organic Synthesis and Medicinal Chemistry*, 2012, John Wiley and Sons, Chichester.
2. (a) X. Shao, P. L. Schnau, M. V. Fawaz, P. J. H. Scott, *Nucl. Med. Biol.*, 2013, 40, 109; (b) X. Shao, M. V. Fawaz, K. Jang, P. J. H. Scott, *Appl. Radiat. Isot.*, 2014, 89, 125.
3. GE standard carbon-11 target with a 75 μm Havar target foil and GEHC standard vacuum foil.
4. Ultra high purity nitrogen gas with 0.5% of oxygen was purchased from Purity Plus.
5. Molecular sieve (4A, 80/100 mesh) was purchased from Alltech, pn: 5624.
6. Shimalite-Ni reduced (80/100 mesh) was purchased from Shimadzu (Kyoto, Japan), pn: 22127719.
7. Phosphorus pentoxide (with moisture indicator) was purchases from Fluka, pn: 79610.
8. Carborsphere (60/80 mesh) was purchased from Alltech (Grace), pn: 5680.
9. Iodine (crystals) was purchases from EMD (Germany), pn: IX01203.
10. Ascarite II (8–20 mesh, 3 in. in TRACERlab standard column) from Thomas Scientific, pn: C049H40.
11. Porapak type Q (0.3 g, 50–80 mesh) was purchases from Alltech, pn: 2700.
12. Silver triflate-Graphpac was prepared by dissolving 5 g of silver trifluoromethanesulfonate (Sigma Aldrich) in 100 ml of acetonitrile (HPLC grade, Fisher) and then mixed with 10 g of Graphpac (80/100 mesh, GC grade, Altech). The solvent was evaporated by rotary vapor under vacuum and 0.35 g of dried powder was used to pack a Tracerlab standard column.
13. X. Shao, R. Hoareau, A. C. Runkle, L. J. M. Tluczek, B. G. Hockley, B. D. Henderson, P. J. H. Scott, *J. Label. Compd. Radiopharm.*, 2011, 54, 819.
14. Sterile water for injection, USP was purchased from Hospira (pn: 0409-4887-50) and used as received.

15. 0.9% Sodium Chloride, USP was purchased from Hospira (pn: 0409-4888-50) and used as received.
16. USP saline is not used for elution, as it co-elutes an unknown chemical impurity off the extraction disk.
17. U.S. Pharmacopeia <823>; Radiopharmaceuticals for Positron Emission Tomography Compounding. USP 32-NF 27. 2009.
18. EMD Chemicals Inc., USA (pn: 9578-3).
19. Capintec, Inc. USA.
20. Charles River Laboratories, USA.
21. U. S. Pharmacopeia <85>; Bacterial Endotoxins Test. USP 32-NF 27. 2009.
22. Becton, Dickinson and Company, USA.
23. U. S. Pharmacopeia <71>; Sterility Test. USP 32-NF 27. 2009.
24. U. S. Pharmacopeia <61>; Microbial Enumeration Tests. USP 32-NF 27. 2009.

CHAPTER 22

SYNTHESIS AND APPLICATIONS OF [^{11}C]HYDROGEN CYANIDE

XIA SHAO, MELISSA E. RODNICK, ALLEN F. BROOKS, AND PETER J. H. SCOTT

Department of Radiology, University of Michigan School of Medicine, Ann Arbor, Michigan, USA

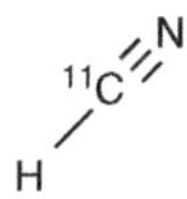

[^{11}C]hydrogen cyanide ([^{11}C]HCN)

1 INTRODUCTION

Carbon-11 is one of the most useful radioisotopes to label bioactive molecules for positron emission tomography (PET) imaging because of the ease with which it can be incorporated into many molecules without a significant effect on biological activity (for reviews or carbon-11 radiochemistry, see Ref. [1]). Radiochemical reactions have to be very fast (must typically occur within 5 min), high yielding, and clean enough that crude reaction mixtures can be purified rapidly (by semi-preparative high-performance liquid chromatography, HPLC or solid-phase extraction (SPE) techniques) to provide pure radiopharmaceuticals as sterile, pyrogen-free isotonic solutions suitable for i.v. administration to a patient.

Carbon-11 is prepared in our facility as [^{11}C]carbon dioxide (see Section 2.1) and can be used as is by reacting it with, for example, Grignard reagents [2], or converted into some other reactive intermediate. Typically, this is a methylating agent such as [^{11}C]CH_3I or [^{11}C]CH_3OTf, and this is the most widely used radiolabeling strategy in our carbon-11 radiopharmaceutical manufacturing program [2, 3]. However, not all target radiotracers can be prepared using these labeling strategies, and it is important to have other methods available. One such strategy that is being increasingly used in our group is carbon-11 cyanation chemistry

Radiochemical Syntheses, Volume 2: Further Radiopharmaceuticals for Positron Emission Tomography and New Strategies for Their Production, First Edition. Edited by Peter J. H. Scott.

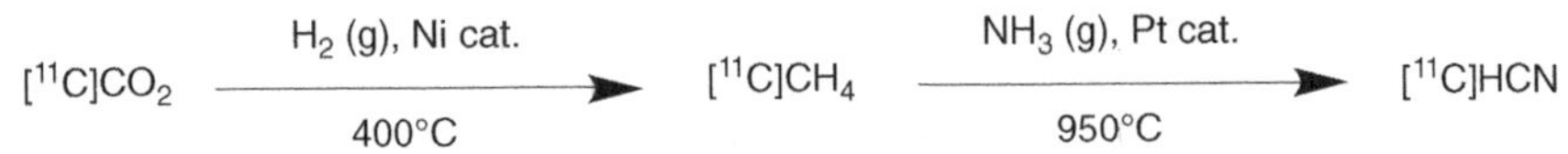

FIGURE 1 Synthesis of [^{11}C]HCN.

(for a review of the scope of such chemistry, see Ref. [4]). Herein, we describe the production of [^{11}C]HCN (Fig. 1), and use of the reagent in the synthesis of [^{11}C]LY2795050 and [^{11}C]gluatime, radiotracers of interest to our research program.

2 SYNTHESIS PROCEDURES

CAUTION: All radiochemical syntheses must be carried out using appropriate equipment in a facility authorized for the use of radioactive materials. Personal protective equipment must be worn and all local radiation safety laws followed.

2.1 Production of [^{11}C]CO_2

[^{11}C]CO_2 was produced by General Electric (GE) PETTrace cyclotron [5]. The target [6] was loaded with [^{14}N]N_2 gas [7] and bombarded with a proton beam (40 μA beam for 30 min) to generate approximately 3 Ci of [^{11}C]CO_2 by the $^{14}N(p,\alpha)^{11}C$ nuclear reaction.

2.2 Production of [^{11}C]HCN

[^{11}C]HCN is prepared in the University of Michigan PET Center using a GE Process Cabinet (ProCab, Fig. 2) [5]. Initially, [^{11}C]CO_2 was transferred from the cyclotron to the ProCab, where it was trapped on molecular sieves [8], mixed with a nickel catalyst [9], and reacted with hydrogen at 400°C to afford [^{11}C]CH_4. [^{11}C]CH_4 was then converted to [^{11}C]HCN by reaction with ammonia [10] over a platinum catalyst [11] at 950°C. Typical yields of [^{11}C]HCN were 1 Ci (from ~3 Ci of [^{11}C]CO_2).

3 REACTIONS WITH HYDROGEN CYANIDE

CAUTION: All radiochemical syntheses must be carried out using appropriate equipment in a facility authorized for the use of radioactive materials. Personal protective equipment must be worn and all local radiation safety laws followed.

[^{11}C]HCN was transferred from the ProCab to either a homemade synthesis module or a GE TRACERlab FXm synthesis module [5]. *Note: [^{11}C]HCN adsorbs to stainless steel lines, and so Teflon transfer lines should be employed to move [^{11}C]HCN around the laboratory.*

(a)

(b)

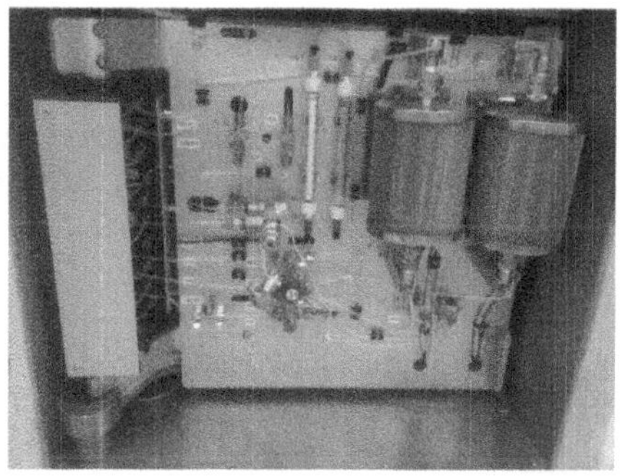

FIGURE 2 GE ProCab (a) containing system for the production of [^{11}C]HCN (b).

3.1 Synthesis of [^{11}C]LY2795050

3.1.1 Introduction

LY2795050 is a potent and selective antagonist of the kappa opioid receptor, and in carbon-11 labeled form has been successfully used in both preclinical and clinical positron emission tomography (PET) imaging studies [12]. The recently reported radiochemical synthesis of [^{11}C]LY2795050 involves a palladium-mediated ^{11}C-cyanation reaction [12], and we were interested in determining if that process could be simplified for routine clinical production in our laboratory. Here, we present initial approaches toward a radiochemical synthesis of [^{11}C]LY2795050 using a semi-automated two-step metal-free labeling procedure (Fig. 3).

3.1.2 Radiosynthesis

No-carrier-added [^{11}C]HCN (produced per Section 2.2 above) was bubbled into a solution of potassium carbonate (3 mg) and kryptofix-2.2.2 (9 mg) in DMF (300 µl) to generate [^{11}C]KCN. Iodo-precursor **1** (2–3 mg in 150 µl DMF) was added, and the reaction was heated at 120°C for 5 min to form the aryl[^{11}C]cyanide labeled intermediate **2** *in situ*. 30% H_2O_2 and 7.5 M NaOH were then added, and the reaction was maintained at 80°C for an additional 5 min to oxidize the cyano group and generate amide **3**. Final purification by semipreparative HPLC (Phenomenex Luna C18(2), 250×10 mm [13], 25:75 MeCN:H_2O, 10 mM NH_4OAc, pH 4.6, 3.0 ml/min) provided [^{11}C]LY2795050 (3.7% EOS yield,)

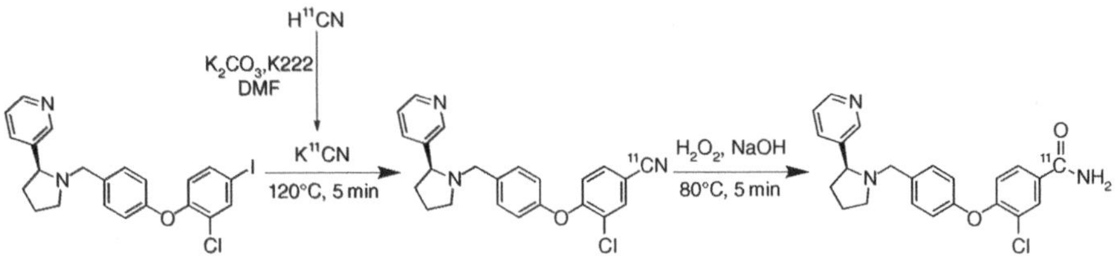

FIGURE 3 Synthesis of [^{11}C]LY2795050.

3.1.3 Quality Control

CAUTION: All radiochemicals produced for clinical use must have local regulatory approval (e.g., FDA, EMEA, MHRA, PFSB, etc.) prior to human use. Quality control procedures must be carried out by trained personnel, and each dose must meet all established quality control criteria before release to the clinic.

HPLC analysis of radiochemical purity was conducted using a Shimadzu LC-2010A$_{HT}$ Liquid Chromatograph [14] fitted with a UV detector and a Bioscan γ-detector [15] (column: Phenomenex Luna C18(2), 150×4.6mm [13], mobile phase: 25:75 MeCN:H_2O, 10mM NH_4OAc, pH 4.6, flow rate: 2.0ml/min, wavelength: 254nm, column temperature: 30°C. Radiochemical purity was greater than 99% and specific activity greater than 15,000Ci/mmol—for a typical analytical HPLC trace, see Fig. 4).

3.2 Synthesis of [^{11}C]Glutamine

3.2.1 Introduction

In recent years, there has been renewed interest in the synthesis of [^{11}C]glutamine as a tumor imaging agent because glutamine is known to play a key role in cancer cell proliferation [16]. A number of syntheses of [^{11}C]glutamine have been reported [17, 18], and we have adapted the method reported by Qu and co-workers [18] for use in our PET Center (Fig. 5).

3.2.2 Radiosynthesis

[^{11}C]HCN (~1000mCi prepared as described in Section 2.2) was bubbled into a mixture of K_2CO_3 (3mg) and kryptofix-222 (9mg) in DMF (300μl) at room temperature, and then (*S*)-*tert*-butyl 2-((*tert*-butoxycarbonyl)amino)-4-iodobutanoate **4** (3.5mg) in DMF (150μl) was added. The mixture was heated for 5min at 120°C to generate **5**. After cooling to room temperature, the reaction mixture was transferred into water (15ml) and passed through C-18 Plus cartridge [19]. The cartridge was further washed with water (10ml). The intermediate product (~500mCi) was eluted with acetonitrile (2ml). The solvent was removed with nitrogen bubbling at 120°C. Excess acetonitrile (2×1 ml) was added to aid in the removal of residual water azeotropically. To the dried residue, TFA/H_2SO_4 (5/1 ratio, 0.6ml) was added and the solution was heated at 110°C for 10min to generate [^{11}C]glutamine **6**. After cooling

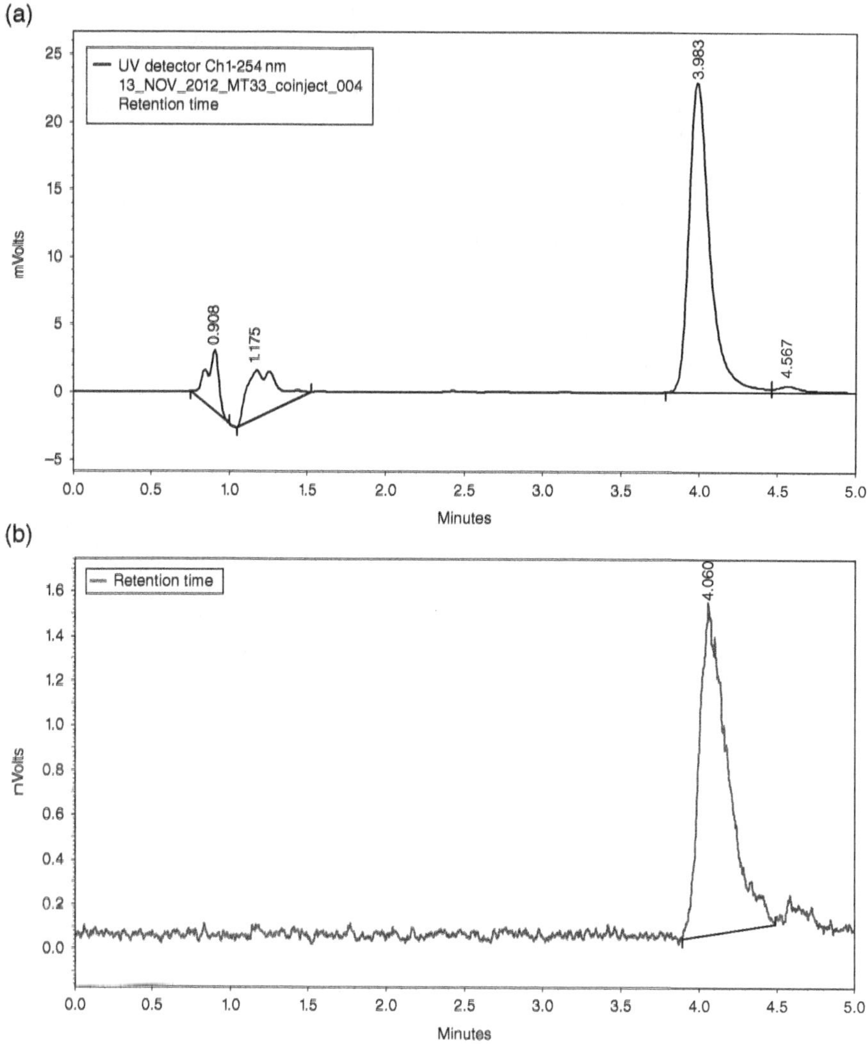

FIGURE 4 Analytical HPLC Trace of [^{11}C]LY2795050 (a: UV, b: Radioactivity).

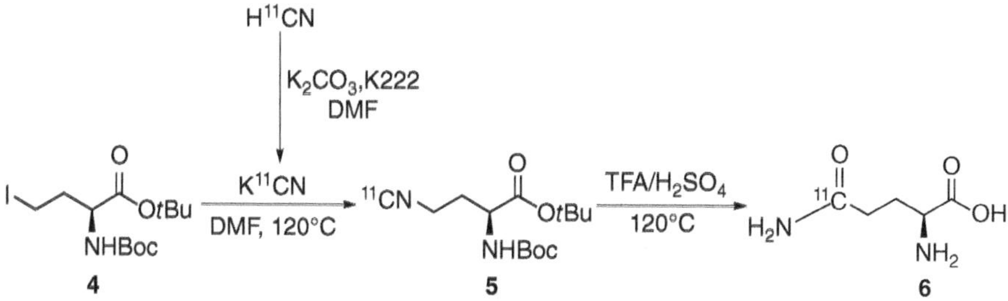

FIGURE 5 Synthesis of [^{11}C]Glutamine.

to room temperature, saturated Na_2CO_3 solution (1.75 ml) was slowly added and the mixture (~100 mCi) was passed through a C18 plus cartridge [19]. The eluent was further passed through a column of Ag11-A8 resin (10 g) [20], followed by a 0.22 μm Millex-GS sterile filter [21]. The column and filter were washed with excess water (3 ml) to recover the activity. The final product, L-[5-^{11}C]-glutamine (47 mCi), was obtained as a clear solution in water (5 ml, pH=6). The total synthesis time was 70 min. Uncorrected radiochemical yield was 1.6% based on $^{11}CO_2$ (corresponding to 17.8% decay-corrected to end-of-bombardment).

3.2.3 Quality Control

CAUTION: All radiochemicals produced for clinical use must have local regulatory approval (e.g., FDA, EMEA, MHRA, PFSB, etc.) prior to human use. Quality control procedures must be carried out by trained personnel, and each dose must meet all established quality control criteria before release to the clinic.

HPLC analysis of radiochemical purity was conducted using a Shimadzu LC-2010A_{HT} Liquid Chromatograph [14] fitted with a UV detector and a Bioscan γ-detector [15] (column: Phenomenex Chirex 3126, 150×4.6 mm [13], mobile phase: 1 mM $CuSO_4$, flow rate: 1.0 ml/min, wavelength: 254 nm, column oven: 30°C. Radiochemical purity was 97%. Optical purity was 95% (D-Gln 2%). Specific activity was greater than 2359 mCi/μmol (end-of-synthesis)—for a typical analytical HPLC trace, see Fig. 6).

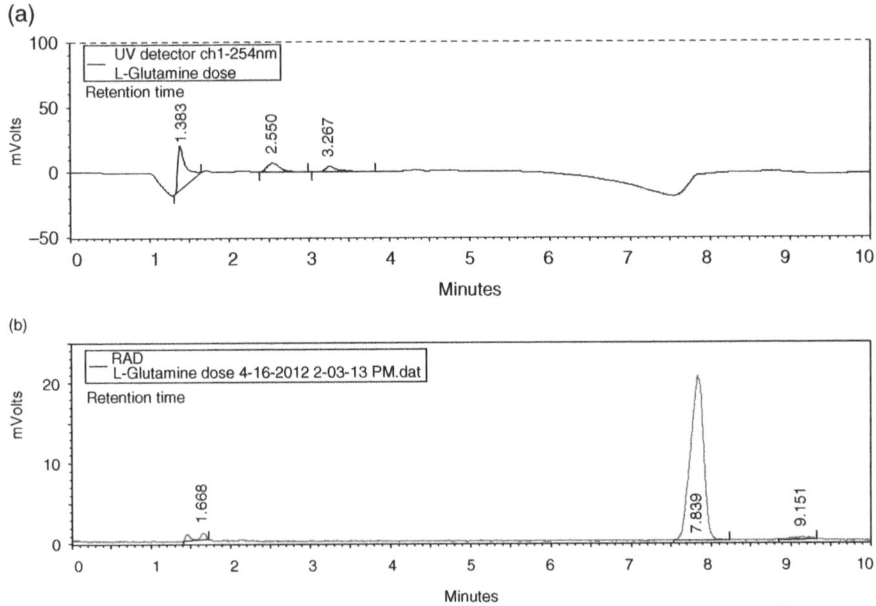

FIGURE 6 Analytical HPLC Trace of [^{11}C]L-Glutamine: (a) UV, (b) Radioactivity.

WASTE DISPOSAL INFORMATION

All hazardous chemicals and toxic materials were disposed of according to Prudent Practices in the Laboratory (Washington, DC: National Academy Press, 1995).

CHEMICAL ABSTRACTS NOMENCLATURE (REGISTRY NUMBER)

Acetonitrile (75-05-8)
Ammonia (7664-41-7)
(*R*)-2-(3-Pyridinyl)-1-pyrrolidinyl]methyl]phenoxy]-(1346133-09-2)
Carbonic acid, potassium salt (1:2) (584-08-7)
Cyanide-^{11}C (33942-38-0)
N,N-Dimethylformamide (68-12-2)
4,7,13,16,21,24-Hexaoxa-1,10-diazabicyclo[8.8.8]hexacosane (23978-09-8)
Hydrocyanic-^{11}C acid (14904-70-2)
Hydrogen peroxide (7722-84-1)
Sodium carbonate (497-19-8)
Sodium hydroxide (1310-73-2)
Sulfuric acid (7664-93-9)
Trifluoroacetic acid (76-05-1)

REFERENCES AND NOTES

For detailed supplier information, see Appendix 1.

1. (a) P. W. Miller, N. J. Long, R. Vilar, A. D. Gee, *Angew. Chem. Int. Ed.* 2008, 47, 8998; (b) P. J. H. Scott, *Angew. Chem. Int. Ed.* 2009, 48, 6001; (c) F. Wuest, M. Berndt, T. Kniess, *Ernst Schering Res. Found. Workshop* 2007, 64, 183; (d) B. Långström, O. Itsenko, O. Rahman, *J. Label. Compd. Radiopharm.* 2007, 50, 794; (e) G. Antonin, B. Långström, Progress in ^{11}C Radiochemistry in *Positron Emission Tomography: Basic Sciences* (Eds: D. L. Bailey, D. W. Townsend, P. E. Valk, M. N. Maisey), Springer, London, 2005, p. 223; (f) M. Allard, E. Fouquet, D. James, M. Szlosek-Pinaud, *Curr. Med. Chem.* 2008, 15, 235.
2. X. Shao, R. Hoareau, A. C. Runkle, J. J. M. Tluczek, B. G. Hockley, B. D. Henderson, P. J. H. Scott, *J. Label. Compd. Radiopharm.* 2011, 54, 819.
3. (a) X. Shao, P. L. Schnau, M. V. Fawaz, P. J. H. Scott, *Nucl. Med. Biol.* 2013, 40, 109; (b) X. Shao, M. V. Fawaz, K. Jang, P. J. H. Scott, *Appl. Radiat. Isotop.* 2014, 89, 125.
4. Y. Andersson, B. Långström, J. Chem. Soc., Perkin Trans. 1 1994, 1395.
5. GE Healthcare, USA.
6. GE standard Carbon-11 target with a 75 μm Havar target foil and standard vacuum foil.
7. Ultra high purity Nitrogen gas with 1% of Oxygen was purchased from Purity Plus.
8. Molecular sieve (4A, 80/100 mesh) was purchased from Alltech, pn: 5624.
9. Shimalite-Ni reduced (80/100 mesh) was purchased from Shimadzu (Kyoto, Japan), pn: 22127719.
10. Ammonia Ultra Pure, Matheson Gases.
11. Platinum gauze, 100 mesh, 99.9% trace metals basis was purchased from Sigma-Aldrich, pn: 298093-1.7G.
12. (a) C. H. Mitch, S. J. Quimby, N. Diaz, C. Pedregal, M. G. de la Torre, A. Jimenez, Q. Shi, E. J. Canada, S. D. Kahl, M. A. Statnick, D. L. McKinzie, D. R. Benesh, K. S. Rash, V. N. Barth, *J. Med.*

Chem. 2011, 54, 8000; (b) M-Q. Zheng, N. Nabulsi, S. J. Kim, G. Tomasi, S-F. Lin, C. Mitch, S. Quimby, V. Barth, K. Rash, J. Masters, A. Navarro, E. Seest, E. D. Morris, R. E. Carson, Y. Huang, *J. Nucl. Med.* 2013, 54, 455; (c) S. J. Kim, M-Q. Zheng, N. Nabulsi, D. Labaree, J. Ropchan, S. Najafzadeh, R. E. Carson, Y. Huang, E. D. Morris, *J. Nucl. Med.* 2013, 54, 1668.
13. Phenomenex, USA.
14. Shimadzu Corporation, USA.
15 Bioscan, Inc., USA.
16. (a) C. V. Dang, *Cell Cycle* 2010, 9, 3884; (b) M. J. Seltzer, B. D. Bennett, A. D. Joshi, P. Gao, A. G. Thomas, D. V. Ferraris, T. Tsukamoto, C. J. Rojas, B. S. Slusher, J. D. Rabinowitz, C. V. Dang, G. J. Riggins, *Cancer Res.* 2010, 70, 8981.
17. W. Fu, H. Orlefors, M. Bergström, G. Antoni, H. Omura, B. Eriksson, Y. Watanabe, B. Långström, *Anticancer Res.* 2000, 20, 251.
18. W. Qu, S. Oya, B. P. Lieberman, K. Ploessl, L. Wang, D. R. Wise, C. R. Divgi, L. P. Chodosh, C. B. Thompson, H. F. Kung, *J. Nucl. Med.* 2012, 53, 98.
19. C-18 Plus cartridges were purchased from Waters (pn: WAT020515) and conditioned with ethanol (10 ml) and water (10 ml) prior to use.
20. Ag11-A8 resin was purchased from Bio-Rad, pn: 142-7834.
21. Millex-GS 0.22 μm filters were purchased from Millipore (USA), pn:SLGSV 255F.

CHAPTER 23

DETERMINATION OF RADIOCHEMICAL PURITY AND RADIOCHEMICAL IDENTITY OF [^{13}N]NH_3 USING THIN LAYER CHROMATOGRAPHY

PETER A. RICE AND DANIEL L. YOKELL

Division of Nuclear Medicine and Molecular Imaging, Massachusetts General Hospital, Boston, Massachusetts, USA

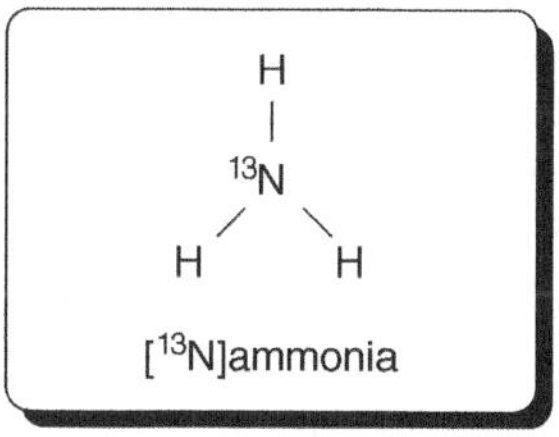

1 INTRODUCTION

The US Pharmacopoeia (USP) and European Pharmacopoeia (EP) methods for radiochemical purity and identity determination of N-13 ammonia injection currently require using a high-performance liquid chromatography (HPLC) system configured with a conductivity detector. This chapter describes a rapid, robust method for determining radiochemical purity and identity of N-13 ammonia using thin-layer chromatography (TLC).

Current compendial EP and USP HPLC conductivity methods are resource-intensive, requiring lengthy mobile phase preparation, and system suitability determination, while occupying valuable instrument time. Also, HPLC conductivity methods can have a higher incidence of system suitability failures in comparison with other HPLC detection methods. System suitability failures can lead to extensive delays in the production schedule of PET radiopharmaceuticals.

An ideal alternative method would rapidly and reproducibly determine the radiochemical purity and radiochemical identity of N-13 ammonia injection, while

Radiochemical Syntheses, Volume 2: Further Radiopharmaceuticals for Positron Emission Tomography and New Strategies for Their Production, First Edition. Edited by Peter J. H. Scott.

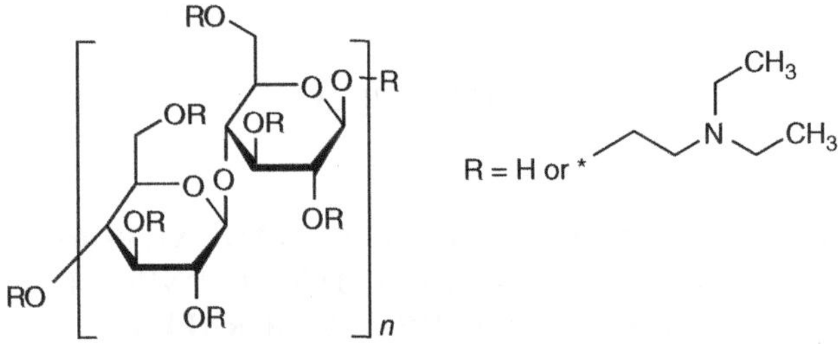

FIGURE 1 Diethylaminoethyl cellulose (DEAE-C).

requiring less time to complete and fewer resources to maintain than compendial methods. A TLC method for determining radiochemical purity and identity eliminates the need to use HPLC for N-13 ammonia quality control. This is highly desirable in a PET production facility that produces multiple radiopharmaceuticals per day.

A TLC system was developed that uses a diethylaminoethyl cellulose (DEAE-C) stationary phase (Fig. 1) The DEAE-C stationary phase of the chromatography system was chosen due to its ability to attract anionic species. The mobile phase comprises methanol:water 75:25.

The known radiochemical impurities of the in-target ammonia synthesis, with the exception of O-15 water, are anionic species: F-18 fluoride and N-13-labeled nitrates and nitrites (NOx). F-18 fluoride and N-13 NOx are chromatographically separated using this TLC method.

2 PROCEDURES

CAUTION: All radiochemical syntheses must be carried out using appropriate equipment in a facility authorized for the use of radioactive materials. Personal protective equipment must be worn and all local radiation safety laws followed.

At the MGH PET Center, N-13 ammonia is produced onsite with a GE PetTRACE cyclotron using the Wieland et al. method [1] of irradiation of 5 mM ethanol in water in-target synthesis method. Using an automated process, N-13 ammonia is then purified by passing through an anion exchange column (Waters, QMA Chloride [2]) and trapped on a cation exchange column (Waters, Accell CM [3]). The N-13 ammonia is released from the cation exchange column using 8 ml of 0.9% sodium chloride for injection, USP. The formulated N-13 ammonia in 0.9% sodium chloride for injection is then transferred through a Teflon® line via nitrogen overpressure to a ISO class 5 isolator for sterile filtration through a vented 0.22 μ polyethersulfone (PES) membrane filter (B Braun [4]) into a vented 30 ml sterile empty vial (Allergy Labs [5]).

In TLC method validation, N-13 ammonia was analyzed using a DEAE-C stationary phase [6] and a mobile phase composition of methanol:water 75:25. For radiochemical impurity testing, F-18 fluoride and N-13 NOx were also produced onsite. Testing was performed by spotting DEAE-C TLC media with analyte sample along with ammonium chloride USP reference standard. The sample consisted of either N-13 ammonia, or N-13 NOx and F-18 fluoride impurities. For radiochemical purity (RCP) analysis, TLC media were analyzed with a Bioscan AR-2000 TLC scanner. N-13 ammonia samples were also compared with a validated HPLC method [8] for RCP. The RCP specification for both HPLC and TLC was greater than or equal to 95%.

For radiochemical identity, the product radioactive peak was compared with the ammonium chloride TLC reference standard, which was visualized using iodoplatinate spray reagent [7]. The radiochemical identity specification for TLC was the Rf of the ammonium chloride standard must equal the Rf of N-13 ammonia. N-13 ammonia samples were also compared with a validated HPLC method for radiochemical identity. The radiochemical identity specification for HPLC was ±5% from the retention time of the ammonium chloride standard.

3 DISCUSSION

TLC strip is developed in methanol:water 75:25 mobile phase. Run time is approximately 8–10 min. After drying, the strip is analyzed on TLC scanner to determine RCP. In TLC analysis, the N-13 ammonia product Rf range was approximately 0.7–0.9, matching the ammonium chloride reference standard (Fig. 2). N-13 NOx (Fig. 3) and F-18 fluoride (Fig. 4) were each retained at the origin (Rf = 0).

The TLC stationary phase following RCP analysis is then developed with iodoplatinate spray reagent. After 10 min, the strip is placed in an iodine vapor

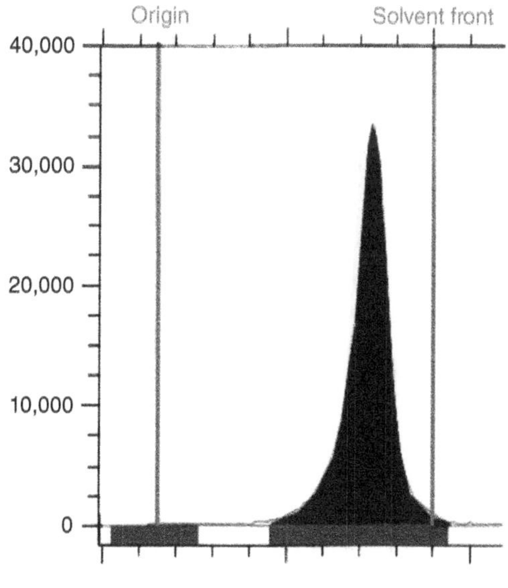

FIGURE 2 N-13 Ammonia TLC chromatogram.

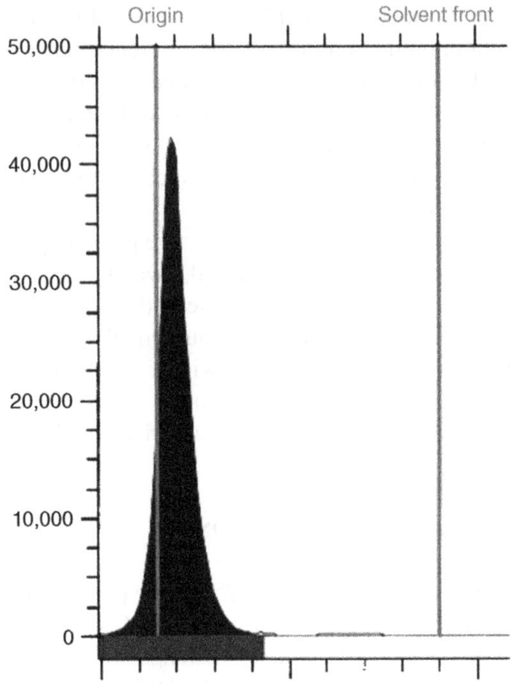

FIGURE 3 N-13 NOx TLC chromatogram.

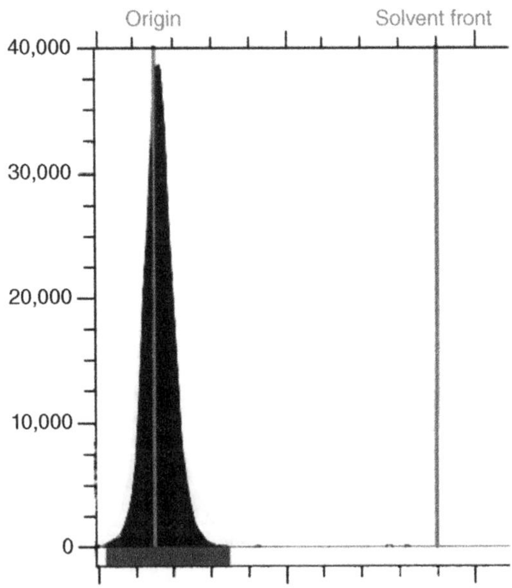

FIGURE 4 F-18 Fluoride TLC chromatogram.

chamber filled with iodine pellets for approximately 10 s for ammonia standard visualization to determine radiochemical identity (Fig. 5).

In comparison between the TLC and HPLC methods on N-13 ammonia validation runs, the mean RCP was 97.9% (range 97.3–98.2%) for TLC and greater than 99% for HPLC ($n=9$). In each test, the standard was visualized with an Rf factor of 0.7–0.9. All radiochemical identity tests passed (Table 1).

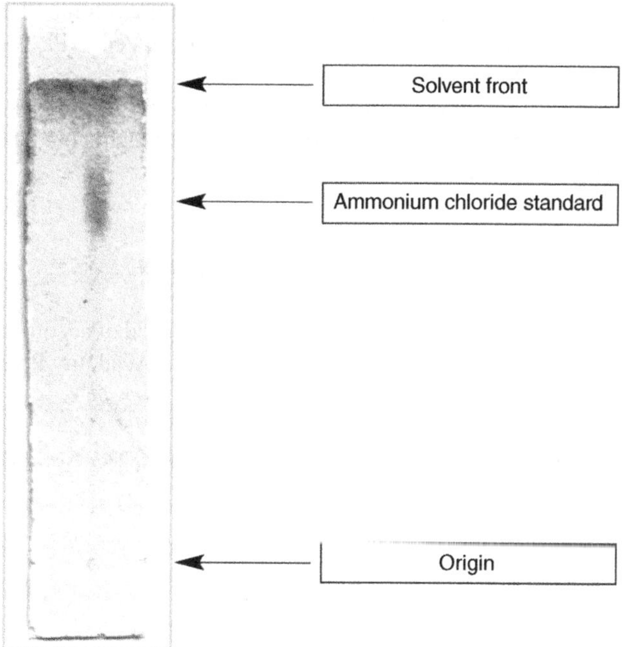

FIGURE 5 DEAE-C TLC after development of ammonium chloride standard in iodoplatinate spray reagent.

TABLE 1 Summary of Ammonia TLC and HPLC Validation Study

Run	TLC		HPLC	
	Radiochemical Purity (%)	Radiochemical Identity	Radiochemical Purity (%)	Radiochemical Identity
1	98.21	Pass	>99	Pass
2	97.54	Pass	>99	Pass
3	98.03	Pass	>99	Pass
4	98.21	Pass	>99	Pass
5	98.03	Pass	>99	Pass
6	97.54	Pass	>99	Pass
7	97.33	Pass	>99	Pass
8	97.90	Pass	>99	Pass
9	97.98	Pass	>99	Pass

4 SUMMARY

In this chapter, we describe a newly developed TLC method for determining the radiochemical purity and identity of N-13 ammonia injection is rapid and reproducible, and the results were comparable with a known, validated HPLC method.

The use of a TLC system for N-13 ammonia quality control eliminates the need for using an HPLC system, improving workflow in a high-volume PET production facility, as well as enabling facilities without specialized HPLC conductivity detectors to produce N-13 ammonia injection.

This novel TLC method using DEAE-C stationary phase developed in methanol:water 75:25 is valid as a stand-alone test for radiochemical purity and radiochemical identity determination.

WASTE DISPOSAL INFORMATION

All hazardous chemicals and toxic materials were disposed of according to Prudent Practices in the Laboratory (Washington, DC: National Academy Press, 1995).

CHEMICAL ABSTRACTS NOMENCLATURE (REGISTRY NUMBER)

Ammonia N 13 (34819-78-8)

REFERENCES AND NOTES

For detailed supplier information, see Appendix 1.

1. B. Wieland, G. Bida, H. Padgett, G. Hendry, E. Zippi, G. Kabalka, J.-L. Morelle, R. Verbruggen, M. Ghyoot, *Appl. Radiat. Isot.*, 1991, 42, 1095–1098.
2. QMA Chloride Light Sep-Pak Plus Cartridge, Waters Part WAT023525.
3. Accell CM Short Sep-Pak Plus Cartridge, Waters Part WAT020550.
4. 0.22 μ PES vented filer, B Braun Part PFE2000.
5. 30 ml Sterile Empty Vial, Allergy Labs Part SV30S.
6. Cellulose DEAE TLC Plates, J.T. Baker Part 4477-00.
7. Iodoplatinate Spray Reagent, Sigma Alrich Part I9157.
8. Waters IC-Pak Cation M/D Column Care and Use Manual, Waters Corporation; 2007, Manual ID WAT036549.

APPENDIX

SUPPLIER INFORMATION

ABX Advanced Biochemical Compounds
www.abx.de

Acros
www.acros.com

Advion
www.advion.com

Agilent Corporation
www.agilent.com

Akorn
www.akorn.com

Alltech (A Division of Grace Davison Discovery Sciences)
www.discoverysciences.com

American Regent
www.americanregent.com

Apollo Scientific Ltd.
www.apolloscientific.co.uk

Atomic Products Corp
See Biodex Medical Systems, Inc.

B. Braun
www.bbraun.com

Baxter
www.baxter.com

Radiochemical Syntheses, Volume 2: Further Radiopharmaceuticals for Positron Emission Tomography and New Strategies for Their Production, First Edition. Edited by Peter J. H. Scott.

Beckman, Inc., USA
www.beckmancoulter.com

Becton, Dickinson and Company
www.bd.com

Biodex Medical Systems, Inc.
www.biodex.com

Bio-Rad
www.bio-rad.com

Bioscan, Inc.
www.bioscan.com and http://www.trifoilimaging.com

Burdick & Jackson
http://www.honeywell-burdickandjackson.com

Canberra
http://www.canberra.com

Capintec
www.capintec.com

Carroll & Ramsey Associates
http://www.carroll-ramsey.com

Charles River Laboratories International, Inc.
www.criver.com

Cole Parmer
www.coleparmer.com

Dionex
www.dionex.com

Eckert and Ziegler
www.ezag.com

EMD Chemicals, Inc., USA
www.emsl.com

EMD Chemicals, Inc., USA
www.emdchemicals.com

EMD Millipore
www.emdmillipore.com

Fisher Scientific
www.fishersci.com

Fluka
See Sigma-Aldrich

GE Healthcare
www.gehealthcare.com/worldwide.html

Gilson
www.gilson.com

Hollister-Stier Laboratories LLC
www.hollister-stier.com

Hospira
www.hospira.com

Huayi Isotopes
www.hic.cc

ILMVAC GmbH
www.ilmvac.de

ISOFLEX USA, Inc.
www.isoflex.com

ISO-MED
www.iso-med.com

Jenway
www.jenway.com

Knauer
www.knauer.net

LabLogic Systems Ltd.
www.lablogic.com

Macherey-Nagel
www.mn-net.com

Merck
www.merck.de

Millipore
www.millipore.com

National Instruments
www.ni.com

NovaBioChem
See EMD Millipore

Oakwood Products, Inc.
www.oakwoodchemical.com

ORTG
http://www.ortg.com

Phenomenex
www.phenomenex.com

Purity Plus
www.purityplusgases.com

Ray Test Corporation
www.raytest.com

Restek
www.restek.com

Rheodyne
http://www.upchurch.com/Rheodyne.aspx

Rotem Industries Ltd.
www.rotemi.co.il

Shimadzu
www.shimadzu.com

Siemens Medical Solutions USA, Inc.
www.healthcare.siemens.com

Sigma-Aldrich
www.sigmaaldrich.com

Sonntek Inc.
www.sonntek.com

Swagelok
www.swagelok.com

TCI America
http://www.tcichemicals.com/en/us/

Thomas Scientific
www.thomassci.com

Varian
www.varian.com

VWR
www.vwr.com

Waters
www.waters.com

W. R. Grace & Company
https://grace.com/en-us

INDEX

Radiochemical Syntheses, Volume 2: Further Radiopharmaceuticals for Positron Emission Tomography and New Strategies for Their Production, First Edition. Edited by Peter J. H. Scott.